S.L. Hopp · M.J. Owren · C.S. Evans (Eds.)

Animal Acoustic Communication

Springer

Berlin
Heidelberg
New York
Barcelona
Budapest
Hongkong
London
Milan
Paris
Santa Clara
Singapore
Tokyo

S.L. Hopp   ·   M.J. Owren   ·   C.S. Evans (Eds.)

Animal Acoustic Communication

Sound Analysis and Research Methods

With 115 Figures

 Springer

DR. STEVEN L. HOPP
University of Arizona
Dept. of Ecology and Evolutionary Biology
Tucson, AZ 85721
USA

DR. MICHAEL J. OWREN
Cornell University
Dept. of Psychology
Ithaca, NY 14853
USA

DR. CHRISTOPHER S. EVANS
Macquarie University
Dept of Psychology
Sydney, NSW 2109
Australia

Library of Congress
Animal acoustic communication : sound analysis and research methods /
Steven L. Hopp, Michael J. Owren, and Christopher S. Evans (eds.).
p. cm.
Includes bibliograhical references and index.

ISBN-13: 978-3-642-76222-2   e-ISBN-13: 978-3-642-76220-8
DOI: 10.1007/978-3-642-76220-8

1. Animal sounds. 2. Sound production by animals. 3. Animal
communication. I. Hopp, Steven L., 1954- . II. Owren Michael
J., 1955- . III. Evans, Christopher Stuart, 1959.
QL765.A.645 1977
591.59#4—dc21   97-10402

© Springer-Verlag Berlin Heidelberg
1998
Softcover reprint of the hardcover 1st edition 1998

Camera ready by Claudia Seelinger, Springer-Verlag Heidelberg
Cover design: D&P, Heidelberg
Cover photograph: courtesy of J.M. Whitehead
SPIN 10014776 31/3137 5 4 3 2 1 0 - Printed on acid free paper

The last decades have brought a significant increase in research on acoustic communication in animals. Publication of scientific papers on both empirical and theoretical aspects of this topic has greatly increased, and a new journal, Bioacoustics, is entirely devoted to such articles. Coupled with this proliferation of work is a recognition that many of the current issues are best approached with an interdisciplinary perspective, requiring technical and theoretical contributions from a number of areas of inquiry that have traditionally been separated. With the notable exception of a collection edited by Lewis (1983), there have been few volumes predominately focused on technical issues in comparative bioacoustics to follow up the early works edited by Lanyon and Tavolga (1960) and Busnel (1963). It was the tremendous growth of expertise concerning this topic in particular that provided the initial impetus to organize this volume, which attempts to present fundamental information from both theoretical and applied aspects of current bioacoustics research. While a completely comprehensive review would be impractical, this volume offers a basic treatment of a wide variety of topics aimed at providing a conceptual framework within which researchers can address their own questions. Each presentation is designed to be useful to the broadest possible spectrum of researchers, including both those currently working in any of the many and diverse disciplines of bioacoustics, and others that may be new to such studies.

Several previous collections have addressed communication-related topics in particular taxonomic groups, such as fish (Tavolga et al. 1981), birds (Kroodsma et al. 1982; Catchpole and Slater 1995; Kroodsma and Miller 1996), and nonhuman primates (Snowdon et al. 1982; Todt et al. 1988; Zimmermann et al. 1995). Typically these treatments have provided a mix of technical, empirical, and theoretical contributions. Other compiled volumes have been devoted to specific subtopics in animal communication including animal perception (e.g., Berkley and Stebbins 1990; Stebbins and Berkley 1990) and sound-broadcast, or playback, studies (McGregor 1992). This book differs from these earlier volumes in that it specifically responds to recent developments in digital computing, and its applications. In addition, we include several chapters outlining the current technological and methodological approaches to studies in particular areas of animal acoustic communication.

A large part of the growth in animal acoustic communication research can be attributed to the development of accessible computer hardware and software that allows sophisticated digital signal processing (DSP) algorithms to be applied to sound-derived data. A technological revolution of the last decades has produced a wide variety of new approaches, new ways of thinking about problems, and much greater capacity for processing and manipulating information. However, the extent to which new developments have been transferred across disciplines has, in part, determined some of the advances in a given area. For research in acoustics, these advances include the development of processing hardware, the discovery of efficient methods for the imaging and analysis of signals, and development of applications for these procedures across disciplines and even across taxa. Many of these techniques have been used widely in engineering fields but have only recently been employed routinely by comparative researchers. Theoretical discussions of digital signal processing have been published primarily in the engineering literature, and have consequently been relatively inaccessible to comparative researchers. Similar difficulties also occur within the discipline of comparative bioacoustics. Here, important factors include the great diversity of backgrounds found among animal-oriented investigators, as well as significant variation in both the nature of the sound signals of interest and the circumstances of the animal species that produce them. The result has been that barriers to mutually beneficial sharing of knowledge have arisen due to rather arbitrary divergence of terminology and methods. One overriding purpose in presenting this volume, then, is to introduce common techniques in DSP in a manner that is comprehensible to both students and professionals conducting bioacoustics research in diverse subdisciplines. In addition, the various authors here have tried to provide clear examples of both typical and possible applications of given methods or techniques, and outlining practical suggestions wherever possible.

The first portion of the book is dedicated to sound analysis and computer applications. In the first chapter, Gerhardt outlines the basic techniques of recording and analyzing animal sounds. The methods discussed therein provide a solid background for applications in diverse areas of research. This presentation is followed by a pair of complementary chapters discussing the applications of digital technology to sound signal representation and processing in the time- and frequency-domains. The first, by Clements, outlines the general principles of DSP in both mathematical and verbal form. The second, by Beeman, explicitly explains and demonstrates these principles in the context of the computing equipment and research topics that might be typical of scientists working in bioacoustics. The next chapter, by Stoddard, offers a practical discussion of the crucial role of electronic filters in bioacoustics research, along with explanations of related terminology, design, and application. The final chapter in this section, by Owren and Bernacki, addresses the conceptually narrower topic of how to characterize energy patterning in the frequency spectrum of sound. The approach described, based on linear prediction, has been used in analogous applications in other disciplines, particularly in the analysis and synthesis of human speech.

The second section addresses four general areas of study in which recent technological advances have allowed researchers both to address long-standing questions of interest with greatly increased power and to pioneer entirely new lines of inquiry. Each of these chapters combines the presentation of important theoretical and methodological topics with both an overview of the associated empirical literature and examples of studies in which new approaches have been used. The first two chapters discuss three realms of acoustics that fall outside the typical human auditory experience. The chapter by Pye and Langbauer on infrasound and ultrasound, airborne acoustic signals that fall below and above the range of human sensitivity, respectively, is complemented by the discussion by Tyack, on underwater sound communication. Each of these signaling domains presents unique technical challenges and have given rise to quite unexpected research findings. In the third chapter, Rubin and Vatikiotis-Bateson discuss the acoustics of species-typical communication in human speech, an extensively studied communication system. The techniques and theoretical models outlined therein provide a rich base for studies with a comparative focus. This area of research is typically not considered in a comparative perspective, although it has historically been immensely important to the investigation of sound signals in nonhuman animals. The development of much of the hardware and software currently in use in comparative bioacoustics can be directly linked to studies of speech production or perception. For example, ideas about the source-filter theory of voice production in humans have provided models for understanding some aspects of sound production by both primates and birds. In the final chapter in this section, Gaunt and Nowicki specifically examine the latter topic, i.e. sound production by birds, and in doing so demonstrate the power of combining technological advances with interdisciplinary thinking.

The three chapters in the final section focus on different aspects of the question of how communicative signals affect the receiver. The methods are diverse, ranging from how animals overtly or covertly respond to signals, to basic questions such as whether organisms can discriminate among similar signals. Hopp and Morton discuss sound playback studies, which are used to assess the reaction of animals to the broadcast of particular sounds. Such studies are typically conducted in the field with free-ranging animals. Where naturally occurring responses to sounds are limited, training animals to respond differentially to signals allows researchers to ask questions about sound perception and categorization of signals. These techniques are discussed by Cynx and Clark. The last chapter is a contrast to the other contributions in that its focus is not primarily on issues related to measurement and analysis of acoustic signals. In this chapter, Ball and Dufty provide detailed, practical methodological information about the interplay between acoustic signals and the neuroendocrine systems of animals. As the chapter demonstrates, hormonal responses to auditory stimulation play an important and immediate role in many aspects of acoustic communication.

As we prepare to enter a new century, it is nearly impossible to anticipate the inevitable advances in technology, or the profound effects that these advances will have on studies of animal acoustic communication. We hope this volume will provide readers with a framework for understanding the dramatic changes that have taken place in animal communication research. It is meant as a stepping stone to our shared future in comparative bioacoustics.

Steven L. Hopp *Tucson, Arizona, USA*
Michael J. Owren *Ithaca, NY, USA*
Christopher S. Evans *Sydney, Australia*

References

Berkley MA, Stebbins WC (eds) (1990) Comparative perception, vol I. Basic mechanisms. Wiley, New York

Busnel RG (ed) (1963) Acoustic behavior of animals. Elsevier, Amsterdam

Catchpole CK, Slater PJB (1995) Bird song. Cambridge New York

Kroodsma DE, Miller EH, Ouellet H (eds) (1982) Acoustic communication in birds, vols I and II. Academic Press, New York

Kroodsma DE, Miller EH (eds)(1996) Ecology and evolution of acoustic communication in birds. Cornell University Press, Ithaca

Lanyon WE, Tavolga WN (1960) Symposium on animal sounds and communication. Am Inst Biol Sci, Washington

Lewis B (ed) (1983) Bioacoustics: a comparative approach. Academic Press, New York

McGregor PK (ed) (1992) Playback and studies of animal communication. Plenum, New York

Snowdon CT, Brown CH, Petersen MR (1982) Primate communication. Cambridge University Press, New York

Stebbins WC, Berkley MA (eds) (1990) Comparative perception, vol II. Complex signals. Wiley, New York

Tavolga WN, Popper AN, Fay RR (eds) (1981) Hearing and sound communication in fishes. Springer Berlin Heidelberg New York

Todt D, Goedeking P, Symmes D (eds) (1988) Primate vocal communication. Springer Berlin Heidelberg New York

Zimmerman E, Newman JD, Jürgens, U (eds) (1995) Current topics in primate vocal communication. Plenum, New York

Acknowledgments

Over the years there were many people who offered suggestions, support, discussion, inspiration, time, and chapter reviews. We thank the following people for their contributions: W. Au, C.A. Boone, B. Burgess, E. Carterette, F. Cheever, D. Czeschlik, V.J. DeGhett, U. Gramm, P.S. Kaplan, B.E. Kingsolver, A. Kirby, T. Krammer, D.E. Kroodsma, C.A. Logan, A. Popper, C. Seelinger, J. Spiesberger, R.A. Suthers. In addition we thank the authors of this volume for their dedication and endurance.

GREGORY F. BALL
Department of Psychology, Johns Hopkins University, 3400 N. Charles Street, Baltimore, Maryland 21218-2686 USA

KIM BEEMAN
Engineering Design, 43 Newton Street, Belmont, Massachusetts 02178, USA

ROBERT BERNACKI
P.O. Box 3188, Bloomington Scientific, Bloomington, Indiana 47402 USA

STEPHEN CLARK
Psychology Department, 124 Raymond Ave, Vassar College, Poughkeepsie, New York 12601, USA

MARK CLEMENTS
School of Electrical Engineering, Georgia Institute of Technology, Atlanta, Georgia, 30332-0250, USA

JEFFERY CYNX
Psychology Department, 124 Raymond Ave, Vassar College, Poughkeepsie, New York , 12601, USA

ALFRED M. DUFTY, JR.
Department of Biology, Boise State University, 1910 University Drive, Boise, Idaho, 83725, USA

CHRISTOPHER S. EVANS
School of Behavioural Sciences, Macquarie University, Sydney NSW 2109, Australia

ABBOT S. (TOBY) GAUNT
Department of Zoology, Ohio State University, 1735 Neil Av., Columbus, Ohio, 43210, USA

H. CARL GERHARDT
Division of Biological Sciences, 105 Tucker Hall, University of Missouri at Columbia, Columbia, Missouri, 65211, USA

Steven L. Hopp
Department of Ecology and Evolutionary Biology, University of Arizona, Tucson, Arizona, 85721, USA

WILLIAM R. LANGBAUER, JR.
Pittsburgh Zoo, P.O. Box 5250, Pittsburgh, Pennsylvania, 15206, USA

EUGENE S. MORTON
Conservation and Research Center, Smithonian Institution, National Zoological Park, 1500 Remount Road, Front Royal, Virginia 22630, USA

STEPHEN NOWICKI
Department of Zoology, Duke University, Durham, North Carolina, 27706, USA

MICHAEL J. OWREN
Department of Psychology, Cornell University, 224 Uris Hall, Ithaca, New York 1485,3 USA

J. DAVID PYE
Department of Zoology and Cpmparartive Physiology, Queen Mary College, University of London, Mile End Road, London E1 4NS, UK

PHILIP RUBIN
Haskins Laboratories and Yale University School of Medicine, Department of Surgery, Otolaryngology, 270 Crown Street, New Haven, Conneticut, 06511, USA

PHILIP K. STODDARD
Department of Biological Sciences, Florida International University, University Park, Miami, Florida, 33199, USA

PETER L. TYACK
Woods Hole Oceanographic Institution, Woods Hole, Massachusetts, 02543, USA

ERIC VATIKIOTIS-BATESON
ATR Human Information Processing Research Laboratories, 2-2 Hikaridai/Seika-cho, Sorkau-gun,Kyoto 619-02, Japan

CHAPTER 3
Digital Signal Analysis, Editing, and Synthesis
K. BEEMAN 59

CHAPTER 8
Measuring and Modeling Speech Production
P. RUBIN AND E. VATIKIOTIS-BATESON

CHAPTER 9
Sound Production in Birds: Acoustics and Physiology Revisited
A. S. GAUNT AND S. NOWICKI

SECTION III
ASSESSING BIOLOGICALLY IMPORTANT RESPONSES

CHAPTER 10
Sound Playback Studies
S. L. HOPP AND E. S. MORTON

CHAPTER 11
The Laboratory Use of Conditional and Natural Responses in the Study
of Avian Auditory Perception
J. CYNX AND S. J. CLARK

Section I:

Processing and Analysis of Acoustic Signals

Acoustic Signals of Animals: Recording, Field Measurements, Analysis and Description

H. C. GERHARDT

1
Introduction

The main aim of this chapter is to outline techniques for recording and characterizing animal communication signals. Signal analysis provides the basis for assessing the repertoires of individuals and species, and for relating variation in signal structure to variation in other phenotypic attributes of the signaler. Correlations are also usually found between signal structure and both social and ecological contexts of signal production (e.g., Falls 1982; Wells 1988). Hypotheses arising from such correlational data can be tested experimentally by playbacks of synthetic sounds, and acoustic analyses of natural sounds provide the information needed to generate such stimuli (e.g., Gerhardt 1991; Wagner 1992). Signal analysis may also provide important hints for studying mechanisms of sound production (e.g., Elsner 1994). Finally, in phylogenetic analyses, acoustic characters can be used along with other traits to generate hypotheses about the evolutionary history of signal structure within a group of animals (e.g., Heller 1990; Crocroft and Ryan 1995).

The order of topics in this chapter mirrors the usual order of acquiring, analyzing, and describing acoustic signals. The first section deals with recording animal sounds under natural conditions and with measurements of acoustic properties that can only be made at the site rather than being extracted from recordings. The second section provides an overview of the kinds of acoustic properties that are usually analyzed and outlines how the results of these analyses are usually presented. The third section provides examples of descriptions of animal sounds. These descriptions are divided into time- and frequency-based analyses, and, where appropriate, I show how these two approaches can estimate values of the same acoustic property. Many of the examples are based on signals produced by amphibians, but are nevertheless representative of the range of complexity observed in the basic acoustic units produced by a wide range of species, including invertebrates, birds, and mammals. The next level of complexity involves the production of a rich variety of such basic acoustic units, e.g., the syllables of bird song. Sophisticated methods for efficiently classifying different syllable types are treated in detail by Beeman (this Volume).

2
Field Recordings and Measurements

2.1
Equipment

The expected frequency range of the animal signals to be recorded dictates the selection of the audio recorders, microphones, and parabolic microphones with particular frequency response characteristics. Moderately priced professional audio recorders (both analog and digital; e.g., Wahlström 1990) and microphones are available for the frequency range of about 50 Hz to 15 kHz. Equipment capable of high-fidelity recording of very low frequencies, such as the infrasonic signals of elephants (Langbauer et al. 1991), high but audible frequencies produced by some birds and insects, and the ultrasonic signals of various insects and mammals is less commonly available and more expensive (see Pye and Langbauer, this volume, for specific suggestions).

Two other important factors in equipment selection are the speed variability (usually expressed as *wow* and *flutter* percentages) associated with the recorder's tape transport system, and the distortion introduced by the recorder and microphone. For the former, stability in tape speed is generally correlated with the price of the recorder—except in the case of digital recorders, where stability is generally not a concern. With respect to distortion, my experience is that overloading the recorder's input amplifiers (thereby generating harmonic and intermodulation distortion) is the most common error in bioacoustics. This problem usually arises when using a VU (volume unit) meter to set the input level when recording very short sounds. Because the time-constant of such meters is relatively long, the meter does not register the peak values (see Wickstrom 1982 for a detailed and practical discussion). Fortunately, most recorders now have peak-reading devices that respond adequately to short sounds.

2.2
On-Site Measurements

Although most acoustic properties of animal sounds are determined from tape recordings, some relevant measurements are normally made outdoors at the time and place the animal is signaling. These include estimates of the absolute amplitude of an animal's signal, the pattern of radiation of the signal and its propagation, the amplitudes of the signals of nearby conspecifics, and the level of ambient background noise.

Estimates of the amplitude of the signal and of directional properties of sound production by an animal in the field have several important applications. First, differences among individuals in their acoustic output may be important in determining their relative success in attracting a mate or repelling a competitor (e.g., Forest and Raspert 1994). Second, the amplitude of a nearest neighbor's signals at the receiver's position is likely to affect the probability and type of response an individual may make. For example, in anurans, some of the variation in particular properties of the advertisement call, and the probability of a switch to aggressive calls, is predictable from measurement of the amplitude of the signals of the nearest neighbour (e.g., Wells 1988; Wagner 1989;

Brenowitz and Rose 1994). Third, estimates of signal amplitude can guide the choice of playback levels in behavioural experiments in the field and laboratory. Too many studies have chosen playback levels arbitrarily or even adjusted the levels subjectively. Finally, measurements of signal amplitude and directional properties provide a starting point for estimating the maximum communication distance for a species (e.g., Gerhardt and Klump 1988a; Römer 1993).

2.3
Signal Amplitude, Directionality, and Background Noise Levels

Because they are usually easy to approach, acoustically signalling insects and anurans have been the subjects of most field measurements of amplitude [usually measured as sound pressure level (SPL) in decibels (dB) re 20 µPa] and directionality (e.g., Loftus-Hills and Littlejohn 1971; Gerhardt 1975; Bennet-Clark 1989; Prestwich et al. 1989). There are a number of factors to consider in making these measurements.

First, the type of microphone will dictate its optimal orientation relative to the signaling animal. Manufacturers of *random-incidence* microphones suggest that the microphone be oriented so that it forms an angle of about 70-90 with the sound source; a microphone designed for *perpendicular incidence* is pointed directly at the animal (Broch 1971; Peterson and Gross 1972; Hassall 1991). In these cases, if the microphone is not aimed directly at the animal, the amplitudes of high-frequency components will be reduced relative to low-frequency ones in the recording, a result that is termed *off-axis coloration* (Woram 1989). Indeed, even when appropriately aimed, parabolic reflectors have non-linear frequency responses that should be measured and taken into account during acoustic analyses (e.g., Wickstrom 1982).

Second, animals themselves differ in the extent to which they beam acoustic signals. Bats, for example, emit echolocation signals in a narrow frontal beam (Simmons 1969), whereas some species of frogs (Gerhardt 1975; Prestwich et al. 1989) and crickets (e.g., Forrest) approximate omnidirectional sound sources with hemispherical radiation patterns. More complex, dumbbell-shaped patterns of sound radiation have been observed in toads (Gerhardt 1975) and mole crickets (Forrest 1991). In the latter, the burrow from which an individual cricket sings can affect its sound radiation pattern (e.g., Bennet-Clark 1989; Forrest 1991). Ideally, simultaneous readings from multiple positions around a signaling animal are desirable.

Third, the weighting network (for frequency filtering) of the sound-level meter must match the signal and acoustic environment. In low-noise situations, or where the frequency spectrum of the animal's sounds is unknown, the *unweighted* (*flat*, or *linear*) setting should be used. If low-frequency background noise is high and the animal's signals do not contain acoustic energy below about 500 Hz or above about 10 kHz, then the researcher can use one of the weighting networks that are standard in most sound-level meters. The A-weighting network attenuates frequencies below about 1 kHz (by about 3 dB at 0.5 kHz and 11 dB at 0.2 kHz, re the amplitude at 1.0 kHz), whereas C-weighting only affects frequencies below about 50 Hz (Harris 1991). I recommend using the flat- or the C-weighted setting whenever possible. If the sound-level meter has a built-in or attachable filter set, then rough (i.e., octave or one-third octave) measurements of the animal's signals and the background noise can be made in the field.

Fourth, the time-constant of the sound-level meter must be taken into account. The time constant is defined as the time required for a quantity that varies exponentially with time to increase by the factor 1-1/e, where e is the base of the natural logarithm (Yeager and Marsh 1991). Nearly all sound-level meters have a so-called *fast* setting. This setting provides a root-mean-square (RMS) sound pressure level and has a time-constant of 125 milliseconds (ms). However, because many animals produce very short signals that fluctuate rapidly in amplitude, the readings obtained with the fast setting are influenced by temporal properties of a signal as well as by its amplitude.

One solution is to use the *peak-reading* mode, with a time-constant of less than about 50 μs. Unfortunately, readings of peak amplitude are sensitive to the phase (relative timing relationships) of the components of complex signals or direct and indirect sound waves. Therefore, even small changes in microphone position may result in considerable variation in readings. This problem is accentuated in reverberant environments or where frequency-dependent absorption is prominent (Yeager and Marsh 1991). An intermediate value is provided by the *impulse* setting, which has an exponential time-constant of about 35 ms. Overall, the best policy is to use the fast setting for amplitude measurements and then to provide oscillograms that show the duration, rise-fall characteristics, and repetition patterns of signals that are less than about 100 ms long when reporting the measurement results. Interpretation of the biological significance of SPL measurements of short signals can be aided by psychophysical or neurophysiological estimates of the auditory integration times for the species being studied (e.g., Dooling 1982; Dunia and Narins 1989).

The last consideration to be discussed is the distance at which the microphone is placed from the animal. In general, the optimum distance is one that insures that the microphone is in both the *free-field* and the *far-field* (see Figure 1). "Free-field" refers to an idealized area free of sound reflections. The far-field is an idealized region in an idealized environment in which, assuming that the sound generator approximates a point (monopole) source, decrements in the amplitude of sound waves with distance will be attributable to spherical (geometric) spreading alone. Specifically, SPL will decrease by a factor of 2 (6 dB) for each doubling of distance.

Whereas being too close to a vocalizing bird or mammal in the field is seldom a problem, a researcher can often place a microphone within a few centimeters of a frog or an insect. Areas very close to the source are in the *near-field* (or *reactive-field*; see Figure 1), where the contribution of particle velocity (movements of the molecules in the medium) to sound energy is greater than that of sound pressure and where these components are not in phase (as they are at greater distances). Ewing (1989) provides a practical introduction to this topic in relation to some insects, in which the particle-velocity component dominates acoustic communication (see also Michelsen et al. 1987). Crocker (1991), Harris (1991), and Skudrzyk (1971) present more technical and mathematical treatments. Particle velocity decreases much more rapidly with distance than does sound pressure. However, there is no generally accepted rule for estimating the extent of the near-field, which depends not only on wavelength (frequency), but also on size, shape, and vibration-mode of the sound source. The most conservative recommendation (Broch 1971; Peterson and Gross 1972) is that the microphone and acoustic source should be separated by a distance corresponding to at least one wavelength of the lowest frequency in the sound.

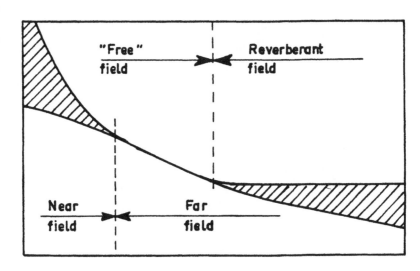

Log of Distance from Sound Source

Fig 1. Variability in sound pressure level as a function of distance from the source. The hatched areas indicate in a very general way the extent of variability that may be encountered in the near field and far field for the reasons discussed in the text. (Modified from Broch 1971)

When recording near an animal, the *proximity effect*, an increase in the low-frequency sensitivity of the microphone, may also occur. The proximity effect is a function of microphone design and is especially pronounced in directional microphones whenever the path length from the front to the rear of the diaphragm becomes large relative to the distance between the source and the diaphragm. Woram (1989) provides a detailed discussion, and examples are shown in Figure 2.

2.4
Patterns of Sound Propagation in Natural Habitats

Beyond the free-field, in the *reverberant-field*, objects in the environment, including the substrate may affect sound waves in diverse ways that cause departures from the inverse square law. Piercy and Daigle (1991) provide a detailed technical and practical discussion in terms of general outdoor acoustics. Several authors cover many of the same topics in the context of animal communication (Wiley and Richards 1978; Michelsen 1978, 1983; Wickstrom 1982, see also Hopp and Morton, this Volume). If the animal is at, or near, ground level and the substrate is absorptive, then there will be a selective attenuation of high-frequency components. If the substrate is reflective, measurements may be affected by indirect waves. Indeed, the effects of indirect waves can be especially

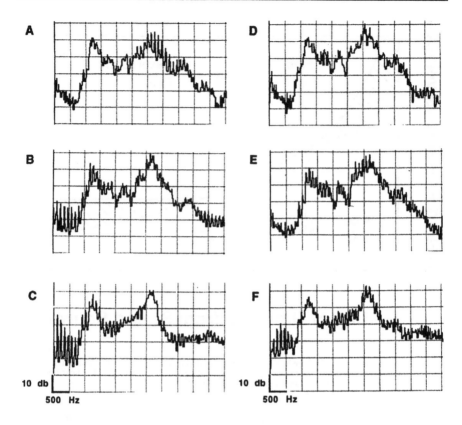

Fig 2 A-F. Power spectra of recordings of single calls of a squirrel treefrog (*Hyla squirella*) made at different distances from the calling male. Each division along the x-axis equals 500 Hz; each division along the y-axis equals 10 dB. The spectra in A and D were of calls recorded at a distance of 50 cm with a directional and omnidirectional microphone, respectively. In B and E, the microphones were 30 cm from the frog. Notice that a series of components below about 800 Hz are prominent in the recording made with the directional microphone but not in the recording made with the omnidirectional microphone. The relative emphasis of these low-frequency components is a result of the proximity effect. In C and F, the microphones were within 3 cm of the frog. Notice that the proximity effect is now evident in recordings made with both microphones

severe and complex if both the animal and the microphone are elevated. The sound waves that propagate directly between the animal and microphone will interact with reflected waves in a destructive or constructive fashion, depending on the dominant wavelength(s) of the sound waves and the distances separating the animal, the microphone, and the substrate. Hence, the spectral structure (relative amplitudes of different frequency components) of broad-band signals may vary considerably as a function of the distance from the animal at which measurements and recordings are made.

Weather conditions and time of day must also be considered because these factors also affect sound propagation. For instance, measurements or recordings should not be attempted if wind velocity is high. Even on days when there is little wind, turbulence will arise from the heating of the ground, especially in open habitats (Wiley and Richards 1978). This effect provides another reason for obtaining measurements as close to an animal as is practical. At distances of greater than about 30 m, fluctuations in meter readings can occur even if the signal being measured does not vary at its source (e.g., Embleton et al. 1976). Conditions for signal propagation may change dramatically at night, and there may even be channels for highly efficient propagation close to the ground when temperature inversions cause cool air to be trapped near the surface (Wiley and Richards 1978).

Several studies have described patterns of attenuation of various animal sounds and artificial signals over relatively long distances in natural environments, but there have been some contradictory results (see discussions in Michelsen 1978, 1983). One reason is that investigators have often used loudspeakers to broadcast animal sounds or artificial test signals through natural environments without calibrating the sound output of a speaker at a given distance in uniform (anechoic) conditions. Computing attenuation patterns by comparing the output of a distant microphone with that of a closer, reference microphone may result in errors because frequency-, distance-, and elevation-dependent variation will also affect acoustic structure recorded by the reference microphone in the natural environment (Michelsen 1978).

A conservative approach is to make measurements of sound pressure levels and recordings at various distances from signaling animals. Estimating the magnitude and characteristics of the background noise is also valuable (see below). If this general strategy is impracticable, then multiple recordings or measurements of signals from the same or different individuals should be made from a standard distance to provide meaningful comparisons. Playbacks of such recordings can show whether the suite of complex variables that affect signal amplitude and quality during propagation through natural environments alter the effectiveness of the signal in eliciting biologically meaningful responses. For example, females of the barking treefrog (*Hyla gratiosa*) were attracted to playbacks of recordings of chorus sounds made at a distance of 160 m from the breeding pond even though the high-frequency components of male calls were strongly attenuated and background noise levels were high (Gerhardt and Klump 1988a).

A highly innovative approach has been to use a portable, neurophysiological preparation as a *biological microphone* (see Römer 1993). For example, recordings of spike activity from an auditory interneuron, which receives inputs from all auditory receptors, at various distances from singing males can be used to estimate the maximum communication distance (i.e., the distance at which the response threshold of the interneuron was exceeded). Moreover, the distances at which biologically pertinent temporal properties are still encoded in the spike activity can also be assessed.

However, the maximum communication distance is very often not well predicted by comparing the minimum auditory threshold with the pattern of attenuation of signals, because many animals communicate in aggregations that generate high levels of background noise. In such cases, detection of the pertinent properties of the signals of other conspecific individuals may be limited to a few close neighbors, even though the signals of many other individuals are well above auditory threshold (e.g., Gerhardt and Klump 1988b; Römer 1993). Therefore, investigators who conduct field playbacks should characterize both the acoustic properties of experimental stimuli and the background noise levels at the point where the target animal was situated at the time of the experimental trial.

3
Laboratory Analysis of Animal Sounds

By taking two simple precautions, two basic errors can be avoided when analyzing bioacoustic signals. First, the investigator should check the transport speeds of both the recorder used to make the original recordings and the recorder used to play back the signals into the instrument to be used for analysis. Obviously, errors in measurements of frequency and time will result from any discrepancies. Second, it is important to remember that both harmonic and other kinds of distortion can be introduced if the input circuitry of the analysis instrumentation is overloaded. The slow response time of the VU meter of the analog Kay Sona-Graph, for example, has frequently resulted in overloading and the production of spurious harmonics in sonograms of animal sounds of short duration (Watkins 1967).

3.1
Terminology

The spectacular diversity of the acoustic signals of animals dooms any attempt to create a definitive, uniform scheme for classifying all important acoustic properties of these sounds. Some animals produce single, short sounds of a few milliseconds duration at very long, highly irregular intervals. Other animals produce nearly continuous sounds that may last for minutes or even hours. Moreover, the sounds that animals produce over long time-periods may consist of uniform, repeated acoustic elements or highly diverse elements arranged in complex temporal sequences (both cases are well represented in bird song, for instance; see Figure 3 for examples from insects). These extremes exemplify the difficulty of deciding what elements constitute basic gross-temporal properties of animal signals such as duration and repetition rate. Thus, most authors who have proposed some form of universal scheme for describing animal sounds have started with the shortest acoustic units.

Broughton (1963), who primarily studied grasshoppers and katydids, suggested that the fundamental acoustic unit of animal sounds is the *pulse*, which he defined as a simple *wave-train*. Workers describing the vocalizations of frogs and toads have also frequently labeled short, repeated wave-trains as pulses (e.g., Duellman and Trueb 1986), although relatively long (i.e., > 30 ms) pulses are often referred to as *notes* or *calls*. The basic unit of birdsong is usually referred to as a note or syllable (e.g., Thompson et al. 1994).

Broughton (1963) proposed a hierarchy of pulsatile structures: (1) a simple pulse, (2) a simple pulse-train, and (3) complex pulse-trains. Complex pulse-trains were further classified as being either straight (the simple pulse-trains are uniform), or figured (the simple pulse trains are patterned). Each of these basic units, from a simple pulse to a figured, complex pulse-train, potentially constituted a chirp, which Broughton (1963, p. 12) defined as the "shortest unitary rhythm-element that can be readily distinguished as such by (the) unaided human ear." However, in my opinion, classifications based on subjective impressions should be avoided. Indeed, Alexander (1962) classified the pulse patterns of cricket songs solely on the basis of the physical properties (time patterns) of the songs (see also Greenfield and Shaw 1982). He recognized about 22 patterns that were grouped under two arbitrary categories: trilled songs and chirping songs. Chirps were defined as short groups of pulses produced in sequence (Figure 3). Pulse-trains (or trills) produced by frogs and toads have typically been labeled as calls or notes (e.g., Blair 1964) or simply as notes (e.g., Littlejohn 1965).

Alternative schemes have classified acoustic units in terms of basic mechanisms of sound production. For example, the *syllables* of insect song (not to be confused with syllables in birdsong) have been defined as single stridulatory (to-and-fro) movements

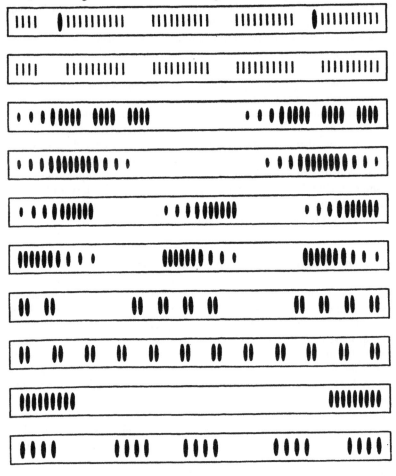

Fig 3. Schematic representation of the temporal patterning of pulses in the songs of chirping crickets. The relative intensity of pulses is indicated by the thickness of the vertical lines. Modified from Alexander (1962)

of the apparatus. Elsner and his colleagues have developed sophisticated ways of relating such movements not only to basic sound patterns in insect song, but also to the activation of single muscles and motoneurons (review in Elsner 1994). More recently, McLister et al. (1995) used the term *note* to label any acoustic unit or units that are produced by a single cycle of airflow back and forth from the lungs and vocal sacs of anuran amphibians.

Thompson et al. (1994) surveyed more than 100 papers that describe bird vocalizations, and found a surprising amount of agreement about the methods to distinguish between different sound elements. In a manner similar to that adopted by students of insect and frog sounds, a common first step is to define units in the time-domain on the basis of the duration of signals and time intervals between signals (the *temporal method*).

Morphological methods involve a spectral (i.e., frequency) analysis of units generated by the temporal analysis (illustrated in Figure 4, and discussed in detail by Beeman, this Volume). Because birds typically produce numerous syllable- or note-types, these sounds are differentiated most readily by the pattern of frequency modulation evident in sonograms. Another level of description involves characterizing sequences of such units derived from the sonographic analysis. This task is not a trivial one when it involves the acoustic output of songbirds, such as mockingbirds and wood thrushes, or marine mammals, such as humpback whales.

In summary, some researchers relate particular acoustic units to mechanisms of sound production, while others define acoustic properties in terms of perceptually relevant patterns. Still other investigators use acoustic characters for phylogenetic analyses. These diverse goals inevitably lead to different ways of defining, describing, and labeling particular acoustic structures. Thus, a description of typical exemplars of each animal signal should include labeled oscillograms, sonograms, or both. The use of such figures avoids the confusion and errors that can arise from the assumption that a particular term always refers to the same acoustic unit.

3.2
Temporal and Spectral Analysis: Some General Principles

Extracting information about the physical properties of sounds usually involves visual representations either in the temporal or spectral domains. In a temporal analysis, usually accomplished with an oscilloscope or its digital equivalent, the amplitude of the signal is displayed as a function of time. The acoustic properties to be measured from such a display include: duration, rise-fall characteristics (shapes, times), patterns of amplitude modulation (if any), and repetition rate. Such measurements are repeated at each level of temporal organization, as in the pulse trains (discrete groups of pulses) commonly produced by insects and frogs.

In a spectral analysis, the relative amplitudes of the frequency components of a signal are displayed as a function of frequency. A continuous signal having a sinusoidal waveform (i.e., a pure tone) will have a spectrum consisting of a single component, the frequency of which is equal to the reciprocal of its period. If a nonsinusoidal waveform is periodic or quasi-periodic, the spectrum will consist of a series of discrete frequency components (harmonics, or harmonics plus sidebands-see below) and these may

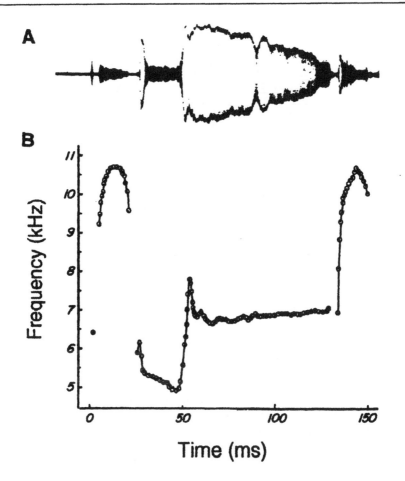

Fig 4. A Oscillogram of a song phrase from a brown-headed cowbird (*Molothrus ater*); B a display of instantaneous frequency on the same time base. (Modified from Greenewalt 1968).

change in frequency and relative amplitude with time. Such complex patterns can be visualized with so-called *waterfall* displays, such as the example in Figure 5, which depicts a bat's echolocation pulse. White noise and impulses are idealized signals that have a continuous spectrum, i.e., their long-term, averaged spectrum is uniform, with equal energy present at every frequency. Animal sounds only approximate these ideal signals, which can be readily characterized mathematically. As the examples below show, the acoustic terminology applied to idealized, artificial signals can still be used profitably to describe the temporal and spectral patterning of animal signals.

The spectra of sounds that are rapidly and repetitively amplitude or frequency modulated (AM or FM, respectively) are also complex. In the simplest case, a single component of fixed frequency (the *carrier* or *dominant frequency*) is sinusoidally amplitude-modulated at a constant rate (Figure 6A). The spectrum then consists of three compo-

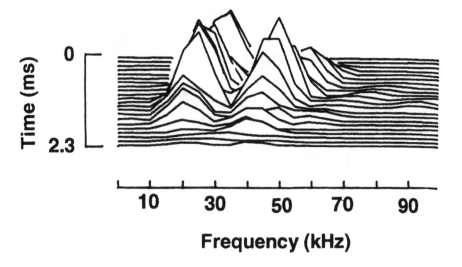

Fig 5. Waterfall display of an echolocation pulse of a bat, *Plecotus phyllotis*. Notice that the amplitude-time profiles of the two harmonics change differently with time. Waterfall displays are especially useful in the analyses of signals with complex, time-varying spectra. Modified from Simmons and O'Farrell (1977)

nents: the carrier frequency and two *sidebands*, the latter with frequencies above and below that of the carrier. The frequency interval between the carrier and each sideband corresponds to the rate of modulation. If the frequency of the carrier remains constant but the rate of amplitude modulation changes with time, the interval between the carrier and frequency and the sidebands changes accordingly (Figure 6B). If the carrier frequency changes but the rate of modulation stays constant, the sideband interval remains constant and all three components rise and fall in a parallel fashion (see Figure 6C).

Sidebands are easily distinguished from harmonics when either the rate of amplitude modulation or the carrier frequency varies with time. By definition, harmonics must have frequencies that are integer multiples or fractions of the frequency of the carrier wave. In the example of Figure 6a, where the frequencies of the carrier and sidebands remain constant, the sidebands could be confused with harmonics. However, the carrier frequency is 1000 Hz and the rate of amplitude modulation is 150 Hz. To interpret these components as three harmonics, it would be necessary to assume that the fundamental frequency of this sound is 50 Hz (the largest common denominator of 850, 1000, and 1150 Hz) and that all other harmonics of the 50-Hz fundamental frequency have been eliminated by some filtering mechanism. Such filtering of the fundamental and lower harmonics occurs in some songbirds (Greenwalt 1968).

Although it is possible for an amplitude-modulated sound to have a suppressed carrier frequency, in simpler cases the magnitude of the sidebands relative to that of the carrier depends directly on the degree (depth) of AM, i.e., the percentage difference between the amplitude peaks and troughs in the overall temporal waveform. Moreover,

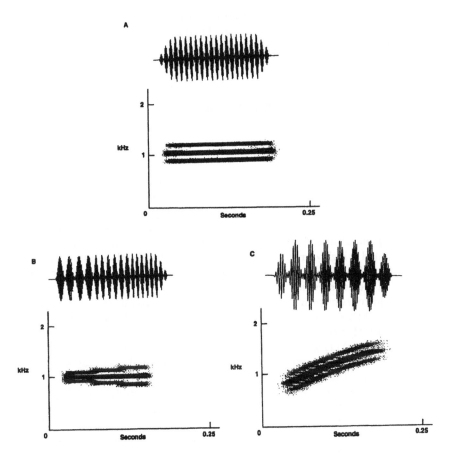

Fig 6. A Oscillogram (*above*) and sonogram (*below*) of a sinusoidally amplitude-modulated tone burst. The carrier frequency was 1000 Hz, and the rate of amplitude modulation was 150 Hz. The sonogram shows the carrier frequency of 1000 Hz and two sidebands, one at 850 Hz and the other, at 1150 Hz. Note that the frequency interval between the carrier frequency and the sideband components is 150 Hz. The three-component spectrum can be mathematically derived as a multiplication of two sinusoids (see Greenewalt 1968, p. 89); in an equivalent fashion, the addition of the three components shown with appropriate amplitudes and phase relationships yield the amplitude-time envelope shown in the oscillogram. Although the sonogram superficially resembles part of a harmonic series, the three frequencies are not integral multiples of 150 Hz. See the text for additional discussion. B Oscillogram (*above*) and sonogram (below) of a sinusoidally amplitude-modulated tone burst in which the rate of amplitude modulation was changed from 80 Hz to 200 Hz. The frequency interval between the carrier frequency (1000 Hz) and the sidebands increases accordingly. C Oscillogram (*above*) and sonogram (*below*) of a sinusoidally amplitude-modulated tone burst in which the carrier frequency was swept from 900 Hz to 1200 Hz. Because the rate of amplitude-modulation was held constant at 150 Hz, the intervals between the two sideband frequencies and the carrier frequency remain constant

non-sinusoidal amplitude-modulation results in multiple sidebands, as does both sinusoidal and non-sinusoidal frequency-modulation of a repetitive kind, like vibrato. Greenewalt (1968) and Beecher (1988) provide useful discussions of spectral analyses and modulation phenomena in a bioacoustic context.

4
Examples of Descriptions and Analyses

4.1
Temporal Properties of Pulsatile Calls

The spring peeper (*Pseudacris crucifer*) produces single, nearly pure-tone pulses at irregular intervals (Figure 7); each pulse or call sweeps upward in frequency by about 400 Hz from beginning to end. The bird-voiced treefrog (*H. avivoca*) also produces tonal pulses, but organizes them into discrete trains (as in Figure 8). The pine woods treefrog (*H. femoralis*) produces short pulses in groups of four to six; each of these units is then repeated at highly irregular intervals (Figure 9). In all of these species, the basic unit, which can be defined as a pulse, has a temporal (and spectral) morphology that is uniform and species-specific.

Interval histograms provide a useful summary of the complexity of the temporal organization of basic acoustic units, defined in terms of their duration, spectral properties, or both. Moreover, plotting intervals as a frequency distribution on a log time scale (see Figure 10) makes it possible to visualize the modes and the variability of each kind of interval, and to compare quantitatively the temporal organization of a wide variety of signals. The spring peeper's calls are repeated at irregular intervals, and an inter pulse histogram based on 45 such calls shows a single modal peak (Figure 10A). The bird-voiced treefrog's organization of pulses into trains that are repeated at relatively longer intervals results in a histogram with two modes (Figure 10B). The first corresponds to the intervals between pulses and the second corresponds to the intervals between pulse-trains. The intervals between the short pulses in the calls of the pine woods treefrog are highly regular. However, in contrast to those of the bird-voiced treefrog, these pulse-trains are repeated at highly variable intervals (Figure 10C).

4.2
Amplitude-Time Envelopes

Each acoustic unit considered in the preceding section has a distinctive amplitude-time envelope (the overall shape of the waveform over time). A common way of characterizing this envelope is to determine its *rise-fall characteristics*. For example, rise-time is usually defined as the interval between the time when the amplitude of the sound reaches 10 % of its maximum value and the time when it reaches 90 % of its maximum amplitude. The shape or form of the envelope is also of interest. A linearly rising envelope is one in which the amplitude reaches 50 % of its peak value at one-half of the time that it reaches maximum peak amplitude (also true for a sigmoidally shaped rising envelope). The second pulse of the bird-voiced treefrog's call in Figure 8c has an approximately linear onset. For an exponentially rising envelope, the amplitude reaches half of the maximum at a point in time greater than 50 % of the time required to reach maximum amplitude (as shown in the first pulse of Figure 8C). In an inverse exponential (logarithmic) form of rise, the half-maximum amplitude occurs at a time less than 50 % of the time of peak amplitude (illustrated by the pine-woods treefrog pulse shown in Figure 9D).

The amplitude-time envelopes of the pulses and longer acoustic units of many, but not all, animal signals can be approximated by one or the other of these shapes. Envelopes may, for instance, be combinations of linear, exponential, or logarithmic functions, or they may just be highly irregular. Moreover, as in the bird-voiced treefrog's calls, the envelope shape (function) may even change from one pulse or pulse-train to the next. Finally, envelope shapes and rise times can vary significantly with distance because of changes in phase relationships among components in spectrally complex signals. Such changes, in turn, arise through interactions between direct and reflected sound waves.

4.3
Relationships between Fine-Scale Temporal and Spectral Properties

The approximately sinusoidal nature of the pulses of some of the species considered above makes it possible to measure the periods (durations) of successive cycles, for instance using a *zero-crossing* protocol (e.g., Figure 4B). The reciprocal of these periods is the *instantaneous frequency*. An analysis of a single pulse of the pine woods treefrog (*H. femoralis*) resulted in a mean period of 0.45 ms, or 2226 Hz, with instantaneous frequencies ranging from 2045 to 3030 Hz. Spectral analysis of one pulse produced an estimate of 2190 Hz as the carrier frequency (Figure 9F). However, the frequency resolution of spectral analysis is typically limited by signal duration (the so-called *uncertainty principle*). Moreover, the practical resolution of frequency will generally be somewhat less, depending on the effective bandwidth of the analysis and other factors (see discussions in Beecher 1988, and both Beeman and Owren and Bernacki, this Volume).

Greenewalt (1968) and Beecher (1988) recommend a period-by-period temporal analysis for short signals that are rapidly modulated in frequency. However, the perceptual significance of such a fine-grained analysis is dubious because both auditory systems and spectrum analyzers perform frequency analyses by averaging over some finite time interval. Very short sounds often do not have a very well-defined pitch. For a single pulse of the pine woods tree frog, for example, the bandwidth (-10 dB) is about 600 Hz (Figure 9F). This value is obtained by determining the frequencies above and below the peak frequency (2190 Hz, 0 dB) that have a relative amplitude of -10 dB, and subtracting the lower frequency from the higher one.

The pulses in the calls of the pine woods treefrog are grouped into trains of four to six pulses, with a mean period of about 8.4 ms (Figure 9C). When a spectral analysis is performed to average over an entire pulse group, this periodicity is obvious in the spectrum (Figure 9E). Here, the spectral peaks on either side of the carrier frequency are separated by intervals of about 110 to 120 Hz, close to the reciprocal of the mean pulse period (8.4 ms) of 119 pulses/s.

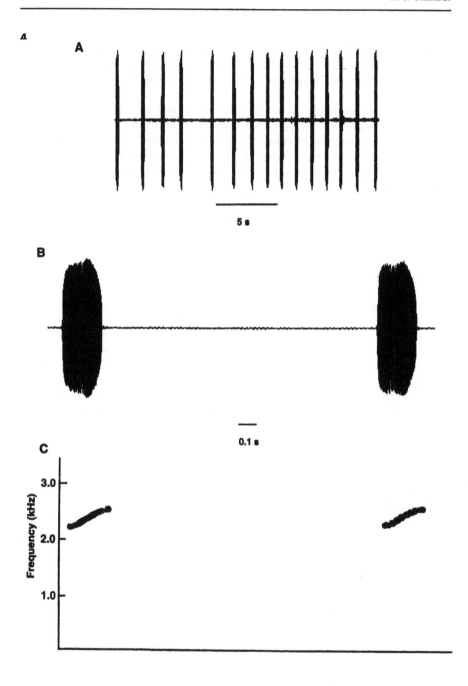

Fig 7. A Oscillogram of long pulses produced by the spring peeper (*Pseudacris crucifer*) over a period of 20 s B Oscillogram with an expanded time base to show the amplitude-time envelopes of two pulses. C Narrow band (45 Hz) sonograms of the two pulses of B; the time-scale is the same. Notice that the pulse is frequency-modulated, rising in frequency by about 380 Hz from beginning to end

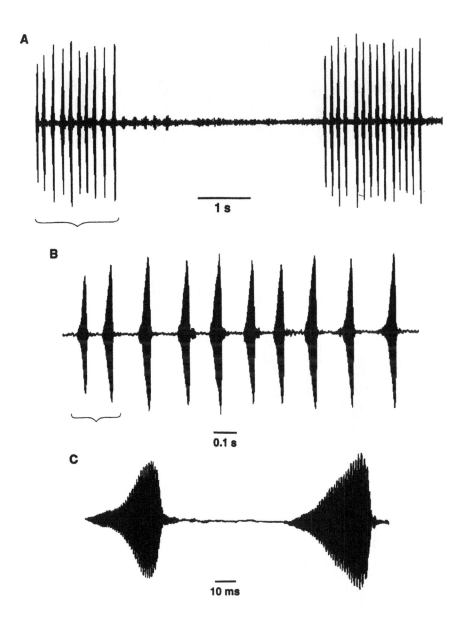

Fig 8. A Oscillogram showing two pulse trains produced by the bird-voiced treefrog (*Hyla avivoca*). B Oscillogram with an expanded time-base to show the first of the two pulse trains in A. C Oscillogram with an expanded time-base to show the first two pulses of the pulse train in B Notice that the onset of the first pulse is distinctly exponential, whereas that of the second pulse is more nearly linear

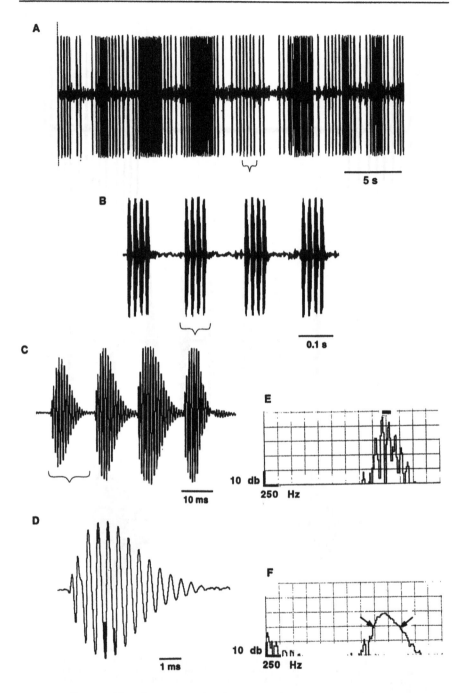

Fig 9. A Oscillogram of the pulse trains produced over a period of about 20 seconds by a pine woods treefrog (*Hyla femoralis*). B Oscillogram with an expanded time-base to show four pulse trains from A. C Oscillogram with an expanded time-base to show one pulse train from A. D Oscillogram with an expanded time-base to show the first pulse from C. E Spectrum generated by averaging over the entire pulse train of C. The interval between the frequency components (about 120 Hz; indicated by the horizontal bar) is equal to the reciprocal of the pulse period. F Spectrum of a the first pulse of the train. Notice that there is only a single, broad spectral peak. The arrows indicate the extent (bandwidth) of the spectrum within 10 dB of its peak amplitude

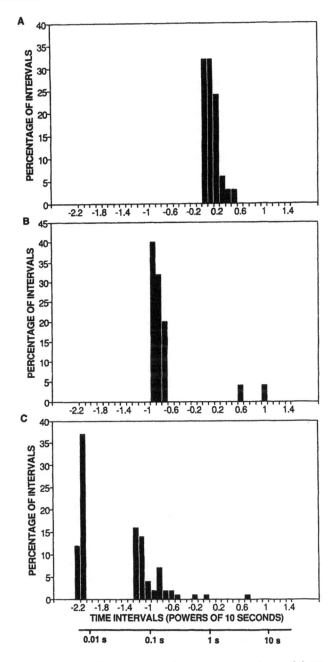

Fig 10. A Interval histogram for 45 pulses produced by a spring peeper (*P. crucifer*). B Interval histogram for eight pulse trains produced by a bird-voiced treefrog (*H. avivoca*). C Interval histogram showing pulse patterning during one minute of calling by a pine woods treefrog (*H. femoralis*). The X-axis for each panel is shown as powers of 10 seconds; a more conventional scale of time intervals is shown at the bottom of the figure

4.4
Spectrally Complex Calls

The calls of three closely – related species of treefrogs — the green (*H. cinerea*), barking (*H. gratiosa*) and pine barrens treefrogs (*H. andersonii*) — range in duration from about 120 to 200 ms, and are repeated at rates of one to two calls per second (Figure 11). The main differences among the species are in the amplitude-time envelopes and spectral properties of the calls. The spectral properties, in turn, can be related to the fine-temporal analysis of the waveform. The amplitude-time envelopes of representative calls of each species are shown in the left panel of Figure 11. While the rise-fall characteristics of each are species-specific, there are also differences within and between individuals of the same species (e.g., Figure 11C and 11D). The pulsatile beginning of the call is a relevant characteristic for female barking treefrogs, for instance, but not for female green treefrogs (Gerhardt 1981a). Indeed, female green treefrogs respond as well to playback of natural calls presented backwards (so that the pulsatile part of the call ends the signal) as they do to natural calls played normally.

Oscillograms with an expanded time base are also included in Figure 11, in order to show fine details of the waveforms. In each case, two cycles of the waveform from the last one-third of the call are shown. Although slight cycle-to-cycle differences occur in these waveforms (see Capranica 1965, for a similar analysis of bullfrog croaks), the (quasi-)periodicity of each is evident in the corresponding spectra in the even spacing (in Hz) of their frequency components. For example, the periods of the repeating waveforms for the calls of the barking and green treefrogs are about 2.4 and 3.15 ms, respectively, and the spectral components in their calls are spaced at frequency intervals equal to the reciprocals of these periods, i.e., at about 420 and 317 Hz, respectively.

A comparison of the spectra of these signals (shown both in the rightmost panel of Figure 11 and in Figure 12) indicates that in barking and green treefrogs there are two bands of emphasis: a low-frequency band, represented by a single component, and a high-frequency band, represented by one to two components. In the barking treefrog, the spectrum of most of the signal comprises a harmonic series. The fundamental frequency (or first harmonic) is about 420 Hz, and it changes slightly from beginning to end of the call (Figure 11A). As all of the components of higher frequencies are integer multiples of the fundamental frequency, the slight frequency modulation of the fundamental is multiplied such that the higher harmonics have a higher absolute degree of frequency modulation.

In the calls of green and pine barrens treefrogs, the spectra could also be described as harmonic series. For instance, in green treefrogs there is a fundamental frequency at about 950 Hz, a second harmonic at 1900 Hz, and a strongly emphasized third harmonic at about 2850 Hz. However, there are other components in the spectrum that arise through modulation. This effect is clearly seen in the calls of the pine barrens treefrog. Here, the emphasized components are the fundamental and second harmonic (arrows in Figure 12E). In the calls of this species, changes in overall rate of modulation of the waveform often occur from call to call (Figures 11C and 11D), or even within a call (Figures 12D and 12E). The mismatch in the frequency-time profiles of the sidebands and those of the fundamental and second harmonics are obvious. The fact that the rate of modulation is usually an integer fraction of the fundamental frequency in the calls of

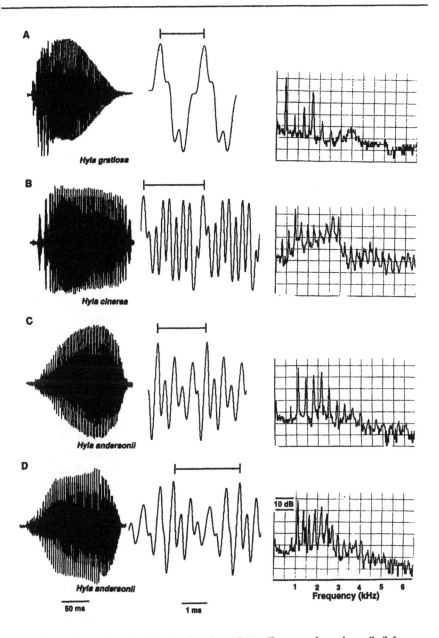

Fig 11. Temporal and spectral properties of treefrog calls. Oscillograms of complete calls (left panels), oscillograms with the time-base expanded (middle panels) to show two cycles of the repeating waveform (the period of one cycle is indicated by the horizontal bars) from the middle of the last third of each call, and amplitude-frequency spectra (right panels) based on averaging over about 10 cycles of the repeating waveform from the middle of the last third of the call. A A representative call of a barking treefrog (*H. gratiosa*). B A representative call of a green treefrog (*H. cinerea*). C A representative call of a pine barrens treefrog (*H. andersonii*). D Another call of the pine barrens treefrog, produced immediately after the call illustrated in C. Note the difference in the amplitude-time waveform, repeating waveform, and spectrum between the two successive signals.

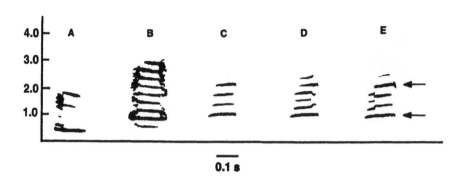

Fig 12. Narrow band sonagrams of representative calls of A *H. gratiosa, H.* B *cinerea;* and *H. andersonii* (C-E). The differences in the frequency-time profiles of harmonics (indicated by arrows in E) and sidebands are especially evident in the call variants of *H. andersonii;* the calls labeled C and D have the same letter designations in Fig 11. See the text for details

both green and pine barrens treefrogs suggests that whatever structures are responsible for modulation are likely to be physically coupled to the structures that generate the components making up the complex carrier waveform (Oldham and Gerhardt 1975).

5
Summary

This chapter provides some general guidelines for the measurement and recording of animal sounds in the field and laboratory, beginning with frequency-response characteristics of the recording equipment, considerations about tape transport speed, and the common mistake of overloading of input amplifiers (thereby generating distortion in the form of spurious frequency components). There are a number of compelling reasons to measure amplitudes of animal sounds under natural conditions, but such measurements must take into account microphone orientation and inherent directionality of the sound source itself, the particular weighting-network setting and time constant of the sound-level meter used, and the distance of the microphone from the animal. Patterns of sound propagation in natural habitats produce frequency-dependent effects on animal signals, suggesting that a conservative approach involving multiple recordings is usually warranted.

Laboratory analyses of animal sounds usually include description of both temporal and spectral patterning. Although adoption of a universal system of terminology for such descriptions of animal sounds is unlikely, a first-order description should be based on the physical properties of the signals rather than perceptual or (presumed) functional criteria. Where this has been done, there is wide agreement about the temporal and spectral structures that should be measured and described, even if the terminology has varied, especially across taxa. In the temporal domain, the acoustic properties measured typically include duration of units, rise-fall characteristics of the waveform, patterns of amplitude modulation, and repetition rates. In the spectral domain, periodicities in the underlying waveform are shown by the concentration of energy in the carrier frequency or frequencies. Harmonic components are integer multiples of the fundamental frequency of vibration of the sound source, and if the waveform is modulated, sidebands occur around the carrier frequency or frequencies at intervals equal to the rate of modulation. In addition, some animals produce noisy (aperiodic) or very short (impulsive) sounds, which show more continuous frequency spectra. Each of these effects can be demonstrated in the calls of insect, frogs, and birds.

Acknowledgments I thank S. Hopp and C. Evans for inviting me to contribute to this volume. M. Owren made many useful editorial changes, and G. Klump, C. Hopp, F. Huber M. Jørgensen, and C. Murphy also made helpful comments on the content and style of the manuscript. The National Science Foundation and National Institutes of Mental Health have provided generous research support to my research program.

References

Alexander RD (1962) Evolutionary change in cricket acoustical communication. Evolution 16: 443–467

Beecher MD (1988) Spectrographic analysis of animal vocalizations: implications of the "uncertainty principle". Bioacoustics 1: 187–208

Bennet-Clark H (1989) Songs and the physics of sound production. In: Huber F, Moore TE, Loher W (eds) Cricket behavior and neurobiology. Cornell University Press, Ithaca p 227

Blair WF (1964) Evolution at populational and interpopulational levels: isolating mechanisms and interspecies interactions in anuran amphibians. Q Rev Biol 39: 77–89

Brenowitz EA, Rose GJ (1994) Behavioural plasticity mediates aggression in choruses of the Pacific treefrog. Anim Behav 47: 633–641

Broch JT (1971) The application of Brül and Kjaer measuring systems to acoustic noise measurements, 2nd edn, Larsen Borg, Denmark

Broughton WB (1963) Method in bio-acoustic terminology. In: Bushnel R-G (ed) Acoustic behaviour of animals. Elsevier, Amsterdam, p 3

Capranica RR (1965) The evoked vocal response of the bullfrog. MIT Press, Cambridge, Massachusetts.

Crocker MJ (1991) Measurement of sound intensity. In: Harris CM (ed) Handbook of acoustical measurements and noise control. McGraw-Hill, New York, p 14.1

Crocroft RB, Ryan MJ (1995) Patterns of advertisement call evolution in toads and chorus frogs. Anim Behav 49: 283–303

Dooling RJ (1982) Auditory perception in birds. In: Kroodsma DE, Miller EH, Ouellet H (eds) Acoustic communication in birds. vol 1. Academic Press, New York, p 95

Duellman WE, Trueb L (1986) Biology of amphibians. McGraw-Hill, New York

Dunia R, Narins PM (1989) Temporal integration in an anuran auditory nerve. Hear Res 39: 287–298

Elsner N (1994) The search for the neural centers of cricket and grasshopper song. In: Schildberger K, Elsner N (eds) Neural basis of behavioural adaptations. Gustav Fischer, Stuttgart, p 167

Embleton TFW, Piercy JE, Olson N (1976) Outdoor sound propagation over ground of finite impedance. J Acoust Soc Am 59: 267–277

Ewing AW (1989) Arthropod bioacoustics: Neurobiology and behavior. Comstock, Cornell, Ithaca

Falls B (1982) Individual recognition of sound in birds. In: Kroodsman DE, Miller EH, Ouellet H (eds) Acoustic communication in birds, vol. 2. Academic Press, New York, p 237

Forrest TG, DM (1991) Power output and efficiency of sound production by crickets. Behav Ecol 2: 327–338

Forrest TG, Raspert R (1994) Models of female mate choice in acoustic communication. Behav Ecol 5: 293–270

Gerhardt HC (1975) Sound pressure levels and radiation patterns of the vocalizations of some North American frogs and toads. J Comp Physiol 102: 1–12

Gerhardt HC (1981a) Mating call recognition in the barking treefrog (*Hyla gratiosa*): responses to synthetic calls and comparisons with the green treefrog (*Hyla cinerea*). J Comp Physiol 144: 17–25

Gerhardt HC (1981b) Mating call recognition in the green treefrog (*Hyla cinerea*): importance of two frequency bands as a function of sound pressure level. J Comp Physiol 144: 9–16

Gerhardt HC (1991) Female mate choice in treefrogs: static and dynamic acoustic criteria. Anim Behav 42: 615–635

Gerhardt HC, Klump GM (1988a) Phonotactic responses and selectivity of barking treefrogs (*Hyla gratiosa*) to chorus sounds. J Comp Physiol A 163: 795–802

Gerhardt HC, Klump GM (1988b) Masking of acoustic signals by the chorus background noise in the green tree frog: a limitation of mate choice. Anim Behav 36: 1247–1249

Greenewalt CH (1968) Bird song: acoustics and physiology. Smithsonian Institution Press, Washington, DC

Greenfield MD, Shaw, KC (1982) Adaptive significance of chorusing with special reference to the Orthoptera. In: Morris GK, Gwynne DT (eds) Orthopteran mating systems: Sexual competition in a diverse group of insects. Westview Press, Boulder, p 1

Harris CM (1991) Definitions, abbreviations, and symbols. In: Harris CM (ed) Handbook of acoustical measurements and noise control. McGraw-Hill, New York, p 2.1

Hassall JR (1991) Noise measurement techniques. In: Harris CM (ed) Handbook of acoustical measurements and noise control. McGraw-Hill, New York, p 9.1

Heller KG (1990) Evolution of song pattern in east Mediterranean Phaneropterinae: constraints by the communication system. In: Bailey WJ, Rentz DCF (eds) The Tettigoniidae: Biology, systematics and evolution. Crawford House Press, Bathurst, p 130

Langbauer WR, Payne KB, Charif RA, Rapaport L, Osborn F (1991) African elephants respond to distant playbacks of low-frequency conspecific calls. J Exp Biol 157: 35–46

Littlejohn MJ (1965) Premating isolation in the *Hyla ewingi* complex (Anura:Hylidae). Evolution 19: 234–243

Loftus-Hills JJ, Littlejohn MJ (1971) Mating-call sound intensities of anuran amphibians. J Acoust Soc Am 49: 1327–1329

McLister JD, Stevens ED, Bogart JP (1995) Comparative contractile dynamics of calling and locomotion muscles in three hylids frogs. J Exp Biol 198: 1527–1538

Michelsen A (1978) Sound reception in different environments. In: Ali MA (ed) Sensory ecology. Plenum Press, New York, p 345

Michelsen A (1983) Biophysical basis of sound communication. In: Lewis B (ed) Bioacoustics: a comparative approach. Academic Press, London, p 3

Michelsen A, Towne WF, Kirchner WH, Kryger P (1987) The acoustic near field of a dancing honeybee. J Comp Physiol A 161: 633–643

Oldham RS, Gerhardt HC (1975) Behavioral isolation of the treefrogs *Hyla cinerea* and *Hyla gratiosa*. Copeia 1975: 223-231

Peterson AP, Gross P (1972) Handbook of noise measurement, 7th edn. General Radio Company, West Concord, Massachusetts

Piercy JE, Daigle GA (1991) Sound propagation in the open air. In: Harris CM (ed) Handbook of acoustical measurements and noise control. McGraw-Hill, New York, p 3.1

Prestwich KN, Brugger KE, Topping MJ (1989) Energy and communication in three species of hylid frogs: power input, power output and efficiency. J Exp Biol 144: 53–80

Römer H (1993) Environmental and biological constraints for the evolution of long-range signalling and hearing in acoustic insects. Philos Trans R Soc and B Biol Sci 340: 179–185

Simmons JA (1969) Acoustic radiation patterns for the echolocating bats *Chilonycteris rubiginosa* and *Eptesicus fuscus*. J Acoust Soc Am 44: 1054–1056

Simmons JA, O'Farrell MJ (1977) Echolocation by the long-eared bat, *Plecotus phyllotis*. J Comp Physiol A 122: 201–214

Skudrzyk E (1971) The foundati•s of acoustics: basic mathematics and basic acoustics. Springer, Berlin Heidelberg New York

Thompson NS, LeDoux K, Moody K (1994) A system for describing bird song units. Bioacoustics 5: 267–279

Wagner WE Jr. (1989) Social correlates of variation in male calling behavior in Blanchard's cricket frog, *Acris crepitans blanchardi*. Ethology 82: 27–45

Wagner WE Jr. (1992) Deceptive or honest signalling of fighting ability? A test of alternative hypotheses for the function of changes in call dominant frequency by male cricket frogs. Anim Behav 44: 449–462

Wahlström S (1990) DAT- digital audio tape. Bioacoustics2: 344–351

Watkins WA (1967) The harmonic interval: fact or artifact in spectral analysis of pulse trains. In: Tavolga WN (ed) Marine bioacoustics, vol 2, Pergamon Press, New York, p 15

Wells KD (1988) The effect of social interactions on anuran vocal behavior. In: Fritszch B, Wilczynski W, Ryan MJ, Hetherington T, Wakowiak W (eds) The evolution of the amphibian auditory system, Wiley, New York, p 433

Wickstrom DC (1982) Factors to consider in recording avian sounds. In: Kroodsma DE, Miller EH, Ouellet H (eds) Acoustic communication in birds. vol 1. Academic Press, New York, p 1

Wiley RH, Richards DG (1978) Physical constraints on acoustic communication in the atmosphere implications for the evolution of animal vocalizations. Behav Ecol Sociobiol 3: 69–94

Woram JM (1989) Sound recording handbook. Howard Sams, Indianapolis

Yeager DM, Marsh AH (1991) Sound levels and their measurement. In: Harris CM (ed) Handbook of acoustical measurements and noise control. McGraw-Hill, New York, p 11.1

CHAPTER 2

Digital Signal Acquisition and Representation

M. CLEMENTS

1
Introduction

The methods collectively known as *digital signal processing* (DSP) are algorithms that numerically manipulate ordered lists of numbers (*signals*). Twenty-five years ago, these methods were used as expensive simulators of analog electronics. The power and versatility of DSP, coupled with the fast and cheap computation power that is now available, has transformed this approach into the method of choice for a wide range of applications. Strictly speaking, DSP has existed for many years. Many of the techniques used today were well understood by mathematicians such as Gauss and Euler. Numerical methods for solution of differential equations and evaluation of integrals also fall under the DSP umbrella, and many claim that it was the need for solving such problems quickly and efficiently that led to the development of the digital computer in the first place. This chapter presents the simple ideas that form the basis of DSP and explores various interpretations of digital signals from both time- and frequency-related points of view.

2
Digital Signal Processing

The two components present in any DSP application are the signals and the systems. As noted above, signals are ordered lists of numbers. Consider the sequence of average daily temperature readings at a particular locale. Probably, these numbers are written down on a list somewhere, or reside in a form accessible to a computer. An important point is that there is usually some regular spacing between consecutive entries in such a sequence (in this case, 1 day). We can therefore call such a list a *discrete-time signal*. Another critical point is that the entries are of only finite precision, meaning that the written list or a comparable computer file uses only a certain number of digits for each number. This finite-precision representation is the origin of the descriptor "digital." Signals without time-dependency, such as digital images, are no less useful. Here, the ordered lists form two-dimensional matrices, with numbers in particular locations in a matrix representing the brightnesses of corresponding picture elements (pixels) on a video screen.

Acquisition of signals is often performed through a process known as *analog-to-digital (A/D) conversion*, which performs the sampling (i.e., making a regular sequence of measurements) and quantization (i.e., representing the numbers by a specific number of bits) necessary for computer representation. *Digital-to-analog (D/A) conversion* allows signals that have been acquired or generated digitally to be presented or stored in analog form. Hardware exists to perform just such operations.

Systems are objects that accept one or more signals as input and produce one or more signals as output. Ideally, the output should be a transformation of the input such that the former is more useful to the particular application at hand than is the latter. If the signals are digital in nature, then the systems are called *digital systems*. It is easy to see why *digital systems* are popular: any recipe for manipulating the input sequences can be implemented — there are virtually no restrictions.

2.1
Major Applications of DSP

As can be imagined, virtually any acoustic, geophysical, optical, or electrical signal can be processed digitally. Therefore, rather than tying the field to any particular signal class, this section will discuss some the uses of DSP from a general point of view.

A major advantage of the processing approach is that the signals are often available in a random access mode — when large amounts of data are stored, quick and efficient retrieval of the particular portion of interest is possible. In analog recordings, in contrast, the desired signal segment may be in the middle of a tape and its retrieval may be time-consuming or even impossible. As signals are often predictable in structure and include significant redundancy, digital processing algorithms that take advantage of these attributes can also be used to dramatically reduce storage capacity requirements. Further, since copying digital data involves simply transferring bits of information, signals can be copied over and over again with virtually no degradation. Similarly, digital signals can be transmitted across phone lines, via satellite, or through cables without information loss. Successful transfer can be ensured even in environments with high interference simply by slowing down the transmission. The digital nature of the transmitted signals also permits virtually unbreakable encryption of the data.

Another strength of DSP is that, although a given signal sequence may have an implicit time dependency, once it has been recorded, the strict time-precedence relationship need not be maintained in further processing. For example, it is perfectly valid to process a given signal sequence in reverse order, or to "look ahead" in the sample list. This general property is extremely important in separating signals from noise, keeping signal values within acceptable ranges, tracking discrete frequency components, and the like.

Perhaps the most popular DSP tools are those that operate in the frequency domain. While there are a number of mathematical and physical reasons for their usefulness, one of the most compelling is biological – the hearing mechanisms of higher vertebrates perform similar analyses, and therefore extraction of biologically significant information from these signals for scientific purposes requires analogous digital techniques. Examples of frequency-domain analysis tools include Fourier transforms, filter banks, spectrograms, linear prediction, and homomorphic techniques.

2.2
Definition of Digital Systems

In this section, a notation will be adopted for describing the various components of digital systems. Considering only a single-input, single-output digital system, $x(n)$ and $y(n)$ represent the input and output components, respectively. Both $x(n)$ and $y(n)$ are sequences of numbers in which n is a time or sequence index that is always an integer. Since $y(n)$ is a transformation of $x(n)$, we can also refer to it as T $[x(n)]$.

Signals of interest may be inherently digital (e.g., population statistics) or they may come from digitizing analog signals (e.g., fluctuating voltage values recorded from a microphone). To see some of the advantages of digital signal processing, consider a vocalization recorded from the spring peeper frog, *Pseudacris crucifer* (formerly *Hyla crucifer*). If this recording were used as input to an oscilloscope, a display similar to that shown in Figure 1A would be observed. The display shows that distinct bursts of sound occur. In addition, however, listening reveals that the sounds within the bursts are quite distinctive. If the signal is digitized and stored in a computer-accessible form, it can conveniently be segmented and examined in greater detail. For most applications, quantification of important signal properties requires further processing.

Suppose, for example, that we desire to measure the *envelope* of the call, i.e., to the profile of sound energy over time. The analog hardware required for this task would include a *rectifier* (which takes the absolute value of signal voltages) and a *smoother* (which removes the fine structure of the rectified waveform). Both operations are quite easily performed in a digital domain. If $s(n)$ is our digitized recording, the first operation would be to implement a function $x(n)=|s(n)|$, producing the absolute value of each sample in the signal. The second operation would be smoothing [see Eq (1)], for instance by taking a *moving average* of $x(n)$, where each output sample is an average of the

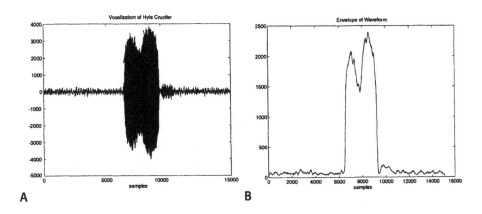

A   B

Fig. 1A, B. The waveform of a call recorded from a aspring peeper, *Pseudacris crucifer*. The amplitude-by-time representation, or oscillogram (A) is rectified and smoothed using a 100-point moving-average filter to produce the amplitude envelope (B) . Waveform energy occurring before and after the call is due to tape-recorder and background noise. The vertical axis depicts arbitrary amplitude units

current input and its previous (M - 1) values (results for the spring peeper vocalization are shown in Figure 1B, with M = 100).

$$y(n) = \frac{1}{M}[x(n) + x(n-1) + ... + x(n-M+1)] .$$

(1)

A fuller discussion of smoothing in terms of frequency units will be presented later.

However, much can be seen from the time response, such as the strength of the primary and secondary bursts. The smoothing operation is one of an important class of systems that are linear-shift-invariant (LSI) and have the following properties. First, if the input is scaled in amplitude, the output is scaled by the same amount. Second, if the input is shifted in time (e.g., delayed), the output is shifted by the same amount. Finally, if two individual input signals are added, the output is the sum of the corresponding individual outputs. All LSI systems can be referred to as filters and can be implemented using the three basic operations of addition of two signal values, multiplication of a signal value by a constant, and delay of a signal value (i.e., storing, or "remembering" its value). Such operations are easily implemented on digital computers.

2.3
Difference Equations

Like all LSI systems, the smoothing filter whose use was illustrated earlier can be described in terms of a difference equation that specifies the implementation of the numerical operations involved. The equation need not explicitly include every term, it can instead be conveniently expressed as a sum where x is the input and y is the output, as shown in Eq. (2).

$$y(n) = \sum_{k=0}^{M-1} b(k)x(n-k).$$

(2)

Here, the filter is of *order* M - 1, meaning that M -1 values must be remembered. In general, the *order* of the filter describes its computational complexity. In this equation, $y(n)$ is the nth (or current) output sample, $x(n - k)$ is the input sample that occurs k units of time (samples) before the current sample, $x(n)$, $b(k)$ is the value by which $x(n-k)$ is multiplied and M - 1 is the order of the filter. In our 100-point smoothing filter, M - 1 was 99 and $b(k)$ was 1 / 100 for all its values. Also note that the smoothing operation required 100 multiplications and 99 summations for every output sample. For the 15,000-sample section depicted, 3,000,000 operations were required. As the original recording was only 2 s in duration, it is clear why digital filtering has only been feasible in recent years. Current technology provides chips costing only a few dollars that are capable of 100 million DSP operations per second.

In general, $b(k)$ can be any set of coefficients. The overall characteristics of such filters are best described in terms of frequency response. Another useful form of the difference equation [shown in Eq (3)] specifies $y(n)$ as a linear combination of the input and its past values, plus a linear combination of the past values of the output. Such filters are recursive, since the output is fed back into the system.

$$y(n) = \sum_{k=0}^{M-1} b(k)x(n-k) - \sum_{k=1}^{N} a(k)y(n-k). \tag{3}$$

Furthermore, consider the smoothing filter implemented by Eq. (4). Here, after an output sample is computed, it is saved for use in computing the next output sample. Applied to the envelope-detection problem described earlier, very similar performance results with only one multiplication and one addition operation per output. Although in this example the recursive approach is much more cost-effective, such is not always the case. The feedback operation, for instance, can sometimes allow the accumulation of small rounding errors to cause problematic inaccuracy in calculation.

$$y(n) = x(n) + 0.98\, y(n-1). \tag{4}$$

3
Digital Filter Frequency Response

These descriptions of recursive and nonrecursive filters have been somewhat qualitative. In both cases, precise specification requires a different set of properties, namely, the frequency responses associated with filter operation.

3.1
Unit-Sample Response Characterization

One common approach to understanding an LSI system is to characterize how it treats an input that consists of a single point. This single point is called a *unit sample* and the output is known as the unit-sample response or sometimes the *impulse response*. An important result is that if this *unit-sample response* is denoted $h(n)$, and the input to the system (or filter) is denoted $x(n)$, then the output $y(n)$ can be computed as shown in Eq. (5).

$$y(n) = \sum_{k=-\infty}^{\infty} x(k)h(n-k). \tag{5}$$

This relation is known as the *convolution* operation and is often abbreviated using the notation $x(n)\, X{*}h(n)$ or $h(n){*}x(n)$, which are equivalent. The essence of convolution is as follows. Each discrete-time signal is comprised of a set of samples. If the system response to a single sample of amplitude (height) 1 at $n = 0$ is known [i.e., $h(n)$], then its response to a sample of height $x(k)$ at $n = k$ is also known (i.e., $x(k)h(n - k)$). In convolution, each of these responses is added. Referring to the difference equation cited above, it can be noted that if all the coefficients $a(k)$ are zero, the unit-sample response will be zero for values of $n > M$. Such filters are known as *finite impulse response* (FIR) or *nonrecursive* filters. In all other cases, the recursive nature of the computation pro-

impulse response (IIR) or *recursive* filters. The terms FIR, IIR, and convolution are important in understanding the descriptions of many common DSP algorithms.

3.2
Frequency-Domain Interpretation of Systems

Another important interpretation of both signals and LSI systems involves their frequency response (which necessarily entails a sequence of samples rather than a single point). First examining filters in this light, an amazing property of LSI filters is that if the input is a *sinusoid* (e.g., a pure tone), the output is a sinusoid of exactly the same frequency. The two aspects of the sinusoid that are changed are its amplitude (height) and its *phase* (e.g., a delay). These two properties depend on the frequency of the input and are collectively described by the filter's *magnitude response* and its *phase response*. Figure 2 shows the frequency response of the recursive smoothing filter described in Eq. (4). As illustrated, the amplitudes of high-frequency components (fast variations) are attenuated. Phase-shifts imposed by the filter are also frequency-dependent, but in a quite different manner than in the case of amplitude.

The variable ω will appear throughout the following discussion, referring to frequency rather than time. The waveform $\cos(\omega n)$ is a discrete-time signal (recall that n is a discrete number) of frequency ω. This frequency has units of radians per sample. The possibly more familiar quantity of cycles per sample is related to radians by a factor of 2π (i.e., radians = cycles $\times$ 2π). In order to describe both magnitude and phase responses efficiently, however, engineers have developed a notation involving complex numbers. Although the notation may seem a bit artificial at first, it can be used to greatly simplify the mathematics involved and therefore appears in almost all DSP-related literature and computer software packages.

In computing the frequency response of a system, it can be shown directly from the convolution sum that an input of the form shown in Eq. (6) produces output as shown

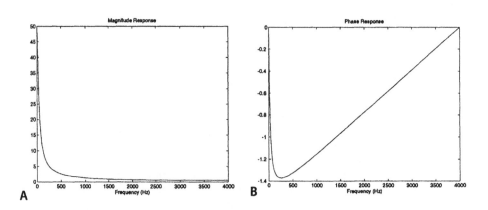

A   B

Fig.2A, B. The magnitude response (A) and phase response (B) of the recursive smoothing filter described in Equation 4. See text for further details

In computing the frequency response of a system, it can be shown directly from the convolution sum that an input of the form shown in Eq. (6) produces output as shown in Eqs. (7) and (8). The complex function $e^{j\omega n}$ is useful because it represents sines and cosines of frequency ω simultaneously through its real part, $\cos(\omega n)$, and its imaginary part, $\sin(\omega n)$. $H(e^{j\omega})$ is a complex value, which like all complex numbers has both a magnitude and an angle.

$$x(n) = e^{j\omega n} = \cos(\omega n) + j\sin(\omega n) \; ; \; j = \sqrt{-1} \; . \tag{6}$$

$$y(n) = e^{j\omega n} \sum_{k=-\infty}^{\infty} h(k)e^{-j\omega k} \; . \tag{7}$$

$$y(n) = e^{j\omega n}H(e^{j\omega}). \tag{8}$$

Several important frequency-domain properties follow. First, if a complex exponential of frequency ω is the input to an LSI system, the output will also be a complex exponential of exactly the same frequency. Its amplitude will be scaled by a magnitude of, $H(e^{j\omega})$,($| H(e^{j\omega}) |$) and it will be phase-shifted by an angle of $H(e^{j\omega})$,($\angle(H(e^{j\omega}))$). In this case, the expression *magnitude* means the length of a complex number in the complex plane and the term *angle* refers to its angle in this plane. The term *complex plane* refers to graphing a complex number in terms of its real part versus its imaginary part. Second, if the input is $\cos(\omega n)$, then the output will be $|H(e^{j\omega})|$ $\cos$ $(\omega n + \angle(H(e^{j\omega})))$. In other words, the filter scales and phase-shifts a sinusoid in a well-defined manner. Note, however, that the frequency is never changed. Third, the frequency response is completely determined by the unit sample response of the system, $h(n)$. Fourth, in the special case where $\angle H(e^{j\omega})$ = - $\alpha\omega$ (known as linear phase), an input of $\cos(\omega n)$ produces an output of $|H(e^{j\omega})|$ $\cos(\omega n - \omega \alpha) = H(e^{j\omega})|$ $\cos(\omega(n - \alpha))$, which is a scaled version of the cosine, delayed by α. Since this delay is independent of ω, all frequency components are delayed by exactly the same amount. Finally, $H(e^{j\omega})$ is periodic in ω with period 2π, since $e^{j\omega n} = e^{jn(\omega + 2\pi)}$.

3.3
Frequency-Domain Interpretation of Signals

It can also be shown that input signals can be expressed in terms of their frequency content. If $x(n)$ is the signal, then its description in terms of frequency is as shown in Eq. (9) which is called the *Fourier transform* of $x(n)$.

$$X(e^{j\omega}) = \sum_{n=-\infty}^{\infty} x(n)e^{-j\omega n} \; . \tag{9}$$

Figure 3A shows the magnitude of the Fourier transform for a segment of the spring peeper call. This plot displays more or less how energy of the acoustic waveform is distributed in the frequency domain. Here, the ordinate range is 0 – 4000 Hz (cycles

over time. Hence, the spike from a different segment may be shifted slightly upward or downward in frequency. In addition, other features are present in the burst. In order to better match the perceptual response of the human auditory system, Fourier transform magnitude can be plotted on a decibel scale, which is logarithmic. Figure 3 B shows such a plot, which reveals the presence of a smaller spike at approximately 2100 Hz.

The most important property of Fourier transforms is shown in Eq. (10 a,b), the latter of which states that the frequency-domain description of the input is multiplied by the frequency response of the filter to obtain the frequency-domain description of the output. Signal processing algorithms are generally designed to perform various manipulations in the frequency domain, such as enhancing certain frequencies, suppressing others, or deriving mathematical models of the observed data. Suppose, for example, that the first frequency peak of the spring-peeper call is to be suppressed, in order to use the resulting signal in a playback experiment. Figure 4(A) shows the magnitude response of a second-order FIR filter that can be used for this purpose, which has the coefficients b(0) = 1.0, b(1) = -0.8415, and b(2) = 1.0 [notation as in Eq. (3)]. Using this filter to process the call, the resulting magnitude output in the frequency domain is as shown in Figure 4(b), demonstrating partial suppression of the highest-amplitude component.

if $y(n) = x(n)*h(n),$ (10a)

then $Y(e^{j\omega} = X(e^{j\omega})H(e^{j\omega}).$ (10b)

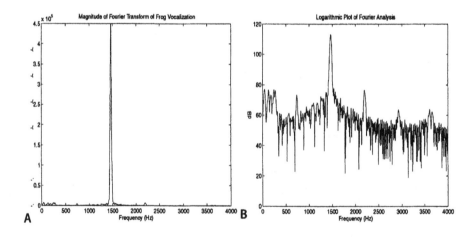

Fig. 3A, B. The frequency content of a short segment of the spring peeper call is shown using Fourier transformation (A), plotting frequency on the abscissa and the relative amplitude of these spectral components on the ordinate. A dominant frequency peak can be clearly seen at about 1400 Hz. Several additional frequency peaks are evident in (B), where amplitude is plotted logarithmically using the decibel scale

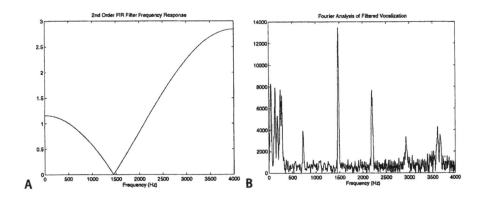

Fig.4 A, B. To remove some of the "noisy" content, the signal used in the previous is first multiplied by a finite impulse response (FIR) filter (A), after which Fourier transformation produces the frequency spectrum shown in (B)

4
Conversion Between Analog and Digital Data Forms

Most digital (discrete-time) signals are produced from analog (continuous-time) versions, through a process known as *sampling*. The sampling process, then, involves converting the signal $x_\alpha(t)$ to $x(n) = x_\alpha(nT)$, a sequence of numbers. The values of the input analog signal (e.g., the voltages at the terminal of a microphone that is recording a spring-peeper call) are only registered at discrete time intervals, T, which is the sampling period. F_s, the sampling frequency, is expressed in samples per second or Hertz. This quantity is the reciprocal of T, or $1/T$. The quantized samples are labeled $\hat{x}(n)$.

Quantization, which is also performed by the A/D converter, is assigning a specific numerical value to each voltage that appears at the input terminal. This process is limited by word width of the converter. A 16-bit A/D board, for instance, can assign a total of 2^{16} or 65,536 different numerical values to time-varying waveform voltages. Although many varieties of A/D-converters exist, just two specifications aptly summarize their performance. The first is the speed of operation, which determines how small T can be (i.e., how high F_s can be). The second is the number of levels available in the quantization process, which determines the overall fidelity of the digitized signal with respect to its analog counterpart.

Ideally, the sampling process would preserve all the input information in the new digital form, allowing inverse conversion to exactly recover the original analog signal. In practice, this ideal is difficult to achieve, although sampling-related degradation can be quite small given sufficient care. The difficulty involved is that gaps of size T now occur between adjacent samples and must be filled in order to return the digitized list to continuous signal form. However, there are many continuous waveforms whose digitally sampled forms would be the same. Since the values of the waveform are only "fixed" at certain places, one could imagine inserting almost any curve between these fixed points. Hence, certain restrictions must be placed on an input signal if it is to be recon-

points. Hence, certain restrictions must be placed on an input signal if it is to be reconstructed later. Heuristically, one reasonable restriction is that the waveform must be fairly smooth between adjacent samples, avoiding oscillations or "wiggles." The exact solution to this problem forms the basis of the *sampling theorem*, which is discussed below.

Another way of looking at the idealized process is as follows. The input to the overall system is $x_\alpha(t)$, an analog signal that appears at the input terminal of an A/D converter, which samples at a constant interval (or period), which is T. The sampled version is $x(n)$, which appears as the input to a digital system or filter, $h(n)$, with the label indicating its unit-sample response. The signal $y(n)$ is the output of the digital system and is necessarily equal to $x(n) * (n)$. To recreate the original signal, a D/A converter (with period T) accepts as input the sequence of numbers $y(n)$ and outputs an analog signal $y_\alpha(t)$. In this last operation, $y_\alpha(t)$ is a signal whose values are exactly equal to $y(n)$ at $t = nT$. An important aspect of the process is, of course, specifying $h(n)$ so as to perform the desired filtering. However, this component can be implemented using a general-purpose computer, special-purpose hardware, an abacus, or anything else that can add, multiply, and store numbers. The key point is actually only that filtering occurs digitally in the overall analog input-output system.

4.1
The Sampling Theorem

Since many signals can have the same sample values, the class of sampled signals must be restricted in order to guarantee a unique D/A inversion. Up to this point, we have only talked about the expression $X(e^{jw})$ designated as the Fourier transform. As discussed, this function describes how the signal $x(n)$ is distributed with respect to frequency, ω, in radians. The continuous-time signal $x_a(t)$ also has a Fourier transform, although defined somewhat differently. We will designate the continuous-time Fourier transform as the CTFT, or $X_a(\Omega)$, where Ω is the continuous-time frequency variable (in radians). Just as with the discrete-time Fourier transform (DTFT), the CTFT describes how the continuous-time signal is distributed with respect to frequency. These definitions will enable us to relate the analog and digital domains and apply the necessary restrictions on the analog signals.

The sampling theorem is as follows. If $x_a(t)$ has no continuous-time frequency components higher in frequency than Ω_M,

$$X_a(\Omega) = 0 \; ; |\Omega| > \Omega_M, \tag{11}$$

then $x_a(t)$ may be recovered from its samples, $x_a(nT)$, provided that

$$T < \frac{\pi}{\Omega_M}, \tag{12}$$

where Ω_M is the maximum frequency (radians) of $x_a(t)$. The condition of Eq. (11) is equivalent to saying $x_a(t)$ is "bandlimited" to Ω_M. Expressed another way, the condition that must be met is

$$\frac{2\pi}{T} > \frac{\Omega_M}{2},$$

(13)

where $\dfrac{2\pi}{T}$ is the sampling frequency (in radians).

This theorem makes clear that the sampling rate must be at least twice that of the highest frequency component in the analog signal. For example, if the analog signal only goes up to 4000 Hz ($2\pi \times 4000$ radians), $x_a(t)$ can be recovered perfectly so long as sampling occurs at a rate of at least 8000 Hz. In a sense, this bandlimiting requirement is one that restricts the waveforms to vary smoothly between the samples. Therefore, the value of T in any A/D operation defines the allowable bandwidth of the analog input.

A complete discussion of the origins and consequences of this important theorem are well beyond the intended scope of discussion. However, some results will be explicitly considered and further information is readily available in standard textbooks on the topic. For the moment, then, it will simply be assumed that a device exists that can record the values of an analog waveform every T time units and convert the analog signal $x_a(t)$ into $x(n)$. In addition, a second device can accept a sequence of numbers (a

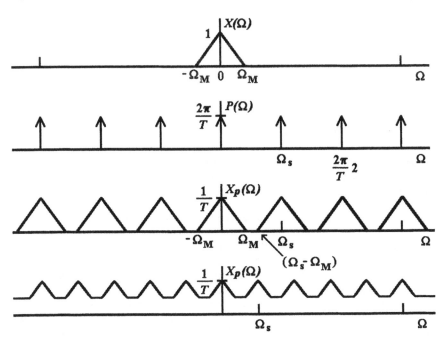

Fig. 5. An analog signal bandlimited between $-W_M$ and W_M (top panel) has a Fourier spectrum with energy at frequency intervals of W_S (upper middle panel). If sampling theory restrictions are met, the spectrum of the digital version is undistorted (lower middle panel). Otherwise, new components are introduced (bottom panel)

digital signal) and output a regularly spaced sequence of pulses of width T and height equal to the sampled values.

Now, consider the same system described in the frequency domain. The Fourier transform of the analog signal is sketched in the top panel of Figure 5. The important point to note is that it is band limited with its highest frequency being Ω_M. The panel just below illustrates the relation between the original spectrum of $x_a(t)$ and that of $x_p(t)$ for the pulses. As can be seen, images of $X_a(\Omega)$ are reproduced every $2\pi / t = \Omega_s$ in frequency. The lower panels illustrate the result for two different values for T. In the first case, the criterion of the sampling theorem is satisfied and the images of $X_\alpha(\Omega)$ are undistorted. The bottom panel shows what happens if t is too large (i.e., the sampling frequency is too low). We can state the Fourier transform relation more precisely using Eq. (14), derived below.

Here, it can be seen that there are an infinite number of images of $X(\Omega)$ in $X_p(\Omega)$. The key condition of the basic sampling theorem is that the lowest frequency image not be distorted.

$$X_p(\Omega) = \frac{1}{2\pi} X(\Omega) * P(\Omega). \tag{14a}$$

$$P(\Omega) = \frac{2\pi}{T} \sum_{k=-\infty}^{\infty} \delta\ (\Omega - k\Omega_S)\ ;\ \Omega_S = \frac{2\Omega}{T}. \tag{14b}$$

$$X_p(\Omega) = \frac{1}{T} \sum_{k=\infty}^{\infty} X(\Omega - k\Omega_S). \tag{14c}$$

4.2
Signal Recovery by Filtering

Now, consider the case where the sampling criterion was satisfied and $x_a(t)$ (or equivalently, $X_a(\Omega)$ is to be recovered from $x_p(t)$ (or $X_p(\Omega)$). Recall that $x_p(t)$ is the output of an ideal system that converts the discrete-time sequence of numbers into a continuous waveform. From the Figure 5, it can be seen that a low-pass filter with cutoff $\frac{\Omega}{2} = \frac{\pi}{T}$ and height T will perform exactly the right processing. On the other hand, if the sampling criterion is not satisfied, $x_a(t)$ cannot generally be recovered.

The distortion that results when an attempt is made to recover such a signal is called *aliasing*. As can be seen in Figure 5 in the case in which the sampling criterion was not met, the higher-frequency components are *folded* into the lower-frequency regions, creating a distortions that are impossible to remove. A further example is illustrated in Figure 6. Here, it is clear that the marked sample points could be used to reconstruct either of these two markedly different the marked sample points could be used to reconstruct either of these two markedly different sinusoids. However, the ambiguity is removed if only sinusoids whose frequency is less than $F_s / 2$ are considered. Similarly, aliasing will be avoided in any signal only if its maximum frequency is less than $F_s / 2$.

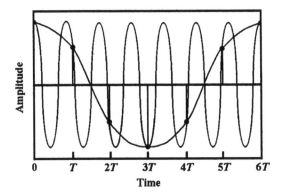

Fig. 6. An illustration of how undersampling can lead to ambiguity in reconstruction of a sinusoid. If the sampling rate is too low, as depicted by the discrete points occurring at time intervals of length T, then the seven cycles occurring in this segment of the original signal may be reconstructed as a single cycle

4.3
Fourier Transform Relations

The DTFT of the sampled signal and of analog signal are shown in Eqs. (15a,b), respectively.

$$x(n) = \frac{1}{2\pi} \int_{-\pi}^{\pi} X(e^{j\omega}) e^{-j\omega n} \, d\omega . \tag{15a}$$

$$x_a(t) = \frac{1}{2\pi} \int_{-\infty}^{\infty} X_a(\Omega) e^{-j\omega t} d\Omega . \tag{15b}$$

Since the sampling process insures that $x(n) = x_a(nT)$, the following result can be derived.

$$s(n) = \sum_{k=1}^{p} a_k s(n-k) + e(n) . \tag{16a}$$

$$X(e^{j\omega}) = \frac{1}{T} \sum_{k=-\infty}^{\infty} X_a\left(\frac{\omega}{T} + \frac{2\pi}{T} k\right) . \tag{16b}$$

In other words, the discrete-time transform is identical to the continuous-time transform when the latter is scaled by $1/T$, compressed in frequency by the factor $1/T$, and repeated every 2π. As Figure 7 shows, the shapes are identical. An extremely important point is that the DTFT, $X(e^{j\omega})$, is only unique over the interval $(-\pi,\pi)$. The fact that the CTFT, $X_a(\Omega)$, is zero outside the frequency range $(-\pi/T, \pi/T)$ is necessary for the correspondence to hold.

One question that is often asked about the DTFT is why the unique range of frequencies is always $(-\pi,\pi)$ irrespective of the sampling interval, T. The answer is that once a signal has been sampled, it consists only of a sequence of numbers and does not include any specific indication of the value of T. Thus, the DTFT processes this sequence of numbers without taking T into account. Since the continuous-time interval T corresponds to exactly one sample in discrete-time format, the mapping of continuous-time frequencies to discrete-time frequencies must be in the ratio 1 / T. Hence, the highest frequency in $X_a(\Omega)$, which by assumption is π / T, always corresponds to $\omega = \pi$ in $X(e^{j\omega})$.

4.4
Effects of Sampling Rates

The effects of different sampling rates on the transform can be seen in Figure 8. In one case, the signal is *undersampled*, creating aliasing. In the other, the signal is *oversampled*, creating large gaps in the frequency domain. Both conditions are undesirable. In the former, the signal is badly distorted. In the latter, many more samples are acquired than are needed, leading to increased storage requirements, increased bit-rate requirements for transmission or copying of the signal, and increased computation in any subsequent digital processing. Overall, to ensure the prevention of aliasing effects, it is generally desirable to reserve a small "guard-band" between the maximum frequency of interest and one-half the sampling rate. To avoid oversampling, however, this band should remain small. Nonetheless, while oversampling does constitute an inefficient use of resources, the signal itself is not degraded.

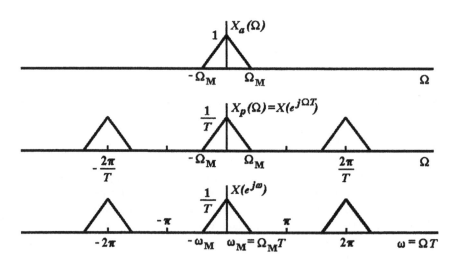

Fig. 7. An analog signal band limited between $-\Omega_M$ and Ω_M (*top panel*) produces identical continuous-time (*middle panel*) and discrete-time (*bottom panel*) Fourier transforms when the former is scaled by 1 / T, compressed in frequency by 1 / T, and repeated every 2π

To ensure that no aliasing takes place, an analog, *anti-aliasing* filter with a cutoff frequency no larger than π / T is usually applied before sampling occurs (see Stoddard, this Volume). Even when the signal to be sampled has a sufficiently narrow bandwidth, one is still well-advised to use such a filter, as high-frequency noise may have entered the system at some stage and will be folded into the low-frequency region if not removed.

4.5
Reconstruction

In D/A conversion, numerical values in binary form are input to a device at a rate of one every per time interval T. Figure 9 illustrates the interpolation that must occur between these values. Exact details of the hardware and associated mathematical analysis used to process the samples $x(n) = x_a(nT)$ to recover $x_a(t)$ are unfortunately beyond scope of this discussion. However, the following description provides a general approximation of the procedure.

In the D/A conversion process, $x(n)$ is the input to a device whose output pulses are spaced at interval T and whose heights are the values $x(n)$. This signal, $x_p(t)$, was introduced earlier and its transform appears between $X_a(\Omega)$ and $X(e^{j\omega})$ in Figure 7. The important step that is still needed is to eliminate any higher-frequency images of the signal. This step can be effected using a low-pass filter, with a cutoff-frequency that is one-half the sampling frequency. Algorithmically, the process simply amounts to a fil

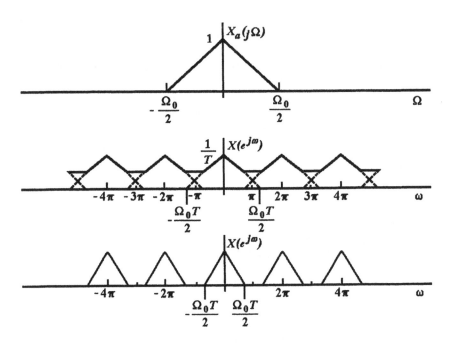

Fig. 8. An analog signal bandlimited between $-W_0 / 2$ and $W_0 / 2$ (top panel) is respectively undersampled, creating aliasing (middle panel), and oversampled, creating gaps in the frequency domain (bottom panel)

tering operation. If the device as a first step creates a set of pulses, spaced by intervals of t and enclosing areas of $x(n)$, the following relationship holds:

$$y_a(t) = \sum_{-\infty}^{\infty} x(n)g(t-nT); \qquad (17)$$

$$Y_a(\Omega) = G(\Omega) \sum_{-\infty}^{\infty} x(n)e^{-j\Omega T} = G(\Omega)X(e^{j\Omega T}). \qquad (18)$$

$G(\Omega)$ therefore only needs to be a low-pass filter with cutoff π / t and height T. $G(\Omega)$ is known as the *reconstruction filter* and has the same cutoff frequency as does the anti-aliasing filter.

Note that this reconstruction procedure is approximate. In electronic devices, there are generally limits on important parameters such as maximum voltage values and speed of transition. D/A converters are used to generate pulses of width T, with appropriate areas. Fortunately, the distortion resulting from this approximation is very small and can be tolerated in most applications. Should one desire, the reconstruction filter can be followed by a *compensating filter* that removes some of these effects, or the reconstruction filter itself can include appropriate frequency shaping.

5
Fundamental Digital Processing Techniques

In the previous sections, ideal operations such as Fourier transforms and filtering were discussed. These kinds of techniques are useful for describing not only the frequency content of a signal, but also the effects that a system has on individual frequency components of the input. Often, however, characteristics of the signal of interest change over time and additional processing can be used to present the data in an accessible and useful form, such as through visual representation. This application is perhaps the most common use of DSP in bioacoustics-related research. If the mechanisms underlying the generation of a signal are known, analysis can be designed to take advantage of production-related information. Human speech is often modeled in this fashion, using techniques such as *linear predictive* coding (LPC) or *homomorphic analysis* that treat the

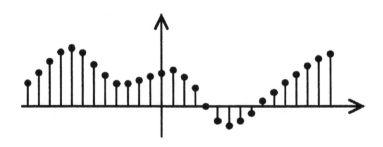

Fig. 9. An illustration of the interpolation needed when an analog signal is reconstructed from discrete sampled values

signal as a linear combination of separable production processes (i.e., quasi-periodic or noisy source energy whose characteristics are modulated by a supralaryngeal vocal tract transfer function; see Rubin and Vatikiotis-Bateson, as well as Owren and Bernacki, this Volume, for further discussion). These techniques, which are very briefly discussed at the end of this chapter, are less often applied to animal acoustic communication signals.

5.1
Power Spectra

The Fourier transform relates two pieces of information concerning a particular frequency component. On the one hand, the transform indicates the density of energy concentration at particular points in the spectrum. As the analysis is performed over a finite period of time, this quantity can be referred to as *power* (energy expended per unit time). The relevant information is obtained by squaring the magnitude of the Fourier transform, a form commonly known as *power spectral density* (PSD; power and energy are always measured in squared units due to the symmetrical, positive-going and negative-going nature of their energy fluctuations). On the other hand, the transform characterizes the phase of the energy component. In many (but not all) cases, phase is considered to be of secondary importance.

5.2
Time and Frequency Resolution

In spectral decomposition of signals, there exists the equivalent of the Heisenberg uncertainty principle in quantum physics. In the acoustic case, it is the time and frequency resolutions that trade off against each other. Although there are various definitions of the concept of "resolution," the basic phenomenon is that the product of the time and frequency resolutions is always equal to 2π. Thus, even if the signal of interest is a pure tone, it will be found to contain energy at components other than its defining, or *fundamental frequency* unless the analysis is carried out over an infinite period of time.

Consider, for example, a very brief sinusoid of a specific frequency — a signal that is equal to the sinusoid over a time interval of width Δ and is zero at all other times. For small Δ-values (i.e., Δ is small in relation to the period of the sinusoid), the signal appears to be a narrow pulse, with energy at all frequencies. As the interval increases, the shape of the sinusoid itself, rather than its onset- and offset-related transients, gradually come to dominate the signal's spectral content. However, this improvement in characterizing the "true" frequency composition of the signal necessarily comes at the cost of decreasing the certainty of its location in time. The same logic holds for any *analysis window* (a *weighting function* or *on/off characteristic*), typically applied to an ongoing signal in order isolate a finite-length, non-zero waveform segment for processing purposes. Overall, increased localization of energy in the time domain implies decreased localization of energy in the frequency domain, and vice versa. This concept is particularly apparent in discrete-time signal processing, where an analysis window of length 1 (i.e., only one sample has a nonzero value) always leads to a uniform power spectrum.

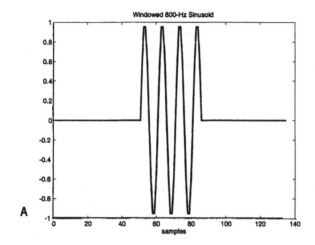

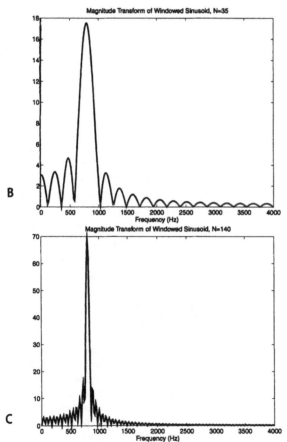

Fig. 10 A-C. Fourier analysis of an 800-Hz sinusoid is affected by the shape and length of the window involved. The first panel (A) shows application of a rectangular window. This window often produces evident distortion in the frequency domain. Distortion is still evident in both (B) and (C). However, the distortion is reduced by increasing the length of the analysis window from 35 (B) to 140 samples (C)

5.3
Windows

Any finite-length transform is subject to the effects of windowing. Representing the windowed signal as a product of the actual signal and the window function, $x_w(n) = x(n) \times w(n)$, the effects of windowing for a sinusoid of frequency 800 Hz and a rectangular window ($w(n) = 1, n = [0,34]$) are shown in Figure 10 A. In Figure 10 B, the Fourier transform of the windowed signal can be seen to have components in addition to the exact frequency of the sinusoid. Two types of distortion can be observed in this spectrum. First, the main, or highest-amplitude, energy *lobe* (occurring at the frequency of the sinusoid) has a nonzero width. This outcome is often termed *smearing*, a loss of frequency resolution or localization. Second, the occurrence of smaller *side lobes* in the spectrum seems to indicate the presence of energy at frequencies substantially removed from that of the original sinusoid. These components can be interpreted as spurious.

Results obtained with a rectangular window of length 140 are shown in Figure 10 C. The main lobe is narrower, meaning that frequency resolution has improved. However, the signal is less well localized in time and the problem of side lobes still exists. Overall, then, one must be very careful when selecting a windowing function for use in frequency analysis. If a window with tapered edges (i.e., gradual onset and offset characteristics, see Figure 11A) is used instead of a rectangular function, the sidelobes are substantially reduced (Figure 11B). The practical result of this demonstration is that a tapered window should be applied to virtually any waveform segment selected from a larger signal for Fourier analysis.

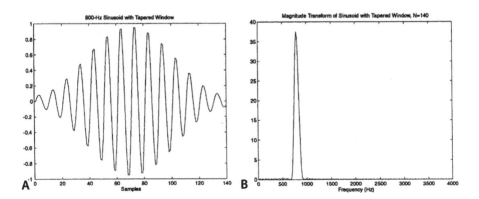

Fig. 11A,B. When a tapered window is used, such as the cosine function whose effects are shown in A, frequency sidelobes are significantly reduced (B)

5.4
Spectral Smoothing

In conducting spectral analyses, an investigator necessarily has to take into account both the nature of the acoustic features of interest and the inevitable effects associated with windowing and other analysis techniques. If the signal is periodic, for instance, it will exhibit energy only at harmonics (frequencies that are integer multiples of the fundamental). In order to localize these energy peaks successfully, spectral smearing cannot be wider in frequency than is the fundamental — the interval between adjacent peaks. In some cases, however, it may be desirable to deliberately smooth these local peaks in order to observe more global spectral structures. Thus, the information made available by a transform can be greatly influenced by length of the analysis window length used to compute it.

In some cases, a signal that is basically random nevertheless exhibits an interesting spectral shape. Using a long analysis window can be undesirable in such instances, as maximizing frequency resolution will also accentuate the noisy nature of the signal. However, using a short analysis window is also problematic in that sufficient frequency resolution may not be attained. Two approaches can be applied to this situation. One technique is to attenuate entirely random noise fluctuations by performing short-window analyses on adjacent signal segments and averaging the resulting power spectra. A second technique is to use longer windows and fit smooth spectral envelopes to the noisy transforms that result. This latter technique, known as the *parametric* approach, includes linear prediction and homomorphic analyses.

5.5
The Discrete Fourier Transform

The bulk of the discussion to this point has concerned discrete-time signals [i.e., $x(n)$] and their Fourier transforms [i.e., $X(e^{j\omega})$]. These transforms are functions of the continuous variable ω, and hence it is impossible to compute the values for all frequencies numerically. Under certain conditions, it is sufficient to compute a finite number of values $X(e^{j\omega})$ sampled over the frequency range. These samples constitute the *discrete Fourier transform* (DFT) or DFT values. Although the DFT is extremely useful in a wide range of applications, great care must be exercised in both the interpretation and application of the results. A basic description of this transform and examples of correct interpretation follow.

The *N-point* DFT of signal $x(n)$ is defined by its Fourier transform, uniformly sampled at n points in frequency.

$$X(k) = X(e^{j\omega})\big|\ \omega = 2\pi k / N. \tag{19}$$

To properly interpret Eq. (19), we must make the following observations. First, the index k is a discrete value that ranges from 0 to $N - 1$. Second, $X(k)$ is a sequence of complex numbers obtained by taking samples of $X(e^{j\omega})$. Finally, the value of the DFT at the index k [i.e., $X(k)$] is equal to the DTFT at frequency $2\pi k / N$.

Equation (20) shows how to compute $X(k)$, the *forward DFT*. Note that only n values of the discrete-time signal are used. The basic assumption of the DFT is that $x(n)$ is only of length N. A sequence of length 128, for example, must use a 128-point DFT. The DFT operation transforms n values in time to n values in frequency. Eq. (21), which is similar to Equation (20), transforms X(k) to $x(n)$. This formulation is the *inverse DFT* (IDFT), which transforms n frequency values to n time values

$$X(k) = \sum_{n=0}^{N-1} x(n)e^{-jk2\pi n/N} \; ; \tag{20}$$

$$x(n) = \frac{1}{N} \sum_{k=0}^{N-1} X(k)e^{jk2\pi n/N} \; . \tag{21}$$

The information content of a digital signal and its DFT are completely equivalent — one can freely move back and forth between the two domains. Since the DFT is a sampled version of the Fourier transform, it consists of complex numbers that are often displayed in terms of magnitude and angle (phase).

A few practical points should be made to assist application and interpretation of the DFT. First, while the DFT can be derived over any signal length, computation of the transform can be performed very efficiently by making use of an algorithm known as the *fast Fourier transform* (FFT). As most FFT algorithms operate only on signals whose lengths are powers of two, frequency analysis often uses window lengths of, for example, 256 or 512 points. Forward and inverse FFTs can be computed equally efficiently.

Second, since the input signals of interest are real functions of time (the imaginary component is ignored), the values of X(k) always exhibit a form of symmetry. Recall that $X(k)$ is complex and must be described in terms of both magnitude and phase. For an N-point DFT, $|X(1)| = |X(N - 1)|$, $|X(2)| = |X(N - 2)|$, and so on. Hence, if we are interested in the magnitude of the spectrum (or the squared-magnitude PSD), we only need to observe $X(K)$ for K = 0,1,...N/ 2. The "upper half" of the DFT magnitude is merely a "reflection" of the lower half.

Third, each component of an N-point DFT represents F_s/N Hz, since the spectrum is sampled at N uniformly spaced frequencies from 0 to 2π and 2π corresponds to the sampling frequency in continuous time. A 1024-point DFT, for example, when applied to a time signal sampled at 10,000 samples per sec has a frequency resolution just under 10 Hz.

Fourth, while the Fourier transform cannot be shorter than the windowed signal segment, it can be longer. For example, it might be desirable to apply an N-point DFT to a signal only M points long (where N is greater than M), if M is not a power of two and N is. Further, frequency resolution greater than F_s / M might be required for a particular analysis. In both cases, DFT-length conversion can be accomplished through *zero-padding*. In this procedure, $N - M$ zero values are added to the end of the signal sample list, producing an N-point signal to which an N-point DFT routine is applied. The outcome of zeropadding may be interpreted as follows. The DTFT of the original signal, $X(e^{j\omega})$, is unaffected by additional zeros at the end of the list since it is based on the inherent assumption that the signal is zero everywhere outside of the range [0,M-1]. However, the M-point DFT samples every $2\pi / M$ Hz, whereas a sample occurs every $2\pi / n$ Hz in the case of the N-point DFT. However, while zero-padding increases the

case of the N-point DFT. However, while zero-padding increases the frequency resolution in the *representation* of $X(e^{j\omega})$, it does not increase the resolution of $X(e^{j\omega})$ itself. The latter quantity depends on the window function.

5.6
Correlation

When signals have a strong random component, it is often useful to characterize them statistically. Beyond simple averages, correlations are the perhaps the most basic of the characterizations used. The concept of correlation is best understood within the realm of probability theory. An attempt will be made here, however, to describe correlation only in terms of averages. Consider one set of numbers, $[x_1, x_2, x_3,..., x_N]$, paired with another set, $[y_1, y_2, y_3,..., y_N]$. The quantity x_i might represent the weight of animal i and y_i its daily food consumption. Further, define m_x and m_y as the averages of x and y, respectively. The correlation between the numbers x and y is then defined as the average of the product $(x_i - m_x)(y_i - m_y)$, as in Eq. (22).

$$R_{xy} = \frac{1}{N} \sum_{i=1}^{N} (x_i - m_x)(y_i - m_y).$$ (22)

One of the fundamental concepts of correlation is that a relationship between two sets of numbers can be quantified. If values of x larger than m_x tend to be paired with values of y that are larger than m_y, the correlation is greater than zero (i.e., x and y are positively correlated). If values of x larger than m_x tend to be paired with values of y that are smaller than m_y, the correlation is negative. If x and y values bear no relation to each other, R_{xy} will be close to zero (and will approach zero as n increases). For digital signals, relationships among sampled values are the quantities of primary interest.

5.7
Autocorrelation

Often, it may be assumed that the average value of a signal to be analyzed is zero. [In many cases, a signal whose average value is not zero (i.e., having an *offset*) is revealing the effects of a consistent measurement error or some electronic miscalibration. Such offsets should typically be removed before analysis]. Any pair of points in the signal $x(n)$, for instance two points separated by m samples, can be correlated. If the signal is of finite duration, N, the definition of the *autocorrelation* as a function of *lag m*, is as shown in Eq. (23).

$$R(m) = \frac{1}{2N+1} \sum_{m=-N}^{N} x(n)x(n+m).$$ (23)

One can easily extend this idea to a case in which N is arbitrarily large. The autocorrelation function has some important properties and applications. First, the maximum value reached by the autocorrelation function is $R(0)$, which is equal to the average power in the signal. Second, if $x(n)$ is periodic with period P [(i.e., $x(n) = x(n + P)$] and N is arbitrarily large (which must be the case if $x(n)$ is truly periodic), then $R(m)$ is periodic with period P. Third, since the maximum value always occurs at $m = 0$, this

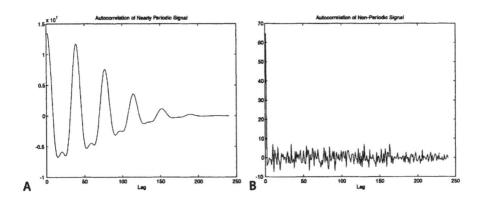

Fig. 12 A,B. Autocorrelation results are plotted as the magnitude of correlation occurring as a function of temporal offset, or lag, when the signal is shifted against itself in time in step-wise fashion. A nearly periodic signal produces a function with pronounced peaks that occur at lag intervals corresponding to its strongest periodic component (A), while no peaks are evident when a non periodic signal is autocorrelated (B)

value repeats every P lags. Fourth, the Fourier transform of $R(m)$ is equal to the power spectral density of $x(n)$. When $R(m)$ is equal to zero for all values of m except $m = 0$, the signal has a flat power spectrum and is referred to as being *white*. In practice, the signals are usually longer than desired for a single analysis. In addition, the signal statistics may be slowly varying, making it desirable to analyze only a portion at a time. Thus a window is needed, preferably one with tapered ends that minimize distortion in the power spectrum.

One application of the autocorrelation function is to detect periodic portions of a waveform. Since few waveforms are perfectly periodic and no windowed waveforms ever are (i.e., they do end at some point), $R(m)$ will never be equal to $R(0)$ for $m \neq 0$.

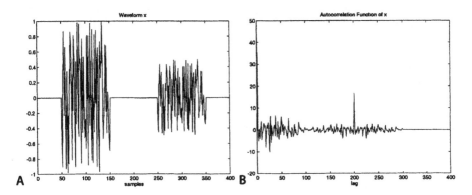

Fig. 13 A,B. Autocorrelation of a two-pulse waveform xl (A) produces a function (B) in which a prominent peak is found at lag 200, indicating that the second energy pulse in waveform x is an echo of the first

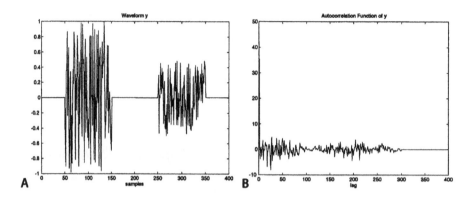

Fig. 14 A,B. Autocorrelation of a two-part waveform depicting unrelated pulses. (A) two-pulse waveform y. (B) The autocorrelation in this case shows no peaks, indicating that the two pulses in waveform y are independent

However, if $x(n)$ repeats its basic form every P samples, $R(m)$ will exhibit a peak at $m = P$. Since even signals with no periodicity exhibit small peaks in $R(m)$, it is advised that unless the ratio $R(P) / R(0)$ exceeds some preselected threshold (often 0.5) no periodicity should be assumed. Figure 12 shows two examples of the autocorrelation function. In one the signal has strong periodicity, while in the other the signal has no periodicity.

Echoes can also be detected using the autocorrelation function. Two pulses of acoustic energy can be seen in the two waveforms depicted in Figures 13 and 14. In each of these cases, it is unknown if the second pulse is an echo of the first or if they are independent in origin. As shown in Figure 13B, the autocorrelation function of $x(n)$ has a distinct peak at lag 200, indicating that the second pulse (seen in Figure 13A) is, in fact, an echo. In contrast, the autocorrelation function of $y(n)$ shown in Figure 14B has no such peak, indicating the absence of a relationship between the two pulses shown in Figure 14A. Another application of the autocorrelation function is estimation of smooth spectra from short data records. The fact that a spectrum can be computed from $R(m)$ forms the basis for a variety of techniques, including the linear prediction approach described briefly in the section on advanced topics.

5.8
Cross-Correlation

Another application of correlation is to relate two different signals. In this case, the cross-correlation between $x(n)$ and $y(n)$ is defined as shown in Eq. (24).

$$R_{xy}(m) = \frac{1}{N} \sum_{n=0}^{N-1} x(n) y(n+m). \tag{24}$$

An important use for the function $R_{xy}(m)$ is in detecting the occurrence of $y(n)$ as a delayed noisy version of $x(n)$, for instance as a component of sonar system operation. If $y(n)$ is, in fact, an echo of $x(n)$, $R_{xy}(m)$ reaches its maximum value at $m = d$, where d is the delay interval. Cross-correlation is extremely effective in these sorts of applications, even in very noisy environments.

5.9
Spectrograms

The concept of spectral analysis for short intervals was described previously. This technique will now be modified to include analyses of a series of intervals, formally designated the *short-time Fourier transform*, or STFT (note that STFTs necessarily include time variables in addition to the frequency dimension associated with the DFT). The simplest way to describe this analysis approach is to picture a window that slides back and forth over the signal. At each time position, a Fourier transform is computed. If the Fourier transform frequency is ω, and the window is positioned at sample n, the value is referred to as $X_n(e^{j\omega})$, shown in Eq. (25).

$$X_n(e^{j\omega}) = \sum_{m=-\infty}^{\infty} x(m)w(n-m)e^{-j\omega m} . \tag{25}$$

In this case then, the function $w(m)$ is a window. The function $w(n - m)$ represents the same window, but reversed in time and positioned at point n. The windowed portion of the signal is $x(m)w(n - m)$, and the overall sum is the Fourier transform of this windowed segment

Two observations concerning the behavior of this function are important in the present context. First, if n is fixed, only one Fourier transform is computed for the windowed signal. The second point is less obvious. If ω is fixed in value, the STFT is $x(n)(e^{j\omega n})$, filtered by a unit-sample response $w(n)$. Unless $w(n)$ is short in duration, time-smearing occurs in accordance with the rise-time associated with $w(n)$. The bandwidth of $W(e^{j\omega})$ determines the amount of frequency smearing. Overall, then, the frequency-versus-time resolution tradeoff in the STFT is entirely determined by the shape of the window used in analysis. When the DFT is used to evaluate the STFT, the density of samples in frequency is a function of DFT size. Note, however, that although the sequence can be zero-padded to obtain arbitrarily high "sampling" of the STFT, the actual frequency resolution is no better than that already determined by window shape. This somewhat surprising result can be used to explain the granular appearance of digital spectrograms.

The two most common spectrogram varieties are designated as *wideband* and *narrowband*, respectively. While these descriptors are, of course, relative and somewhat arbitrary, the two frequency bandwidths typically used do produce vastly different outcomes when applied to both human speech and many other acoustic signals. Wideband spectrograms are based on filters with rapid rise times, which are in turn associated with fairly wide frequency responses. Typical values associated with a wideband filter are a rise-time of 3.3 ms and bandwidth of 300 Hz. The output of the DFT, $X_n(k)$ as a function of n, constitutes a filter "channel" that is centered at its frequency kF_s/N and has this risetime and bandwidth. A filter with these characteristics has the effect of performing spectral smoothing over 300-Hz widths, removing much spectral fine structure. For example, if a periodic acoustic waveform with a fundamental frequency of

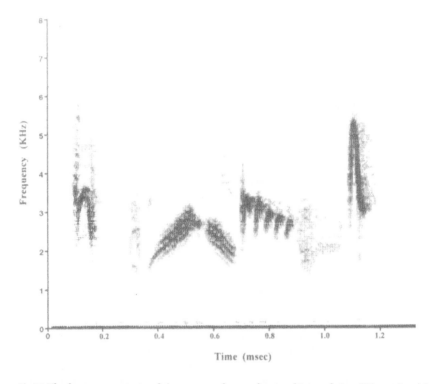

Fig.15. The frequency content and time course of a song from a white-eyed vireo (*Vireo griseus*) is shown in a digital spectrogram that represents frequency on the abscissa, time on the ordinate, and energy amplitude as the darkness of *shading* at each point in the resulting matrix

vals and will therefore not be resolved by the filter. However, the 5-ms periods of the waveform will be resolved in time. In speech analysis, this particular filter is therefore inappropriate for characterization of detailed spectral effects, but is appropriate for fine-grained analysis of periodicities in vocal fold vibration. For a narrowband filter, rise time is typically 20 ms while bandwidth is 45 Hz. Here, individual harmonics of periodic waveforms with fundamental frequencies as low as 45 Hz can be resolved. However, associated periodic variations are smeared in the time domain.

Some of the useful properties of a spectrogram can be illustrated using a song recorded from a white-eyed vireo, *Vireo griseus*, shown in Figure 15. This representation shows time on the ordinate, frequency on the abscissa, and gray-scale shading that represents log magnitude of the energy occurring at any given point. Frequency resolution is determined by the length of the DFT used for each time frame, in this case 512 points. Since the sampling frequency was 20 000 samples per second, each DFT sample therefore represents 20 000 / 512, or roughly 39 Hz. However, since a tapered window was used, the resolution is no better than 78 Hz. Following each DFT computation, the window was shifted 40 points in time, producing significant overlap in the energy represented by immediately adjacent power spectra underlying the spectrographic representation. One might therefore think the time resolution to be 40 / 20 000 or 2 ms. Time

resented by immediately adjacent power spectra underlying the spectrographic representation. One might therefore think the time resolution to be 40 / 20 000 or 2 ms. Time resolution is instead a function of overall DFT window length (i.e., 512), moderated by the effect of window-tapering. Thus, the effective window length is approximately 256, corresponding to a time resolution of roughly 12 ms.

These parameters are probably appropriate in the case of the vireo's song, as the spectrogram captures the most rapid changes in the spectrum, as can be seen in the terminal note. It can be clearly seen in this digital display that there are at least seven components in the song, the fifth being a trill with descending frequency (see both Gerhardt and Beeman, this Volume, for further discussions of spectrograms).

6
An Introduction to Some Advanced Topics

Very brief descriptions of a few more advanced topics are provided below. More extensive discussions of some of these topics are provided elsewhere in this Volume.

6.1
Digital Filtering

Digital filtering is one of the major applications of DSP (see Stoddard, this Volume, for detailed discussion). Using modern signal processing software and hardware packages, many of the difficulties that historically plagued analog filter design have been obviated. Using digital technology, one need do no more than provide a set of filter specifications as input to an automated filter-design program. Unfortunately, no single filter design is ideal in all respects. The ideal transfer function of a lowpass filter, for instance, would have a value of 1 that changed instantaneously to 0 at the designated cut-off frequency. Such a transfer function cannot be achieved.

In filtering applications, there is typically a set of desirable frequencies, the *passband*, that should be passed through the filter with minimal modification and a set of undesirable frequencies, the *stopband*, that should be suppressed or at least attenuated (Figure 16). A *ripple*, or variation in frequency response, is always associated with passbands and stopbands. The following approximations must therefore be made. The passband ripple (δ_1) refers to the maximum deviation from 1 that can be permitted. This factor specifies the maximum distortion occurring in the passband. If $\delta_1 = 0.01$, no more than 1 % ripple occurs at any passband frequency. The stopband ripple (δ_2) refers to the maximum deviation from 0 that can be permitted (sometimes called leakage). This parameter specifies the minimum amount that the stopband is attenuated. If $\delta_2 = 0.01$, then all frequencies in the stopband are attenuated by at least a factor of 100, or 40 dB $[-20\log_{10}(.01)]$. The *transition width* specifies the bandwidth over which the filter response changes from 1 to 0.

FIR filters must be used if a linear phase response is required, meaning that all frequencies are to be delayed by an equal amount. This constraint is critical if we wish to preserve rapidly changing features, or "sharp edges" in the signal waveform. IIR filters can be used if nonuniform phase delays are acceptable. An important advantage of using

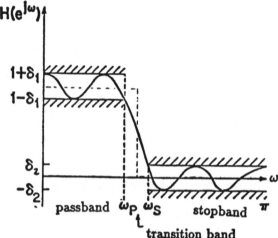

Fig.16. The response characteristic of a lowpass filter, illustrating differential amplitude effects on input components of varying frequencies

IIR filters is that much less computation is required than with many other filter-types. Overall, although digital technology has greatly simplified the task of filter design, determining the minimum performance requirements for the filter itself remains a significant challenge.

6.2
Linear Prediction

While Fourier transformation provides a powerful and very broadly applicable technique for spectral analysis, its results can be significantly enhanced by introducing a specific model of a given signal into the processing algorithm. In LPC, for instance, the signal is assumed to represent the output of a filter based only on peaks, or *poles*. For many acoustical signals, this assumption can approximately be justified. Poles correspond to *resonances*, frequency ranges in which energy passing through the system is amplified. The converse, frequency ranges in which signal energy is attenuated or removed, are referred to as *zeros*, or *antiresonances*. For the human speech signal, unless the nasal cavity is open, only resonances are important.

In a digital system, short segments (e.g., 30 ms) of the output speech s(n) can be approximately represented as in Eq. (26).

$$s(n) = \sum_{k=1}^{p} a_k s(n-k) + e(n) \ . \tag{26}$$

Here, $e(n)$ is a white input, a_k are the filter coefficients, and p is approximately 10. A key feature of LPC is that the coefficients ok are chosen so as to provide the best possible fit to the model. The mathematical technique involved uses the first p lags of the windowed autocorrelation function, then solves for a_k in a computationally efficient manner. A second important aspect is that the smooth spectrum is computed as the transfer function of the recursive filter written above. A third property is that the transfer func-

with associated resolution of frequency peaks that is much superior to that of Fourier-analysis-based spectrograms or DFTs. A bonus is that once a filter transfer function has been computed, it can also be used to synthesize the signal. No naturally occurring signal fits the all-pole model perfectly, although the approach has been shown to be robust to violations of its assumptions in the particular case of human speech production. LPC may also be fruitfully applied to other signals as well, and may be most useful when the spectral feature characteristics of the signal of interest are related to resonance effects occurring during the sound production process (see discussion by Owren and Bernacki, this Volume).

6.3
Homomorphic Analysis

A model less restrictive than LPC is one that merely assumes that the observed signal represents the convolution of one signal with another (i.e., a signal that passes through a filter). In LPC, the input is assumed to have no spectral shape and the filter is assumed to be of the all-pole variety. In assuming only a convolution process, however, one can more generically disentangle rapidly varying from slowly varying waveform components. Convolution involves multiplying the Fourier transforms of two signals. If the logarithms of the Fourier transforms are first computed, the multiplication involved becomes simple addition. Thus, applying an inverse Fourier transfer function to the Fourier transforms combined in this way remains an additive process.

The latter is called the *complex cepstrum*, a representation of spectral components in a novel temporal domain, *quefrency*, in which periodicities occurring in the "spectral waveform" become apparent as individual amplitude spikes. If the two signals that were combined to produce the waveform being analyzed have different properties, then, they may be separable using the complex-cepstrum approach. In the example of human speech, the vocal tract resonances are assumed to be slowly varying in frequency, whereas vocal fold vibration that provides input to the system is assumed to be very "busy" in the frequency domain. Hence, if only the portion of the complex cepstrum corresponding to vocal tract periodicities is preserved and transformed back into the frequency domain, the result will be a smooth spectrum. This technique is called *homomorphic smoothing*. One problem in using homomorphic smoothing, however, is that the Fourier transforms involve complex numbers. Logarithms of complex numbers are not unique, creating a difficulty in selecting particular values for the computations. One compromise solution is to use only Fourier transform magnitudes, producing a representation known as the *cepstrum*.

7
Summary

In this chapter, the basic ideas, philosophies, and capabilities of digital signal processing (DSP) are described. The broad applicability of DSP grows directly from the simplicity of its most basic concepts. In this approach, a signal is any ordered list of numbers representing a time-varying quantity, and a system is any processing algorithm that takes a signal as input and produces a signal as output. The processor involved is often referred to as a filter and produces some useful transformation of the input. Filters are most often of the linear-shift-invariant variety, producing signal outputs though linear processing of input. As many of the signals of interest (particularly in bioacoustics) are originally continuous, conversion between analog and discrete signal representation forms is required. Analog-to-digital signal conversion relies on sampling, which is characterized in terms of rate and quantization. In accordance with the sampling theorem, analog inputs must be sampled at a rate no higher than one-half the maximum frequency of interest in order to avoid aliasing effects. Anti-aliasing filtering is therefore routinely applied upon signal input. Recovering an analog signal from a digital signal requires interpolation of values between adjacent samples, for which filtering is also used, in order to ensure that the maximum frequency of the recovered signal is less than one-half the sampling rate of the digital signal.

Digital signals are typically represented in the time and frequency domains. Time-domain representation involves a series of signal amplitudes recorded over some temporal interval. This waveform representation can be transformed into a new time-based form, like an amplitude envelope, through processes such as rectification and smoothing. The signal's frequency-related characteristics are often of primary interest, and processing in this domain relies on digital systems with known magnitude and phase responses as a function of frequency. The former is considered to be greatest importance in bioacoustics applications, while the latter is typically disregarded. A notation based on complex numbers provides a convenient means of describing and manipulating the frequency-domain characteristics of digital signals.

Depending on its response to a single energy pulse — the unit-sample response — a digital system is often characterized as being either of a finite- or infinite-impulse-response type, each of which has useful frequency-domain properties. The Fourier transform is by far the most commonly used technique for characterizing the spectral content of a signal, and produces a power spectrum that shows the relative magnitudes of a series of sampled frequencies. However, the outcome of Fourier transform computations depends in part on the length the waveform segment used and the relative weighting that is applied to each of its sample points. These parameters are set through selecting a particular analysis window. As the window is finite in length, the resulting discrete Fourier transform is subject to inherent measurement uncertainty. Specifically, increasing resolution in the frequency domain by lengthening the window decreases temporal resolution. Increasing resolution in the time-domain by shortening the window decreases spectral resolution. The effects of this powerful constraint can be alleviated somewhat through techniques like zero-padding, but can never be circumvented. A second implication of the finite nature of Fourier transformation is that spectral distortions are introduced. Such inaccuracies can be reduced by choosing a tapered analysis window, but can also never be entirely eliminated. Restricting window-length to a power of 2 is the basis of the fast Fourier transform, in which a discrete transform is computed with maximal efficiency. By treating a signal as a series of contiguous or overlapping finite segments, its time-varying frequency characteristics can be examined using Fourier-transform-based spectrograms.

Correlation provides a powerful tool for examining certain signal characteristics. In the time domain, for instance, autocorrelation of a given signal or cross-correlation of two signals can be used to reveal waveform periodicities or relationships. Other useful

DSP techniques include digital filtering, in which both magnitude and phase characteristics of signals can be altered, as well as linear prediction and cepstrum-based homomorphic analysis, in which a signal is treated as the product of independent (often production-related) components whose characteristics can be extracted separately.

CHAPTER 3

Digital Signal Analysis, Editing, and Synthesis

K. BEEMAN

1
Introduction

Historically, the quantitative study of sound has been wedded to the development of sound-measurement technology. Researchers have routinely seized on and resourcefully adapted various technological tools, whether intended for sound analysis or not. Sabine (1900), for example, developed acoustical reverberation theory in an empty theater at Harvard University, using an organ pipe, a chronograph, and his own hearing to measure reverberant sound duration. Similarly, Brand (1934) characterized the time-varying frequency of birdsong by recording vocalizations on motion-picture film and measuring spatial line-density on the soundtrack. Successive milestones in sound-measurement technology — notably the microphone, the oscilloscope, and later the sound spectrograph — helped researchers to visualize and measure sounds but not to model them directly. Modeling of acoustic communication was instead typically performed indirectly via statistical analysis, comparison, and classification of individual measured sound features.

The development of digital computers has allowed researchers to extend and automate sound-measurement capabilities in many areas. However, the most important impact of this technology on bioacoustics may ultimately be the resulting accessibility of sophisticated mathematical tools for modeling sound structure, and the integration of these tools into the measurement process. The implications of this merger range from dramatic improvement in efficiency to the opportunity to conduct sophisticated interactive experiments in which the computer, for instance, presents stimuli, receives and analyzes resulting behavioral responses, and selects stimuli on this basis for the next testing cycle. Mathematical tools allow the investigator to measure and manipulate features ranging from discrete sound parameters, such as time and frequency maxima, to more comprehensive sound properties, such as time-varying amplitude and frequency and overall spectrographic similarity.

This chapter describes digital measurement and modeling techniques while providing a practical survey of the tools available to bioacousticians. Topics include measurement of detailed time and frequency parameters over precise intervals, real-time analysis and display of individual spectra and spectrograms, digital sound comparison and statistical classification, precise rearrangement of sound elements in the temporal domain, and sound synthesis based on natural sounds, mathematical functions, or both. Synthesis manipulations such as arithmetic operations, sound combination, time and frequency shifting and rescaling, frequency modulation, harmonic manipulation, and noise addition and removal are covered. Digital sound analysis technology has been applied to a wide variety of animal groups, notably primates (e.g., Owren and Bernacki 1988; Hauser 1991, 1992; Hauser and Schön-Ybarra 1994), anurans (e.g., Gerhardt 1989, 1992; Ryan et al.1992), birds (e.g., Nowicki and Capranica 1986; Nelson and Croner 1991; Suthers et al. 1994; Nelson et al. 1995), insects (e.g., Wells and Henry 1992; Stephen and Hartley 1995), and marine mammals (e.g., Clark 1982; Buck and Tyack 1993).

More advanced topics in signal analysis, measurement theory, and digital technology applicable to sound analysis are discussed in other works. These include mathematical signal analysis (Schafer 1975), engineering measurement theory (Randall 1977), digital hardware principles and guidelines (Stoddard 1990), and signal analysis and bioacoustic instrumentation (Rosen and Howell 1991). Digital sound analysis systems have been developed for various computers, including DEC minicomputer (Beeman 1987), IBM PC (Beeman 1989, 1996b), Amiga (Richard 1991), and Apple Macintosh (Charif et al. 1995).

2
Temporal and Spectral Measurements

Animals and humans respond to both the temporal and spectral structure of sounds. Biologically important sound attributes include temporal properties such as duration, repetition, and sequencing of sound elements, as well as spectral properties such as frequency, bandwidth, harmonic structure, and noisiness. Temporal properties can be measured from the *amplitude-time waveform*, which specifies acoustic pressure as a function of time, and the *amplitude envelope*, which specifies time-averaged acoustic intensity as a function of time. Spectral properties can be derived from the *power spectrum*, which specifies energy distribution as a function of frequency, and the *frequency-time spectrogram*, which specifies energy distribution as a function of both frequency and time. Significant information can be encoded in both the temporal and spectral domains of sound signals, and the following sections provide a survey of the digital techniques used to extract and analyze the associated acoustic properties.

Digital signal analysis typically begins with the measurement of basic time and frequency parameters. Important components include maximum and minimum values and the associated time and frequency coordinates. These extrema can be measured over the entire sound or, by restricting the analysis window to specific segments, over a succession of signal intervals. Temporal and spectral examples include, respectively, the peak level of an amplitude envelope and its time coordinate, and the amplitudes and frequency coordinates of successive peaks within a spectrum. These measurements can be compared statistically over an ensemble of signals to assess a wide variety of features. Important characteristics of the time waveform include its average signal level (which, if different from zero for an acoustic signal, may indicate instrumentation problems), and its *root-mean-square* (RMS) signal level (the square-root of the time-averaged squared amplitude), which is the standard measure of signal energy. Other parameters of interest that are available from spectrographic representations include the overall duration of a signal, its frequency range, and successive maxima and minima in its time-varying pitch contour.

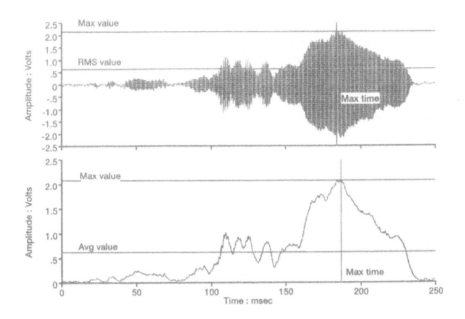

Fig. 1. Measuring duration, amplitude, and root-mean-square (RMS) level of a swamp sparrow (*Melospiza georgiana*) syllable from its waveform *(top panel)* and amplitude envelope *(bottom panel)*

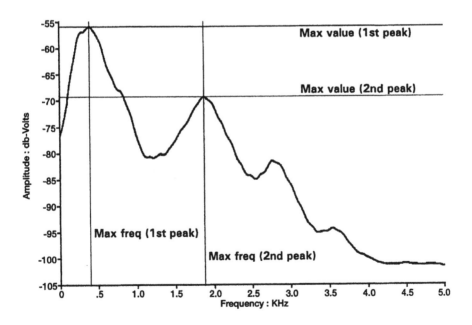

Fig. 2. Measuring the frequency and amplitude of major spectral peaks in a power spectrum

In a digital environment, measurements can be performed on individual notes, entire calls, or average signals derived from an ensemble of sounds (Beeman 1996b). Using the results of such measurements, the sounds can then be manipulated individually or in groups. For example, an ensemble of sounds can be normalized to the same sound energy level before playback, by dividing each sound by its own RMS value. Figure 1 illustrates the use of digital techniques to measure the duration, amplitude, and RMS value of a time waveform and its associated amplitude envelope, while Figure 2 shows frequency and amplitude measurements of peaks in a power spectrum.

3
Time-Varying Amplitude Analysis

3.1
Amplitude Envelopes

Temporal analysis is concerned with the location, duration, and pattern of sound events, such as the sequence and timing of notes within a song, songs within a calling sequence, or the pattern of variation in sound intensity (see Gerhardt, this Volume, for discussion of terminology used to distinguish acoustic units in the temporal domain).

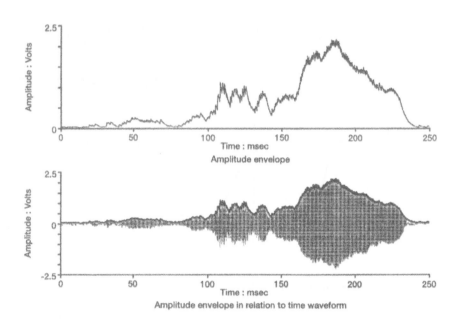

Fig. 3. The amplitude envelope (*top panel*) superimposed on the waveform (*bottom panel*) shown in Fig. 1

Although many parameters of interest can be measured directly from the time-amplitude waveform, it can be more efficient and objective to derive these from the corresponding amplitude envelope.

Visually, the *amplitude envelope* traces the outline of a waveform, as shown in Figure 3. Such tracking is achieved mathematically by jumping from peak to peak and bridging valleys of the signal, with the goal of retaining larger-scale and potentially meaningful amplitude variations while ignoring smaller-scale changes. Before computing the envelope, the time waveform is *rectified* by inverting all negative-going waveform components to positive-going ones. The tracking algorithm is designed to follow the rectified input signal when its amplitude increases, while decaying in accordance with a specified time constant (T) when the input level falls.

Discrimination of larger versus smaller-scale amplitude variations is determined by the decay-time constant used, where different values of T will cause the envelope to reflect signal features of different time scales. With a large T-value, longer, slower variations are tracked and retained while shorter variations are smoothed away. A small T-value, in contrast, produces an envelope that retains detailed aspects of amplitude modulation in the original signal. The optimal T-value for a particular signal type must be determined empirically, and should typically be both small enough to retain significant amplitude-modulation features and large enough to reveal the desired amount of overall envelope shape.

Amplitude envelopes can be used to measure duration, onset and offset patterns, amplitude modulation, intensity variation between notes, and time-varying intensity of waveforms. Envelopes can be used in synthesis to isolate and manipulate a sound's amplitude-time characteristics. Amplitude envelopes can also be used to quantitatively derive the average amplitude behavior of an ensemble of signals (by averaging the entire set of envelope functions), or to compare the similarity of amplitude behavior between notes (by cross-correlating the envelopes). For playback purposes, the amplitude character of a waveform can be removed by dividing the signal by its envelope function, producing a constant amplitude signal. The amplitude modulation of a signal envelope can be derived and analyzed mathematically by deriving a smoothed amplitude envelope (using a larger T) that is subtracted from the original envelope, leaving only the modulation function (see discussion below and in Beeman 1996b).

3.2
Gate Functions

While the amplitude envelope allows the researcher to measure the time-varying characteristics of a signal at various levels of time resolution, one is sometimes interested only in the overall temporal pattern of a sequence of sounds. A useful technique for temporal analysis at this level is to convert the amplitude envelope into a *gate function* (illustrated in Figure 4; also see Beeman 1996b). This continuous time function has only two values, used to distinguish signal events from silences. It is derived by comparing the amplitude envelope to a *threshold level*, which is normally set just above the signal's baseline noise level. The gate function is assigned a value of 1 wherever the signal exceeds the threshold and, zero elsewhere. It is usually also desirable to require the signal to be

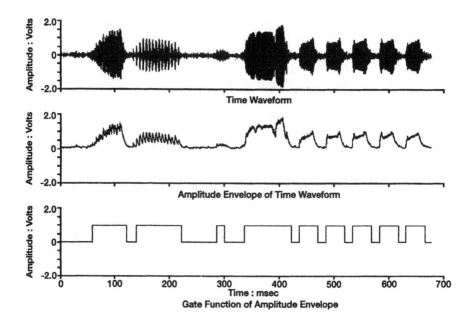

Fig. 4. Using a gate function to measure note duration and spacing. The time waveform (*top panel*) is converted to an amplitude envelope (*middle panel*) which is converted to the binary-valued gate function (*bottom panel*), indicating the presence and locations of sound segments

present or absent for some minimum interval before gate-function polarity can change. This allows the algorithm to respectively reject short noise bursts and bridge brief amplitude gaps within a single coherent event.

Overall, the gate function is a powerful tool that can be applied on a variety of scales in both the time and frequency domains. Applied to temporal features, gate functions can be used to characterize durations of both individual notes and entire songs, producing objective groupings of successive sounds and silences. Temporal variation in repeated note sequences can be characterized by measuring the durations of notes and inter-note silences and collecting the results in two corresponding histograms. These histograms can then be compared among both species and individuals. The similarity between gate functions can also be quantified by calculating their cross-correlation values, for example, to characterize the temporal similarity among trill sequences from different individuals of a species. An example of the application of gate-function analysis to spectral parameters is detection of the bandwidth of frequency peaks.

4
Spectral Analysis

The power spectrum of a signal is most commonly derived by *Fourier transformation* of the time waveform, which is a mathematical integration technique that characterizes the amplitude levels of the various frequencies that are present (see Clements, this Volume, for discussion of Fourier methods). The first digital Fourier technique was the *discrete Fourier transform* (DFT), while an efficient computational algorithm known as the *fast Fourier transform* (FFT) later made this technique routinely available in digital acoustic analysis systems.

4.1
Power Spectrum Features

In generating Fourier-based power spectra, the researcher must consider the time-versus-frequency-resolution trade-off, selection of a particular transform window, application of transform smoothing, and (where absolute energy levels are important) scaling the transform. Various aspects of these issues are discussed elsewhere in this Volume (see Clements, and Owren and Bernacki), and by Beeman (1996b), and only the most important points are summarized here.

The *bandwidth* of a signal is defined as its maximum frequency range, in other words, a signal containing frequencies up to 10 kHz has a 10-kHz bandwidth. The *time resolution* and *frequency resolution* of a spectral measurement express its degree of inherent temporal and spectral uncertainty, respectively. When Fourier transforms are used, resolution is determined solely by the physical duration of the transformed time signal (T), and does not depend on digital sampling rate except as the latter affects analysis duration. Time resolution is then T seconds, while frequency resolution is $1 / T$ Hz.

For measurement purposes, a silent interval can be added at the end of the signal before transforming in order to produce a smoother spectrum, a useful procedure referred to as *zero-padding*. Zero-padding will be beneficial up to an overall segment length of about 4 times that of the original signal (e.g., 1024 data points plus 3072 zeroes). Mathematically, zero-padding increases the density of computed points in the Fourier spectrum, thereby achieving better sampling of the true underlying spectrum. Nonetheless, maximum spectral resolution is limited by the length of the original signal without zero-padding.

In Fourier analysis, time and frequency resolution are mutually constrained by a trade-off known as the *time-bandwidth product*, meaning that the product of time resolution and frequency resolution is always equal to 1. As a result, improving resolution in one domain necessarily degrades resolution in the other. For example, a 100-msec temporal resolution means that the time of occurrence of any acoustic feature of interest can be pinpointed only within a 100-msec interval. The associated spectral resolution of 10 Hz implies that the frequency of any acoustic feature of interest can only be pinpointed within a 10-Hz interval. Improving temporal resolution to 10 msec dilates the bandwidth of spectral uncertainty to 100 Hz, while sharpening the latter to 5 Hz dilates time resolution to 200 msec. This relationship (also known as *temporal-spectral uncertainty*) is discussed and illustrated by Beecher (1988), and Gerhardt (this Volume).

Fourier analysis of a short segment of a continuous sound can produce mathematical artifacts due to the abruptness of the segment's onset and offset. To minimize these edge effects, a *window* function is usually applied to the signal before transformation to taper its beginning and ending amplitudes. A *rectangular window* provides no tapering, while a *Hanning window* varies sinusoidally between a beginning value of 0, a center value of 1, and an ending value of 0. The *Hamming window* is similar to the Hanning, but begins and ends on a value greater than zero. Transform windows differ in the manner and extent to which they blur the true underlying spectrum and the choice involves unavoidable trade-offs. The Hanning window, for instance, separates closely spaced harmonics or sidebands better than does the rectangular window, but also broadens and delocalizes the exact frequency of these bands. Overall, however, the Hanning window is the best choice for biological sound analysis (see also discussion by Clements, this Volume).

Digital power spectra are typically fuzzy, and it is usually desirable to *smooth* them in order to see the underlying contour more clearly. Smoothing is usually performed through the *running-average* technique, in which each point in the power spectrum is recomputed as the mean value of itself and a number of adjacent points (specified as the *width* of the smoothing window). Smoothing should be considered an integral follow-up to the calculation of a digital power spectrum. Optimal smoothing width depends on the scale of the features of interest. Larger smoothing widths, for instance, emphasize overall spectral shape but suppress fine frequency variation, such as closely spaced harmonics. This distinction is illustrated in Figure 5, which shows the power spectrum of a human vowel sound conditioned by two different smoothing widths. The first was smoothed with a 50-Hz window and shows the harmonic energy components produced by periodic vocal-fold vibration. The other spectrum was twice smoothed using a 150-Hz window. The effect is to average across individual harmonics, revealing an overall spectral shape. Frequency resolution of the transform before smoothing was 0.76 Hz, time resolution was 1.32 sec, and a Hanning window was used.

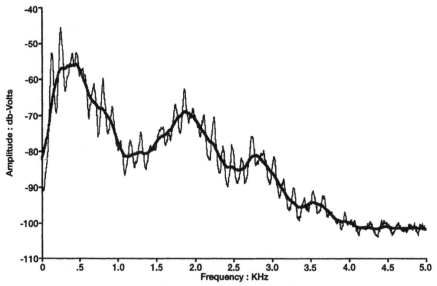

DF: 0.76 Hz T-Lo: 0.ms T-Hi: 800.ms FFT:16384 Wind:HANN Sm: 50.

Fig. 5. Power spectra of a vowel sound from human speech illustrating the effects of smoothing window widths of 50 Hz (*light trace*) and 150 Hz (*bold trace*).

Often, the researcher needs only to measure relative energy levels within a power spectrum, or to compare energy levels between different spectra obtained under constant conditions. Sometimes, however, measurements of absolute sound levels are required. In these cases, spectral energy must be explicitly related to some physical quantity, such as pressure or acceleration. Such relationships are complex; more detailed discussions can be found in Randall (1977) and Beeman (1996b). In brief, power spectra should be displayed in decibel (dB) units relative to 1 Volt-RMS in the source time - signal, so that a 1-Volt-RMS sinewave signal is transformed to a nominal peak spectral level of o dB. This relationship is affected by a variety of factors, including spectral density, the transform window used, and the energy-versus-power units chosen for a particular application. Power spectra can be converted to physical units by applying the calibration factor of the measurement system (e.g., microphone, amplifier, recorder), which is the ratio of the physical input (such as sound pressure) to electrical output (Volts).

4.2
Measuring Similarity Among Power Spectra

As in the temporal domain, patterning in the frequency spectrum can represent a biologically significant aspect of an acoustic signaling process. It is therefore of interest to compare the power spectra derived from various sounds, for instance with respect to overall *sound similarity* and relative *frequency-shift* effects (in which signals show a similar spectral pattern but differ in the absolute frequency position of important spectral peaks). These characteristics can be quantified by cross-correlating pairs of power-

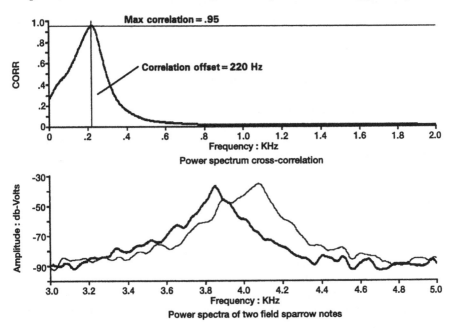

Fig. 6. The power spectra of two field sparrow (*Spizella pusilla*) notes (*bottom panel*), and their cross-correlation function *(top panel)*, showing the maximum correlation value and its frequency offset.

spectrum functions (Beeman 1996b). This computation produces both an overall correlation value and mean frequency-shift for each pair, as illustrated in Figure 6. The results of pairwise similarity analysis over an entire data set can be collected in matrices and analyzed using multivariate statistical techniques like multidimensional scaling to reveal clustering tendencies. Two potential applications of this approach are analyzing the level of spectral similarity among individuals and species, and characterizing spectral changes occurring during ontogeny.

4.3
Other Spectral Analysis Techniques

Spectral analysis can also be conducted using techniques that are not based on the Fourier transform. One technique is *zero-crossing analysis* (ZCA; Greenewalt 1968; Dorrscheidt 1978; Staddon et al. 1978; Margoliash 1983). In this approach, the frequency content of a time-signal segment is calculated by measuring successive time intervals at which the waveform crosses the zero-Volt line, and computing the reciprocal of each such period. The principal advantage of ZCA is that time and frequency resolutions can be as much as ten times greater than in Fourier analysis, because there is no inherent time-frequency resolution trade-off involved. The principal disadvantage of ZCA is that signal components other than the fundamental frequency (e.g., higher harmonics and noise) strongly alter the zero-crossing points in the waveform, degrading the frequency measurements. Fourier transforms, in contrast, produce measurements based on multiple waveform cycles, mitigating the effects of noise, and are unaffected by the presence of higher-order harmonics. Thus, while ZCA avoids the time-bandwidth product limitation, it lacks the performance stability and noise-immunity associated with the Fourier transform.

A statistical technique for spectrum estimation called *linear predictive coding* (LPC; see Markel and Gray 1976, as well as Owren and Bernacki, and Rubin and Vatikiotis-Bateson, this Volume) is widely used for the spectral measurement, modeling, and synthesis of human speech and has also been applied to nonhuman subjects. This technique models the sound spectrum as a sum of spectral resonances, as in the human vocal tract, and consideration should be given to its appropriateness to sound structure and production physiology before application to other species (see Owren and Bernacki for detailed discussion of this issue).

5
Spectrographic Analysis

A frequency-time spectrogram expresses the time-varying spectral content of a signal. It displays energy in three dimensions - amplitude, frequency, and time - where the amplitude of a signal component is expressed visually as the darkness of an area displayed on axes representing time and frequency. Two spectrograms of the human speech utterance "dancing" are shown in Figure 7, produced using approximately the wide and narrow frequency bandwidths traditionally used in the study of acoustic pho-

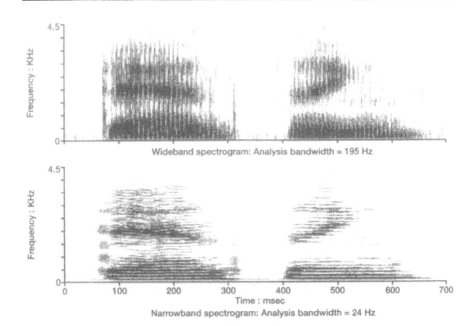

Fig. 7. Narrowband (24 Hz bandwidth, *bottom panel*) and wideband (195 Hz bandwidth, *top panel*) digital spectrograms of the human speech utterance "dancing"

netics. This analysis technique was made widely available with the invention of the analog sound spectrograph (Koenig et al. 1946), an instrument that also became known as the *sonagraph*, the product of which is correspondingly called a *sonagram*. Spectrographic analysis can be enormously useful in discerning spectral-temporal patterns, harmonic structure, and general similarity among sounds, and has historically been the preferred method for investigating bioacoustic signals.

Just as a digitized waveform represents a sampled, numerical version of a physical signal, a digital spectrogram is a sampled, numerical version of the traditional, analog spectrogram. In essence, a digital spectrogram is a two-dimensional, frequency-by-time matrix in which each cell value represents the intensity of a particular frequency component at a given time. Each column in this matrix represents the sound spectrum derived over a short time segment. A digital spectrogram is generated by stepping a short analysis window across the waveform and calculating the discrete Fourier transform of the "windowed" waveform segment at each step. The spectrograms shown in Figure 7 are digital, created using this stepping process, which is illustrated in Figure 8. The spectrogram is displayed on the computer screen by assigning numerical amplitude gradations either to a range of gray shades or to various colors. The matrix can also be shown in a three-dimensional plot, or *waterfall display*, which shows each power spectrum slightly offset from its neighbors on a diagonally oriented time axis.

Digital spectrograms can be used in different ways. First, they can be examined visually, using human judgment to discern patterns and relationships among sounds. Second, unlike the analog case, digital representations can be analyzed using mathematical tools to characterize acoustic properties of interest. This numeric approach allows the researcher to, for instance, measure onset and offset characteristics of notes or syllables both in the time and frequency domains, extract time-varying frequency contours, calculate an average, or template, representation from an ensemble of spectrograms, quantify spectrogram similarity through cross-correlation, and perform pattern recognition and sound classification using a variety of image analysis, similarity, and statistical techniques.

5.1
Spectrogram Generation

Essential considerations in spectrogram generation include time and frequency resolution, the number of transforms used, time *granularity*, and characteristics of the analysis window. As digital spectrograms are based on DFT calculation, the transform parameters used strongly influence the resulting representation. To interpret spectrograms accurately, it is essential to understand these parameters and their effects. These issues are discussed in Beeman (1996b) and only the most important points are reviewed here. Flanagan (1972) and Clements (this Volume) also provide brief mathematical descriptions of DFT-based spectrogram generation.

Time-frequency resolution is constrained by the time-bandwidth product, as discussed earlier. While finer time resolution reveals finer temporal features, the accompanying coarser frequency resolution blurs spectral features. Conversely, achieving finer frequency resolution of these features tends to obscure their temporal locations. These outcomes are clearly illustrated in the two spectrograms shown in Figure 7, which

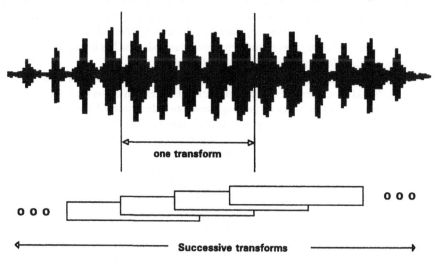

Fig. 8. Generation of a digital spectrogram by applying a transform window to successive segments of the time waveform. The user controls the length of the window and the degree of overlap between adjacent windows

are drawn respectively with wideband (195 Hz, 5 ms) and narrowband (24 Hz, 42 ms) resolutions. Optimum time and frequency resolutions are chosen on the basis of the signal features of interest and the relative importance of temporal and spectral behavior in the analysis (Beecher 1988). Spectrograms with high time and frequency resolution can be produced by non-Fourier-based techniques such as Wigner and Gabor transforms, which avoid the time-bandwidth limitation (Loughlin et al. 1993). These non-linear techniques are promising but can introduce spurious spectral components when applied to non-tonal sounds, and should be used carefully.

Selecting the transform window for spectrographic representation involves essentially the same considerations that apply in calculating a power spectrum. Relative to a rectangular window, the Hanning window produces finer, higher-contrast spectrograms, and is recommended for general use. The Hanning window is useful in separating closely spaced signal harmonics or sidebands, which are blurred by a rectangular window. On the other hand, the Hanning window broadens and delocalizes the bands, making exact frequency determination more difficult.

5.2
Spectrogram Display

As noted earlier, digital spectrograms are displayed by converting the values associated with individual cells in the frequency-time matrix to gray shades or colors on the screen. The *intensity display range* of the spectrogram should be matched to the *dynamic range* of the signal. The display range is the range of screen display levels and varies from darker shades for higher amplitude levels to lighter shades for lower amplitude levels, with white representing the absence of energy. The dynamic range is the difference between the highest and the lowest numerical signal values in the matrix. By varying the display range, the user can emphasize features associated with different intensity levels in the spectrogram. For example, displaying the upper 40 dB of a spectrogram can highlight the signal by causing lower-amplitude, background noise components to lighten or disappear, while narrowing the range to the upper 20 dB may suppress even low-amplitude signal components, producing a *silhouette* spectrogram. Other traditional spectrograph features can also be emulated digitally, such as the *high-frequency shaping filter* used to enhance high-frequency components by increasing their relative amplitude.

In the traditional analog spectrogram, the number of transforms involved is effectively infinite, since the analyzer moves continuously in the temporal domain. In a digital spectrogram, a finite number of transforms are calculated, producing visible time-domain granularity. This granularity can be reduced by computing a larger number of transforms, which increases the processing time required. This trade-off can be ameliorated by interpolating additional columns and rows between the original ones in the spectrogram matrix during the display process. This method is quite effective in increasing the smoothness and visual resolution of the spectrogram display, but does not improve the underlying mathematical time and frequency resolutions.

5.3
Spectrogram Parameter Measurements

Computer-based spectrographic analysis typically involves a spectrogram screen display and a cursor used to measure features of interest. Results of such measurements
are typically stored in a text file for further analysis. Some systems allow the user to
navigate the spectrogram, displaying and measuring waveform and power-spectrum
cross-sections directly. A further enhancement is the ability to view sound data continuously on a real-time, scrolling spectrographic display (Beeman 1991, 1996a), pausing
the display as necessary to make screen measurements.

Computer-based spectrographic screen measurements are faster and more accurate
than traditional approaches involving manual measuring instruments, but are still
manual and thus do not represent a theoretical advance. Truly new approaches include
numerical techniques which analyze the spectrogram matrix to measure sound parameter values, and image analysis techniques for the recognition of sound features. For
example, time and frequency boundaries in the spectrogram can be extracted in a manner analogous to gate-function analysis by locating the coordinates at which matrix
values exceed specified threshold levels. This technique can be used to characterize the
temporal and spectral ranges of the signal, allowing automation of the editing and
analysis process by identifying the temporal segments to be extracted, or the boundaries
of specific frequency bands (see Figure 9).

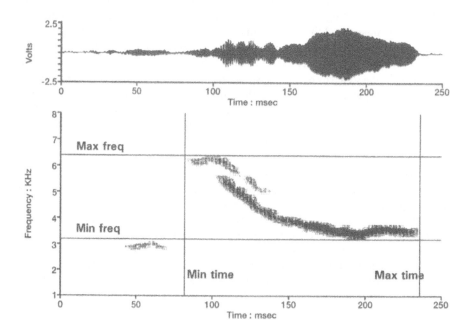

Fig. 9. The onset and offset times and minimum and maximum frequency coordinates of a swamp
sparrow note (*top panel*) are measured from its spectrogram (*bottom panel*)

6
Classification of Naturally Occurring Animal Sounds

In the remainder of this chapter it will be useful to refer to specific sound-types based on their distinctive acoustic features. The terms used for these signals are drawn from the working nomenclature of bioacoustics researchers. Eight basic sound-types are delineated, reflecting both temporal and spectral structures. Each is briefly described below in the context of specific examples from various species, and illustrated by waveforms and spectrograms in Figure 10.

6.1
Properties of Ideal Signals

6.1.1
Periodicity

A brief explanation of the applicable, engineering-based terminology is also helpful (see Gerhardt, this Volume, for a more complete discussion). In this literature a distinction is made between periodic and non-periodic signals. *Periodic* signals are characterized by repeated amplitude cycles in the time-waveform. In a naturally occurring biological signal, such periodicity typically represents the repeated operation of a sound-producing mechanism like vibration of a membrane (e.g., syringeal sound production in birds; see Gaunt and Nowicki, this Volume) or in paired tissue structures (e.g., vocal folds underlying laryngeal sound production in mammals, including nonhuman primates and humans; see Rubin and Vatikiotis-Bateson, and Owren and Bernacki, this Volume).

Periodic sounds are classified as being either *simple* (sinusoidal or pure-tone), containing energy at a single frequency, or *complex*, containing energy at multiple frequencies. In the latter, spectral energy occurs in the *fundamental frequency* (corresponding to the basic rate of vibration in the underlying sound-producing structure), and its *harmonics* (spectral components occurring at integer multiples of the fundamental frequency). Periodic sounds are contrasted with *aperiodic* sounds whose amplitude variation does not repeat cyclically. Aperiodic sounds appear noisy, and their spectra cannot be expressed as a fundamental frequency and its harmonics. Aperiodic sounds can result from a single, sharp impulse like a click, or from inherently unpatterned, turbulent airflow. Note that these sound types are idealized in that no naturally occurring sound would be likely to be a pure tone, nor to repeat exactly from cycle to cycle.

6.1.2
Amplitude Modulation

Amplitude modulation (AM) refers to changes in the overall energy level of a signal occurring over time (see Stremler 1977). As used here, this term refers to any periodic variation in intensity about an average intensity level, or slow changes in intensity that may not be cyclical in nature. The frequency of the energy component whose intensity

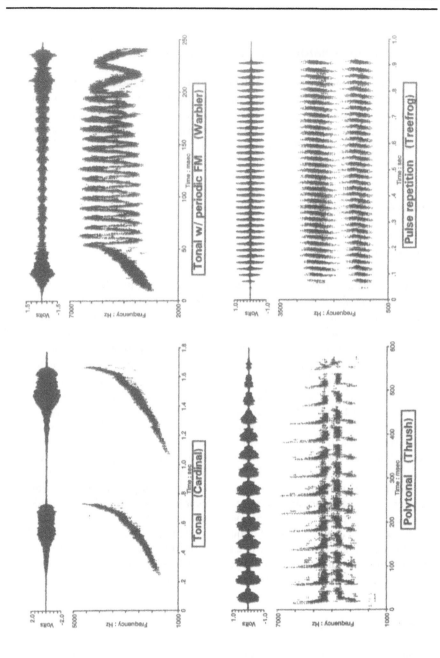

Fig. 10. Classification of eight basic sound types based on biologically relevant temporal and spectral properties. Representative waveforms (*top panels*) and spectrograms (*bottom panels*) are shown for each

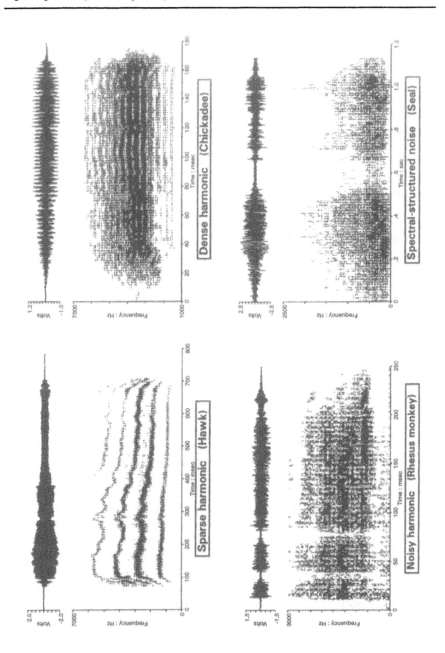

is changing is called the *carrier frequency*, the rate of envelope variation is the *modulation frequency*, and the magnitude of variation is expressed by the *modulation index*. Thus in Figure 20, a 1000-Hz signal whose peak amplitude varies between 0.80 and 1.20 at a rate of 100 times per second has a carrier frequency of 1000 Hz, a modulation frequency of 100 Hz, and a modulation index of 0.20. AM patterning is readily produced and is often of functional significance in bioacoustic signals, for instance as an important component of sound character, or in providing clues about the physical nature and operation of the sound-producing mechanism.

6.1.3
Frequency Modulation

Frequency modulation (FM) refers to changes in the instantaneous frequency of a signal over time (see Stremler 1977). As used here, it refers to any periodic variation of instantaneous frequency about an average frequency value, or overall frequency changes that are not cyclical in nature. In a signal that varies in a periodic fashion, the center frequency of the modulated energy component is called the *carrier frequency*, the rate of variation is the *modulation frequency*, and the frequency magnitude of the variation is the *modulation depth*. Thus, a signal that varies between 980 and 1020 Hz at a rate of ten times per second has a carrier frequency of 1000 Hz, a modulation frequency of 10 Hz, and a modulation depth of 20 Hz.

6.1.4
Biologically Relevant Sound Types

In addition to the above terminology, bioacousticians have devised terms to express the various sound structures in naturally occurring communication signals. These are described here and illustrated in Figure 10 in order of increasing temporal and spectral complexity. *Tonal* sounds contain a single, *dominant* frequency component at each time instant, although frequency and amplitude may vary with time. A dominant frequency is one that is significantly higher in amplitude than any other frequency component present and is often the perceptually salient spectral component. Note, however, that the perceptual significance of any aspect of a bioacoustic signal can only be examined through explicit testing, as described in this Volume by Cynx and Clark, and Hopp and Morton, who examine laboratory- and field-based methods, respectively. Tonal sounds, which often resemble whistles, include many bird songs and signals produced by marine mammals (e.g., dolphins and beluga whales, *Delphinapterus leucas*). *Tonal sounds with periodic FM* are single-frequency sounds that show regular, often rapid pitch variation, whose auditory quality can resemble a harsh buzz rather than a whistle. Because they exhibit only one frequency at a time, they are considered tonal and are commonly observed among birds, especially warblers. The dominant frequency in both types of tonal sounds may represent a fundamental or a higher-level harmonic. For instance, Gaunt and Nowicki (this Volume) discuss how filtering in the vocal tract of songbirds can shape the sound by suppressing energy at the fundamental, leaving the dominant frequency at a higher harmonic.

Polytonal sounds contain two or more spectral components that are not harmonically related, representing the operation of two or more independent periodic sound-producing mechanisms. An example is the two-voice phenomenon of birdsong, also discussed by Gaunt and Nowicki (this Volume).

Pulse-repetition signals consist of a series of energy bursts whose acoustic structure may be constant or may vary between pulses. The pulses themselves may be tonal, show multiple frequency components, or exhibit broadband spectral energy, and may be amplitude- or frequency-modulated. These sounds are produced by a variety of species, including anurans, marine mammals, birds, and nonhuman primates. Their audible quality is often a repetitive buzz and can incorporate underlying pitch variation over the pulse sequence.

Sparse-harmonic sounds contain a relatively small number of harmonically related spectral components, and range in sound quality from a harsh whistle to a light rasp. *Dense-harmonic* signals contain a larger number of spectral components and sound more harsh or nasal than sparse-harmonic sounds. Examples include chickadee *dee* notes, humpback whale (*Megaptera novaeangliae*) cries, nonhuman primate calls, and human speech sounds in which vocal fold vibration (*voicing*) occurs.

Noisy-harmonic sounds consist of a combination of tonal or harmonic components with significant additional noise. These signals have a harsh auditory quality in proportion to their noise content. Examples include nonhuman primate vocalizations, and some calls produced by starlings and crows.

Spectrally structured noise consists of noise energy with one or more time-varying spectral peaks or other coherent spectral structure. Examples include nonhuman primate *screams* and *barks*, Mexican chickadee (*Parus sclateri*) calls, and fur seal vocalizations. Sounds like nonhuman-primate screams that show pronounced spectral peaks can give an audible impression of time-varying pitch. In other cases, the sound quality may be of a grunt, bark, or roar with little pitch impression.

7
Time-varying Frequency Analysis

Many sounds involve frequency variation over time, and this variation is often significant in the communication process. Frequency-time analysis involves the measurement and characterization of such variation, and is concerned with the duration, range, and patterning of spectral changes including, for example, maximum and minimum frequencies and the spectral patterns of different sound types. Tonal and harmonic sounds, in particular, can be characterized effectively by their time-varying dominant frequency, which will be referred to as the *spectral contour* of a signal (Beeman 1987, 1996b). Mathematically, the contour value at each instant is the frequency containing the most energy, and this can be visualized as the frequency "spine" of the sound's spectrogram. Depending on the sound, the spectral contour may represent the fundamental frequency or a higher harmonic.

Spectral contour analysis represents a powerful tool for characterizing tonal and harmonic signals. The resulting function can be measured, manipulated, and compared statistically in the same way as the sound's amplitude envelope. It is more comprehensive than individual sound parameters, and mathematically more tractable than the spectrogram as a whole. Once derived, spectral contours can be used to automatically extract and measure frequency-time features such as contour minima and maxima, rate of frequency change over time, frequency range, and signal duration. Contours can be used as the basis for similarity analysis, which can be advantageous for harmonic or noisy sounds that are less amenable to similarity measurement by cross-correlation of the entire spectrogram (discussed below).

7.1
Deriving Spectral Contours

Any means of determining the instantaneous dominant frequency of a sound can produce a spectral contour function. This section discusses several approaches. More extended discussion and additional references are provided by Beeman (1996b).

Dominant frequency extraction can be performed by determining the frequency at which each column of a digital spectrogram achieves maximum amplitude, then recording these frequency values as a continuous time function – a technique referred to here as *spectrogram contour detection* (SCD; see Beeman 1987, 1996b; Buck and Tyack 1993) and illustrated in Figure 11. Because SCD is based on Fourier analysis, it is inherently frequency-selective and contours can be extracted successfully from time signals containing both multiple harmonics and broadband noise. In fact, SCD is sufficiently immune to noise components that it has been used successfully to extract noise-free playback stimuli from field recordings (see below).

SCD may not function effectively on sounds whose dominant frequency jumps between harmonics [such as human speech, nonhuman primates, or the black-capped chickadee's (*Parus atricapillus*) *dee* note], or which contain noisy components with time-varying spectral peaks. For such sounds, the frequency values returned may jump between harmonics or momentary peaks in the noisy spectrum. In these cases, accurate frequency contours can nonetheless be obtained by *hand-drawing* over a frequency-magnified spectrogram display using a mouse or some other tracing method (Beeman, 1996b). Existing contours can be edited in the same manner for smoothing or error correction.

Zero-crossing analysis (ZCA), can also be used to derive spectral contours, and can yield contours with very high time and frequency resolution on tonal sound material. The limitation of ZCA is that it cannot be used effectively on sounds containing harmonic or noise energy. However, ZCA can be a useful technique, as noted with respect to sound synthesis below. *Hilbert transform analysis* (Dabelsteen and Pedersen 1985; Oppenheim and Schafer 1989) applies the mathematics of complex variable theory to decompose any tonal or harmonic time signal into the product of a frequency function $F(t)$ and an amplitude function $A(t)$. With tonal sounds, Hilbert analysis produces precise and useful results, but like ZCA, it degrades in the presence of harmonic and noise

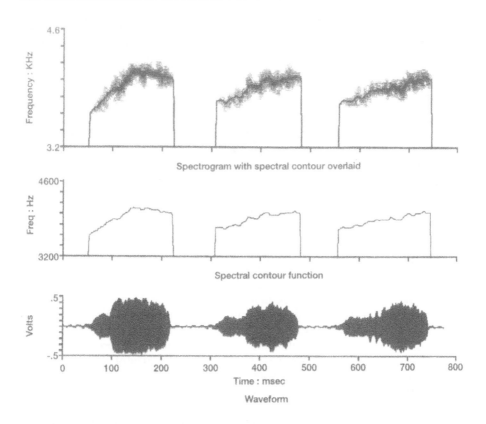

Fig. 11. Spectral contour detection applied to three swamp sparrow notes, showing spectrographic representations with the extracted pitch contours superimposed (*top panel*), the pitch contours alone (*middle panel*), and the waveforms (*bottom panel*).

energy. Like ZCA, Hilbert analysis is not the method of choice in most biological applications, but can perform well on tonal or nearly tonal material

7.2
Sound Similarity Comparison

Researchers have long sought to gauge the overall similarity of various sounds, for instance to relate signal acoustics to ontogeny, individual variation, geographical location, and species identity. Sound comparison has traditionally been performed either by comparing sounds on a qualitative basis (e.g., Borror 1965; Kroodsma 1974; Marler and Pickert 1984), or by reducing sounds to significant parameters that can be compared statistically (e.g., Nelson and Croner 1991). The first approach includes judgments involving auditory comparison of sounds and visual comparison of spectrograms, both of which can include bias related to human perceptual processing, and thereby lack objectivity, repeatability, and mathematical foundation. The second approach is better in these respects, but reducing a complex sound to a limited set of parameters may fail to capture all its salient aspects and parameter choice is subject to bias. Nonetheless, waveforms, power spectra, and spectrograms have all been used as sources for a wide variety of parameters.

Parameter-based sound similarity has been measured in a number of ways. Nelson (1992), for instance, used the simple Euclidean distance between parameter vectors (i.e., the square-root of the sum of the squared differences between corresponding parameter values) as a similarity metric. Parametric and nonparametric analyses of variance can be used to determine which features differ statistically across different sound types. Various multivariate statistical techniques (see Sparling and Williams 1978 for a review) have been applied as well, including discriminant-function analysis (Hauser 1991) and principal-components analysis (Martindale 1980; Clark 1982; Nelson and Croner 1991). These approaches have used the *proportion of variance explained* as a measure of the significance of various sound features and have classified sounds on the basis of statistical clustering on these features. Multidimensional scaling can also be used to portray the clustering tendencies within a dataset, for instance based on pairwise similarities between sounds (Clark et al 1987; Hauser 1991).

Because spectrograms provide an intuitive overview of the spectral and temporal characteristics of a sound, they are a desirable basis for quantitative similarity comparisons between signals. With digital spectrograms, similarity can be measured by comparing the complete numerical matrices underlying the visual representations. This approach has the advantage of being objective and repeatable and of including the whole sound rather than individual, measured parameters. Quantitative similarity can be measured by calculating the normalized covariance between two spectrograms at successive time-offsets (Clark et al. 1987). This process can be visualized as sliding two transparencies containing the spectrograms over each other along the time axis, and measuring the degree of overlap between the images at each point. The resulting correlation function, $R(T)$, is a sequence of correlation values representing spectrogram similarity as a function of time-offset. The peak value of $R(T)$ quantifies the maximum similarity between the two spectrograms.

In this approach, the spectrograms are first normalized for overall amplitude and time-offset becomes a variable (*normalization* refers to the process of removing the variation in one sound feature from an ensemble of sounds, rendering the signals equivalent on that particular parameter). The maximum-similarity value is then independent of overall amplitude differences or temporal offset in the original digital signal files. Figure 12 shows a spectrogram-based cross-correlation comparison between two swamp sparrow (*Melospiza georgiana*) syllables, showing the maximum correlation value and time-offset. Because digital spectrograms are based on Fourier transforms, which have inherent immunity to spectral noise, the spectrogram-similarity technique can be used effectively on quite noisy signals. Furthermore, the comparison can be restricted to a specified bandwidth within the spectrograms, thereby excluding out-of-band extraneous noise.

In general, quantitative spectrogram-similarity comparison provides a sensitive and comprehensive measure for sounds that generally resemble each other. However, the

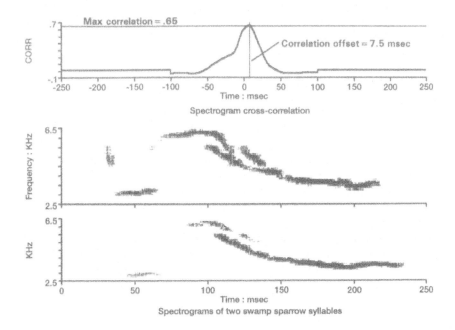

Fig. 12. Quantifying the overall similarity between two swamp sparrow syllables by cross-correlation (*top panel*) of their spectrograms (*middle* and *bottom panels*)

technique can have severe limitations, even for quite similar sounds, unless normalization is performed with respect to characteristics such as mean frequency, frequency range (maximum minus minimum frequency), duration, or harmonic interval. In the absence of normalization, variation in these features may produce low cross-correlation values for apparently similar sounds and mask biologically important similarities. Note, however, that removing a parameter's variation from the similarity calculation also removes that parameter's influence from the analysis. It is therefore important to study clustering tendencies in a dataset with various parameters both included and excluded, as well as specifically testing the behavioral importance of these parameters empirically.

Recently developed tools address some of the limitations noted above (Beeman 1996b). For example, mean frequency differences between individual signals in an ensemble can be removed by cross-correlating their power spectra and frequency shifting the associated spectrograms before cross-correlation. Similarly, differences in signal duration can be removed by uniform linear expansion of the time-scale of each spectrogram as needed before comparison. Alternatively, sounds can be compared on the basis of mathematical functions that are more comprehensive than extracted parameters, but more flexible than spectrograms. Such functions include amplitude-time envelopes, pulse-repetition patterns, spectral contours, and power spectra (Beeman

1996b). For example, differences in frequency range and harmonic interval can be overcome by reducing sounds to spectral contours, performing the necessary normalization, and then calculating similarity.

8
Digital Sound Synthesis

Digital signal processing provides a wide variety of techniques for sound *synthesis*, which will be considered to include any significant creation or manipulation of sound material, whether based on natural or mathematically specified signals. General approaches include digital *editing, arithmetic manipulation*, and *generation* of sound, each of which provides the ability to create precise, repeatable, and incremental signal variations. Editing is manipulation of existing sound material by rearranging waveform segments in the time domain. Arithmetic manipulation includes both combining and altering signals through such operations as addition and multiplicative amplitude scaling using a specific function of interest. Sound generation – creating new waveforms – is a particularly important technique that can combine elements of both editing and arithmetic manipulation with de novo creation of segments or entire signals.

8.1
Editing

Sound editing, as used here, means rearranging existing signal components in time, without generating new physical waveforms. Ideally, a digital time-signal editing system should be visually based, since the process involves moving sound segments from one location to another. Beginning with source and target waveforms on the computer screen, the user graphically selects a signal segment and an editing operation to perform. The system should provide basic cut-and-paste operations, allowing individual signal components to be extracted, inserted, deleted, and concatenated. These processes are illustrated in Figure 13. Editing commands can be used to extract or reorder signals for analysis, or to alter signals for use in playback experiments, for instance in combination with synthesis techniques that can be used to alter the temporal and spectral characteristics of the signal.

Bioacousticians have long desired a digital *spectrogram editor* that would be capable of extracting, moving, or deleting specified segments in *both* the time and frequency domains. Such an editor would be able to shift a signal's energy components in the frequency domain, selectively erase or redraw particular harmonics, and the like (Zoloth et al. 1980), then generate the corresponding time signal. However, such capabilities have not developed far because the approach itself is fundamentally problematical. Frequency-based editing changes can cause severe mathematical artifacts in the corresponding time signals. For example, changes in frequency characteristics can produce amplitude modulation in the waveform, while many frequency-band deletions simply cannot be transformed back to the time domain. More practicable approaches can sometimes be used to produce the desired results, including spectral-filtering and

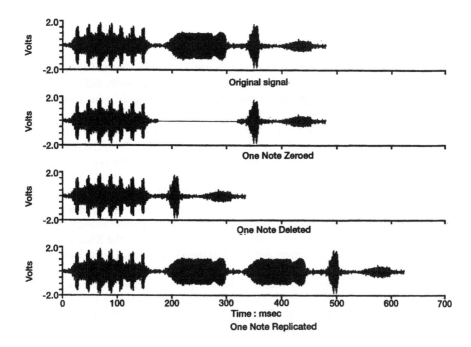

Fig. 13. Using digital signal editing to alter the waveform of a sparrow song (*top panel*) to set one note to zero (*top middle panel*), delete the resulting silent interval (*bottom middle panel*), or replicate that note (*bottom panel*)

waveform-conditioning techniques discussed in the following sections (also see Stoddard, this Volume).

8.2
Arithmetic Manipulation and Generation of Sound

The arithmetic manipulation of existing waveforms and the generation of new sounds are often combined in practice. In fact, many of the functions of the "mythical" spectrogram editor can be implemented using combinations of the techniques described below. Complex signal characteristics like amplitude-envelope shape, spectral composition, AM, FM, and phase relationships of individual components can all be modified with great precision. Furthermore, temporal and spectral features can be independently shifted and expanded. The functions used to alter or synthesize a given signal or signal element can be based on naturally occurring sounds, or created through graphical approaches like mouse-based drawing, or derived from mathematical functions.

8.3
Synthesis Models

There are many synthesis models currently in use. Each embodies different assumptions about sound structure and sound production, and as a result, different models synthesize different sound types better than others. The *tonal model*, which is described in depth below, represents sounds as frequency- and amplitude-modulated sinusoids. Because of its mathematical tractability, this model has received much attention in animal-related studies, and works well for a wide range of natural sound material, including whistles, rapid AM and FM buzzes, harmonic sounds, pulse-repetition signals, and some noisy-harmonic sounds.

Another approach models critical aspects of the physiological processes involved in the production of the sound. For instance, many aspects of speech production have been modeled in this fashion, as described by Rubin and Vatikiotis-Bateson (this Volume). One classic example of this type of *articulatory model* is the *Klatt synthesizer* (Klatt 1980), which has been extensively used in speech research. This software model passes voiced sound (based on periodic energy bursts designed to mimic vocal-fold vibration pulses) and unvoiced sound (noisy energy that mimics turbulent air flow) through a bank of parallel filters that simulates the physical resonances of the human vocal tract. The parameters of these filters are continuously varied, thereby representing the time-varying resonances resulting from the articulatory maneuvers of speech. The Klatt synthesizer has also been used with nonhuman primates to test the auditory processing of both human speech sounds (e.g., Hienz and Brady 1988; Sinnott 1989) and species-specific vocalizations (e.g., May et al. 1988; Hopp et al., 1992).

8.3.1
Tonal Model

A wide variety of bioacoustic signals can be modeled as tonal or harmonic, and as a result can be readily represented, manipulated, and synthesized (e.g., Greenewalt 1968; Dorrscheidt 1978; Margoliash 1983; Beeman 1987, 1996b). Such sounds are mathematically tractable and form versatile building blocks for more complex sounds. The tonal

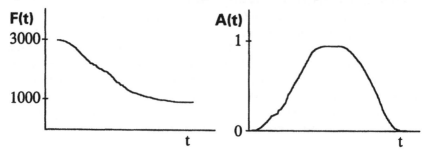

Fig.14. *F(t)* and *A(t)* functions for a tonal sound that descends in frequency from 3000 to 1000 Hz, while first rising and then falling in amplitude

model provides techniques used to synthesize tonal sounds, pulse-repetition sounds, and harmonic sums of tonal sounds, and allows for extensive mathematical manipulation of amplitude, temporal, and spectral sound properties.

Due to the simplicity of its structure, a tonal sound can be completely represented by two time functions – a spectral contour, $F(t)$, which represents its time-varying frequency, and an amplitude envelope, $A(t)$, representing its time-varying intensity. As an example, the functions shown in Figure 14 describe a sound whose frequency descends from 3000 to 1000 Hz, and whose amplitude rises from 0 to a maximum level, remains steady, and then declines to 0.

$F(t)$ and $A(t)$ are used to synthesize sounds as follows. Both functions are represented as time signals, with values specifying instantaneous frequency (in Hz) and intensity (in Volts), respectively. To recover the original signal, $F(t)$ is passed through a voltage-to-frequency sinewave generator, producing a constant amplitude signal of time-varying frequency F(t). This signal is then multiplied by $A(t)$, giving it the appropriate amplitude characteristics. This process is schematically illustrated in Figure 15. Mathematically, the process is justified by Hilbert transform theory, which shows that any signal can be decomposed into frequency and amplitude functions, and then recovered from these functions without loss of information (Oppenheim and Schafer 1989). Because the frequency and amplitude functions are independent, frequency and amplitude features can be independently modified, allowing their communicative significance to be tested separately.

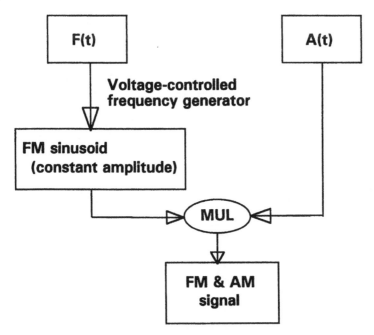

Fig. 15. How $F(t)$ and $A(t)$ functions are used in conjunction with a voltage-to-frequency sinewave generator to produce a synthetic, tonal sound. MUL= Multiply

8.4
Sources of F(t) and A(t) Functions

8.4.1
Mathematically Based Functions

$F(t)$ and $A(t)$ functions can be based on elementary mathematical functions such as *constants, linear ramps, exponential ramps, sinusoids,* and combinations of these. Figure 16 shows five different $F(t)$ *functions,* together with the resulting time waveforms. The constant function produces a single, unchanging frequency component. Linear- and exponential-ramp functions produce signals that change in frequency at a constant rate (in Hz per second) and at an exponential rate (in octaves per second), respectively. A sinusoidal function produces a signal that changes frequency at a sinusoidal rate, for instance ten times per second. Custom functions allow the user to create a wide variety of FM signals through any combination of mathematical functions and natural sound material.

$A(t)$ functions are created in a similar manner. Figure 17 shows *rectangular, trapezoidal taper, exponential taper, cosine taper,* and *sinusoidal* functions, and the resulting

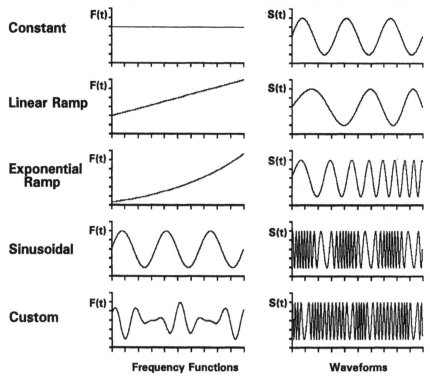

Fig. 16. Mathematical $F(t)$ functions used in sound synthesis to produce frequency-modulated waveforms, $S(t)$

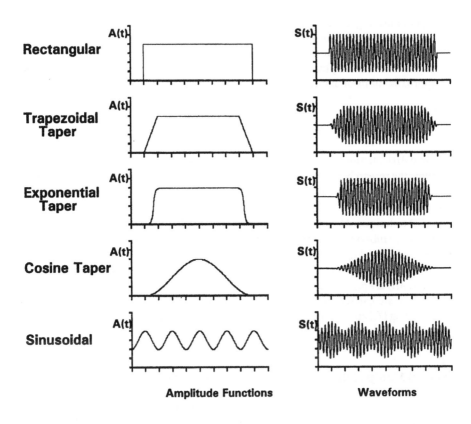

Fig. 17. Mathematical $A(t)$ functions used in sound synthesis to produce amplitude-modulated waveforms, $S(t)$

time waveforms. The rectangular function produces a waveform whose envelope changes amplitude instantaneously at onset and offset, with constant amplitude during the signal. The trapezoidal taper function produces a waveform whose amplitude changes linearly between zero and its maximum value at onset and offset. The exponential taper function produces waveform onset and offsets whose amplitudes change exponentially. The cosine taper function produces a short, heavy taper, with minimum discernible onset or offset. Finally, sinusoidal amplitude modulation is produced by a sinusoidal function, and results in periodic variation in the waveform envelope. Amplitude functions can contribute to sound quality (as in amplitude modulation), or they can simply be used to taper the onsets and offsets of synthesized stimuli before experimental use. Overly sharp onsets and offsets (i.e., rising or falling to full amplitude within in a few milliseconds) generate perceptible pops on playback, in addition to injecting spurious spectral energy, and should be avoided.

$F(t)$ and $A(t)$ functions can be combined and transformed using various mathematical operations to generate sound features like harmonic complexes, modulation, and amplitude scaling. Because they are universal, objective, and readily described, synthetic sounds based on mathematical functions are widely used as stimuli for testing responses at the neurophysiological level. However, they are generally less successful as approximations of naturally occurring biological sounds.

8.4.2
Functions Derived from Natural Sounds

$F(t)$ and $A(t)$ functions can be derived directly from tonal and harmonic natural sounds, to allow the researcher to precisely modify frequency or amplitude features for testing (Margoliash 1983; Beeman 1987, 1996b). $F(t)$ can be derived from the time waveform, using zero-crossing or Hilbert transform techniques, or from the spectrogram, using the SCD technique. $A(t)$ can be derived from the amplitude envelope of the time waveform, or by Hilbert transform techniques. $F(t)$ and $A(t)$ functions of any shape can also be drawn by hand using the mouse.

While synthesis from mathematical functions involves generation and synthesis – $F(t)$ and $A(t)$ are generated then combined – synthesis from natural sounds begins with an analysis step, in which $F(t)$ and $A(t)$ are derived from the source sound. The process of analyzing $F(t)$ and $A(t)$ from natural sounds is summarized in the top panels of Figure 18.

9
Sound Manipulation and Generation Techniques

Once the $F(t)$ and $A(t)$ functions have been selected or derived, manipulation of an existing sound or generation of an entirely new signal can occur. The following sections describe these manipulations, includw'ng frequency shifting, time scaling, amplitude and frequency modulation, intensity manipulation, harmonic removal and rescaling, pulse repetition variation, phase adjustment, noise addition and removal, and the derivation of template sounds. Further details are described by Beeman (1996b). In the simple example shown in Figure 18, the synthesis manipulation consists of time-reversing the sound. Based on the original sound's spectrogram and time waveform (at the top of the figure), respectively, the sound is first analyzed into $F(t)$ and $A(t)$ functions. These functions are then reversed to produce the new, altered functions $F'(t)$ and $A'(t)$. $F'(t)$ is used to synthesize a constant-amplitude sinusoid representing the frequency variation, which is then multiplied by $A'(t)$. The spectrogram and waveform of the synthetic sound are shown at the bottom of the figure.

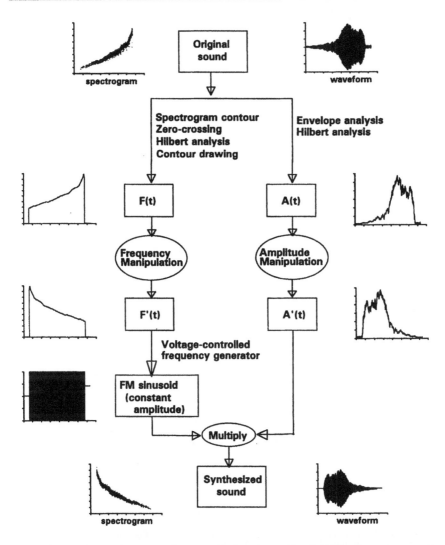

Fig. 18. The complete analysis-manipulation-synthesis process. $F(t)$ and $A(t)$ functions are extracted from a natural sound, reversed in time (in this example), and then resynthesized to produce a time-reversed synthetic version of the original signal

9.1
Duration Scaling

Sound duration can be uniformly expanded or compressed without altering frequency characteristics, by expanding or compressing both the $F(t)$ and $A(t)$ functions via linear interpolation. Attempting to alter temporal features by resampling the time waveform

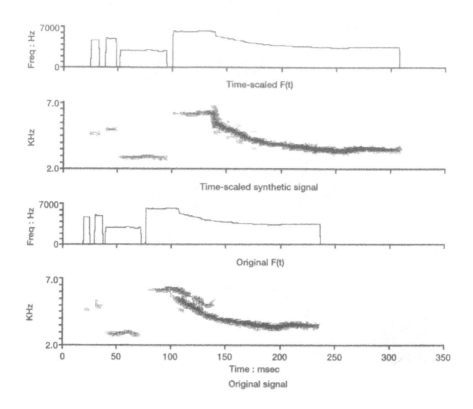

Fig. 19. A synthetic version of the swamp sparrow syllable shown in Fig. 1 is increased in duration by 30 % by time-expanding its *F(t)* and *A(t)* functions

itself would introduce severe artifacts. Instead, all manipulations are performed on $F(t)$ and $A(t)$, and artifacts are avoided. Figure 19 shows a swamp sparrow syllable (the same sound as shown in Figure 1) whose duration has been increased by 30 % by this method, without changing its frequency characteristics. This concept can be extended to non-uniform time expansion or compression, a technique sometimes called *dynamic time warping*, which involves a variable stretching or contracting process.

9.2
Amplitude Envelope Manipulations

A signal's amplitude envelope may include both short-term periodic and long-term non-periodic variation. The former can often be modeled sinusoidally, while the latter may represent signal onset or offset, phonetic emphasis, and other slowly varying effects. As illustrated in Figure 20, a signal's overall amplitude envelope, $A(t)$, can be mathematically modeled as the sum of a slowly varying envelope, $A_s(t)$ (a constant value of unity in the figure), and a rapidly varying modulation function, $A_m(t)$, so that $A(t) = A_{s(t)} + A_{m(t)}$. $A_{s(t)}$ is derived from $A(t)$ by smoothing, to remove the rapid modulation. The smoothing window should be wider than the modulation period, but short enough to retain the time-varying shape of $A_s(t)$. A 15-ms window, for example, might be appropriate for a 100-Hz (i.e., 10-msec period) modulation rate. $A_m(t)$ is then obtained by subtracting $A_s(t)$ from $A(t)$. $A_s(t)$ and $A_m(t)$ can then be analyzed independently.

Amplitude envelopes can be manipulated in other ways as well. As noted earlier, RMS energy can be equalized among a set of signals, for example to equalize the intensity levels of a series of playback signals created from natural sounds. Alternately, amplitude variation can be removed from a signal entirely by dividing the signal by its $A(t)$, leaving a signal of uniform amplitude but time-varying frequency. Note that this ma-

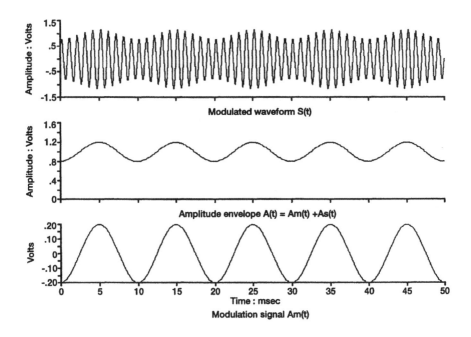

Fig. 20. An amplitude-modulated waveform [$S(t)$, *top panel*] and its complete amplitude envelope [$A(t)$, *middle panel*] and amplitude modulation function [$A_m(t)$, *bottom panel*)]

nipulation will amplify any low-level noise segments in the signal. Finally, all signals in an ensemble can be given the same amplitude envelope $A_{ref}(t)$, by multiplying each signal envelope $A_n(t)$ by the function $A_{ref}(t) / A_n(t)$.

9.3
Spectral Manipulations

9.3.1
Frequency Shifting and Scaling

Tonal synthesis allows spectral changes to be made without altering temporal or amplitude relationships, because $F(t)$ can be modified independently of $A(t)$ and other time characteristics. Uniform frequency shifting is performed by adding a constant positive or negative value to $F(t)$, while leaving $A(t)$ unchanged. Because the adjustment is made to the frequency function rather than the waveform or spectrogram, the output signal is free of artifacts (such as amplitude modulation) that have been a major shortcoming of these other approaches. Figure 21 shows a swamp sparrow syllable whose frequency has been uniformly raised by 500 Hz without altering overall duration. Frequency can be shifted logarithmically as well, for instance by multiplying $F(t)$ by a factor

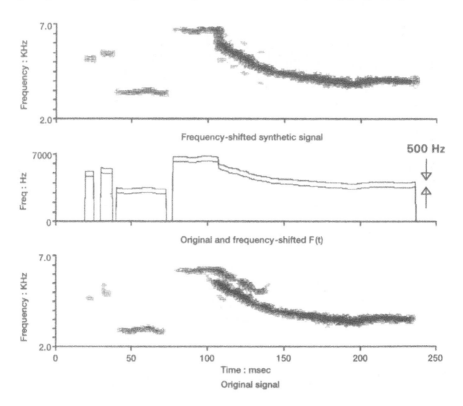

Fig. 21. A synthetic version of the swamp sparrow syllable shown in Fig. 1 is shifted upward in frequency by 500 Hz by adding 500 to its $F(t)$ function

of 2 to produce an increase of one octave. $F(t)$ can also be compressed or expanded in range to alter the signal's frequency variation about its mean frequency.

9.3.2
Frequency Modulation

Any FM characteristics modeled in an $F(t)$ function can be altered, replaced, or even removed. The general approach is to analyze the signal into two components, a slowly varying carrier frequency $F_c(t)$ and rapidly varying frequency modulation $F_m(t)$, so that $F(t) = F_c(t) + F_m(t)$, as described by Beeman (1996b). FM parameters are then determined from these functions. Figure 22 shows an Eastern phoebe (*Sayornis phoebe*) note analyzed in this manner into functions representing the total contour, $F(t)$, its carrier frequency, $F_c(t)$, and its frequency modulation, $F_m(t)$, about the carrier. One can then measure from $F_m(t)$ the signal's mean modulation frequency (about 86 Hz) and mean modulation depth (about 416 Hz). These functions can then be altered to produce a synthetic sound with altered characteristics. Independent changes to $F_c(t)$ and $F_m(t)$ can be used, for example, to increase or decrease the signal's overall frequency change by scaling $F_c(t)$, or to double the magnitude of modulation depth by multiplying $F_m(t)$ by a factor of 2. $F_c(t)$ and $F_m(t)$ are then recombined to produce a new $F(t)$.

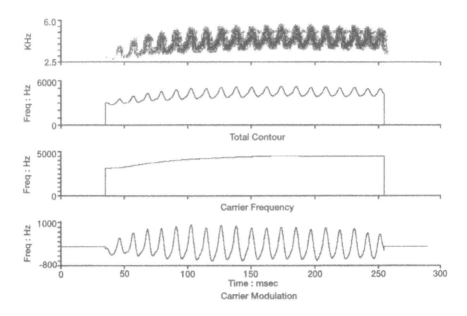

Fig. 22. The spectrogram of a frequency-modulated Eastern phoebe (*Sayornis phoebe*) note (*top panel*) is analyzed into its total spectral contour (*top middle panel*), which is further analyzed into carrier frequency (*bottom middle panel*) and carrier modulation (*bottom panel*)

9.4
Synthesis of Biological Sound Types

9.4.1
Tonal and Polytonal Signals

Tonal signals are the simplest to characterize and synthesize using the methods outlined
in this chapter. Polytonal signals are more difficult to synthesize, and the choice of
modeling approach depends on the original production process. Depending on the
sound and the species, tones may be either added or multiplied during sound production. The former produces tonal sums, while the latter produces AM cross-products
which can assume complex forms, including harmonics at intervals of the amplitude-
modulation frequency, or sums or differences of the input frequencies. Pitch interactions within the polytonal sound can be subtle or quite dissonant, depending on the
degree of coupling occurring between the sound sources and the frequencies involved.
Polytonal sounds can display a great diversity and complexity of spectral characteristics
and a corresponding variation in sound quality.

Synthesis approaches for these more complex signals depend on their structure.
Polytonal sums can sometimes be analyzed by separating the voices via filtering or

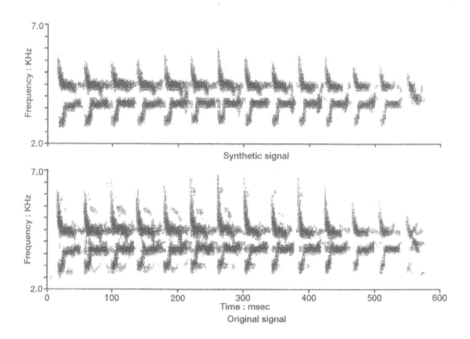

Fig. 23. A polytonal sound produced by a wood thrush (*Hylocichla mustelina; bottom panel*) is
synthesized using polytonal synthesis (*top panel*)

spectrogram contour detection. However, the most complex polytonal sounds, such as some chickadee calls, may involve intermodulation between two harmonic voices, and therefore require modeling the modulation process itself in order to reproduce the complex spectral relationships. Figure 23 shows a segment of wood thrush (*Hylocichla mustelina*) song containing the sum of two distinct nonharmonic voices. Because the voices occupy different frequency ranges, they can be modeled by analyzing the sound into two independent sets of amplitude and frequency functions.

9.4.2
Pulse-Repetition Signals

Pulse-repetition sounds can also arise from a variety of production mechanisms and therefore vary widely in structure, and in the techniques used to synthesize them. One approach is to generate mathematical functions that are used singly or in combination to produce individual pulses, from which a sequence can be built through waveform editing. Pulses can be repeated at constant or variable intervals, and the sequence can be uniform, show patterned variation (such as stepped changes in frequency or intensity), or be randomly changing. The onset and offset amplitudes of each pulse should be tapered, as discussed earlier.

Alternatively, synthetic pulses can be derived from naturally occurring pulse-repetition sounds (such as anuran vocalizations). The pickerel frog (*Rana palustris*) call

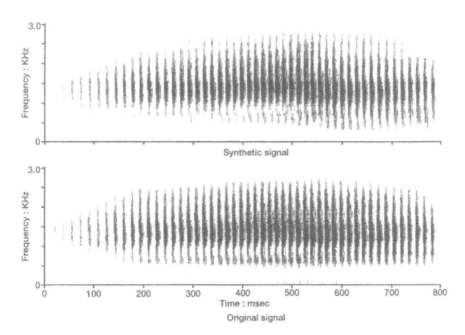

Fig. 24. A pulse-repetition call of a pickerel frog (*Rana palustris*; *bottom panel*) is synthesized using pulse repetition synthesis techniques (*top panel*)

shown in Figure 24, for example, can be modeled as a sequence of broadband FM pulses, further modulated by an AM pulse-repetition envelope. The aural impression of this sound reflects both the pitch characteristics of its FM behavior and the buzziness produced by the rapid pulse repetition. $F(t)$ can be obtained through zero-crossing analysis, and $A(t)$ through amplitude-envelope analysis. FM and pulse-repetition behavior are thus separated. Manipulations include altering or removing the FM component, compressing or expanding the pulse repetition rate, and inserting or rearranging individual pulses.

9.4.3
Harmonic Signals

Harmonic sounds can be generated by synthesizing individual harmonics using the tonal model, then adding these components together. Because individual harmonics are explicitly separated in this approach, they can be individually edited and weighted before being recombined. All harmonics can be synthesized from the fundamental contour, $F(t)$, referred to as $F_0(t)$. Figure 25 shows a chickadee *dee* note in its original form and in a synthetic version based on this technique.

Harmonic synthesis proceeds in five steps. First, the fundamental frequency contour, $F_0(t)$, is extracted using one of the techniques described above. If the energy is concen-

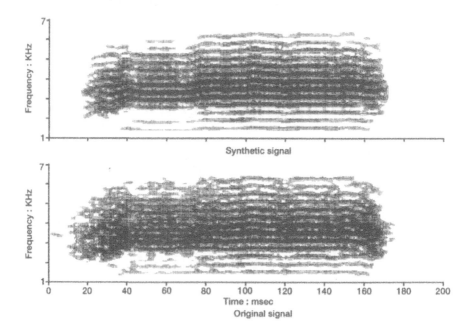

Fig. 25. A black-capped chickadee (*Parus atricapillus*) *dee* note (*bottom panel*) is synthesized using harmonic synthesis techniques (*top panel*)

trated in the upper harmonics, leaving the fundamental undetectable, one of the visible harmonic contours, $F_n(t)$, can be extracted and then scaled by division to recover the fundamental. Second, higher harmonic frequency contours, $F_1(t)$, $F_2(t)$..., $F_n(t)$, are derived from $F_0(t)$ by integer multiplication (i.e., 2, 3...,n + 1). Third, corresponding amplitude functions, $A_0(t)$, $A_1(t)$..., $A_n(t)$, are derived individually from the fundamental and harmonic components of the natural sound. An efficient approach developed by Balaban (described in Beeman 1996b) derives $A_n(t)$ by automatically tracing the contour of the corresponding $F_n(t)$ through the spectrogram and reading successive amplitude values. Fourth, waveforms are generated for the individual harmonics,

(i.e., $S_n(t) = A_n(t) * \sin [2\pi F_n(t)] t$), by applying tonal synthesis to each pair, $F_0(t)$ and $A_0(t)$, $F_1(t)$ and $A_1(t)$, and so on. Finally, these waveforms are summed to produce the composite harmonic signal, $S(t) = S_0(t) + S_1(t) + ...+ S_n(t)$.

Harmonic-synthesis techniques can be used to selectively remove or rescale the amplitude of individual harmonic components, in order to simplify a sound or test the significance of different harmonic components. An individual harmonic is rescaled by multiplying the corresponding $A_n(t)$ function by some factor and then resynthesizing. Similarly, an individual harmonic can be removed by omitting it from the harmonic sum.

9.4.4
Noisy Signals

Noisy signals can be also be synthesized, for instance by adding noise components to existing sounds, or by noise, summing noise bands to create spectral patterns. Noise signals can be based on uniform or Gaussian, both of which have roughly uniform spectral-energy distribution. When adding noise to an existing signal, the noise should be amplitude-scaled for a specific signal-to-noise ratio with respect to the signal. This is accomplished by separately measuring the RMS levels of signal and noise, and scaling these components accordingly. Noise energy also can be limited to a specific frequency band using a bandpass filter, and multiple noise bands can be combined to form a spectrally structured noisy sound. As before, careful amplitude-scaling of energy in different frequency bands can be used to explicitly simulate or alter the spectral energy distribution pattern of a natural sound (see also Owren and Bernacki, this Volume, for discussion of synthesis of noisy bioacoustic signals).

9.5
Miscellaneous Synthesis Topics

9.5.1
Template Sounds

Tonal synthesis techniques allow the user to derive and synthesize an average "template" sound from an ensemble of signals, for instance for morphological analysis or playback experiments (Beeman 1996b). The average sound is derived as follows (see Figure 26). $F(t)$ and $A(t)$ functions are extracted from each sound sample, then the

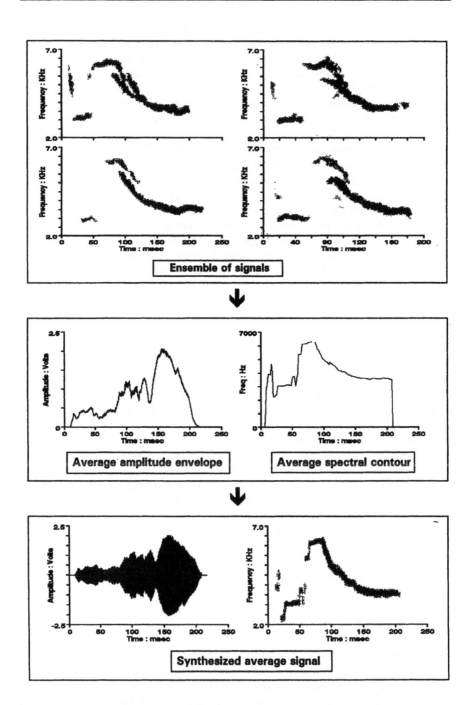

Fig. 26. Sound template derivation. Frequency and amplitude functions are derived from an ensemble of signals (*top panel*), to create averaged *A(t)* and *F(t)* functions (*middle panel*), from which a template sound is synthesized (*bottom panel*)

ensemble of functions is normalized in duration and aligned in time. The $F(t)$ and $A(t)$ functions are then averaged, yielding $F_{avg}(t)$ and $A_{avg}(t)$, which can be used to synthesize a single template waveform representing the average sound for the entire data set. Note that template sounds cannot be obtained by averaging time waveforms (whose random phase components would mutually cancel), from parameter measurements (which cannot be used to regenerate sounds), or from spectrogram averages (which cannot be inverse transformed to an artifact-free time signal).

Template sounds have several important applications. First, they produce an average exemplar for sound types that are too complex to be adequately characterized by measurements of sound parameters alone. Average morphological features of the sound can then be measured from the template version. Second, individual sounds from the data set can be compared to the template sound in order to measure their morphological similarity. Third, because template sounds are generated directly from amplitude and frequency functions, sound features can be altered systematically by manipulating these functions. The researcher can generate an array of test sounds of graduated similarity to the template in order to test, for instance, relationships between functional, perceptual, and acoustical similarity, the perceptual centrality of the template, the perceptual significance of various acoustic features in the template, and the locations and abruptness of functional and perceptual boundaries in relation to acoustic variation.

9.5.2
Noise Removal

Tonal-synthesis techniques can also be used to remove noise from tonal, and even some harmonic signals, by extracting relatively noise-free $F(t)$ and $A(t)$ functions from the natural sounds and using these to produce noise-free synthetic versions. This technique was developed by Nelson (described in Beeman 1996b), and has been used to clean up noisy field recordings for manipulation and subsequent playback. The success of this technique depends directly on the ability to extract accurate $F(t)$ and $A(t)$ representations of the underlying signal in the presence of noise, as well as the adequacy of the tonal model as a represention of the natural sound. This approach is effective because of the noise-insensitivity of the Fourier-transform-based SCD method for deriving $F(t)$, because noise occurring in $A(t)$ can be reduced by smoothing, and because minor artifacts in $F(t)$ and $A(t)$ can be corrected, if necessary, using mouse-drawing techniques. As illustrated in Figure 27, this noise-removal technique can produce quite dramatic results when applied to field-recordings.

10
Summary

Digital signal analysis technology adds many capabilities to the field of biological sound analysis. It provides rapid and flexible ways of visualizing sound signals and a reliable way of recording and storing them. More important, because digitized signals are numerical, digital technology allows researchers to use mathematical analysis and mod-

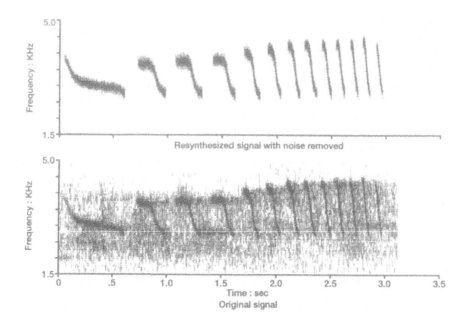

Fig. 27. A natural field sparrow song with a high level of background noise (*bottom panel*) is analyzed and resynthesized (*top panel*) using tonal-sound techniques for noise removal

sound analysis techniques made possible by digital signal analysis, and the underlying theory and practical considerations. Precise and automated measurements can be made on waveforms, power spectra, and spectrograms. Time signals can be converted either to amplitude envelopes to analyze their amplitude behavior, or to gate functions to visualize, measure, and compare their temporal patterning. Frequency spectra and frequency-time spectrograms of specific sound segments can be calculated digitally and measured to characterize sound features. Spectral analysis parameters such as time and frequency resolution, transform window, spectral smoothing, and physical scaling should be selected carefully for accurate analyses. Real-time spectrographic display is also possible.

Other digital techniques carry sound analysis beyond the scope of traditional analog tools. Researchers can digitally extract and calculate spectral contours, which are mathematical functions representing a sound's instantaneous time-varying frequency. Different contour extraction techniques, such as zero-crossing analysis, spectrogram contour detection, or Hilbert transform analysis, can be used depending on sound structure. Spectral contours can form the basis of both sound comparison and sound synthesis operations. Digital analysis also allows the researcher to compare sounds in a variety of ways, based on a variety of sound characteristics. Parameters representing isolated sound features can be measured, often automatically, and compared statistically for similarity. More advanced approaches allow researchers to compare sound functions used to represent a sound's entire behavior over one or more characteristics, such as amplitude envelopes, power spectra, spectral contours, or spectrograms. These

such as amplitude envelopes, power spectra, spectral contours, or spectrograms. These functions can be compared mathematically using cross-correlation techniques to measure quantitative similarity. An ensemble of sounds can be compared pairwise and analyzed statistically to estimate clustering tendencies, or a sequence of individual sounds can be compared to a sound template to measure their graded similarity.

Another important application of digital sound analysis is sound manipulation and synthesis. Digital editing tools can be used to rearrange existing sound material, and these replace traditional cut-and-splice tape techniques with faster and more accurate interactive screen tools. Digital sound generation techniques can create entirely new sounds from mathematical functions for use as auditory stimuli in psychophysical and neurophysiological research. Digital synthesis techniques can decompose natural sounds into their essential components, which can then be manipulated and recombined to form selectively altered sounds. In this manner, natural biological sounds can be varied in detail and then used in playback studies, to explore the behavioral significance of individual sound features. Eight natural sound types were described, along with techniques for synthesizing many of them, including tonal sounds, harmonic sounds, polytonal sounds, pulse repetition sounds, and noise, and combinations and sequences of these sounds. Much natural sound synthesis is based on the tonal sound model, which reduces sounds to one or more pairs of amplitude envelope and spectral contour functions, which can be independently manipulated to alter sound characteristics. Synthesis manipulations include frequency shifting, time scaling, amplitude and frequency modulation, intensity manipulation, harmonic removal and rescaling, pulse repetition variation, and noise addition and removal. Finally, digital analysis can be used to synthesize an average template sound from a sound ensemble, for feature measurement, similarity analysis, or perceptual testing.

References

Beecher MD (1988) Spectrographic analysis of animal vocalizations: implications of the "uncertainty" principle. Bioacoustics 1: 187–208
Beeman K (1987) SIGNAL V2.5 User's Guide [DEC PDP]. Engineering Design, Belmont, Massachusetts
Beeman K (1989) SIGNAL V1.3 User's Guide [IBM PC]. Engineering Design, Belmont, Massachusetts
Beeman K (1991) RTS V1.0 User's Guide [IBM PC]. Engineering Design, Belmont, Massachusetts
Beeman K (1996a) RTS V2.0 User's Guide [IBM PC]. Engineering Design, Belmont, Massachusetts
Beeman K (1996b) SIGNAL V3.0 User's Guide [IBM PC]. Engineering Design, Belmont, Massachusetts
Borror DJ (1965) Song variation in Maine song sparrows. Wilson Bull 77: 5–37
Brand AR (1934) A method for the intensive study of bird song. Auk 52: 40–52
Buck JR, Tyack PL (1993) A quantitative measure of similarity for Tursiops truncatus signature whistles. J Acoust Soc Am 94: 2497–2506
Charif RA, Mitchell S, Clark CW (1995) CANARY 1.2 users manual. Cornell Laboratory of Ornithology, Ithaca New York
Clark CW (1982) The acoustic repertoire of the southern right whale, a quantitative analysis. Anim Behav 30: 1060–1071
Clark CW, Marler P, Beeman K (1987) Quantitative analysis of animal vocal phonology: an application to swamp sparrow song. Ethology 76: 101–115
Dabelsteen T, Pedersen SB (1985) A method for computerized modification of certain natural animal sounds for communication study purposes. Biol Cybern 52: 399–404
Dorrscheidt GJ (1978) A reversible method for sound analysis adapted to bio–acoustical experiments. Int J Bio Med Comput 9: 127–145
Flanagan JL (1972) Speech analysis, synthesis, and perception, 2nd edn. Springer Berlin Heidelberg New York

Gerhardt HC (1989) Acoustic pattern recognition in anuran amphibians. In: Dooling RJ, Hulse SH (eds) Comparative psychology of audition: Perceiving complex sounds. Erlbaum Assoc, Hillsdale, New Jersey, p 175
Gerhardt HC (1992) Multiple messages in acoustic signals. Sem Neurosci 4: 39–400
Greenewalt CH (1968) Bird Song: Acoustics and physiology. Smithsonian Institute Press, Washington, DC
Hauser MD (1991) Sources of acoustic variation in rhesus macaque vocalizations. Ethology 89: 29–46
Hauser MD (1992) Articulatory and social factors influence the acoustic structure of rhesus monkey vocalizations: a learned mode of production? J Acoust Soc Am 91: 2175–2179
Hauser MD, Schön Ybarra M (1994) The role of lip configuration in monkey vocalizations: experiments using xylocaine as a nerve block. Brain Lang 46: 232–244
Hienz RD, Brady JV (1988) The acquisition of vowel discriminations by nonhuman primates. J Acoust Soc Am 84: 186–194.
Hopp SL, Sinnott JM, Owren MJ, Petersen MR (1992) Differential sensitivity of Japanese macaque (*Macaca fuscata*) and humans (*Homo sapiens*) to peak position along a synthetic coo call continuum. J Comp Physiol Psychol 106: 128–136
Klatt DH (1980) Software for a cascade/parallel formant synthesizer. J Acoust Soc Am 67: 971–995
Koenig W, Dunn HK, Lacy LY (1946) The sound spectrograph. J Acoust Soc Am 18: 19–49
Kroodsma DE (1974) Song learning, dialects and dispersal in Bewick's wren. Z Tierpsychol 35: 353–380
Loughlin PJ, Atlas LE, Pitton JW (1993) Advanced time-frequency representations for speech processing. In: Cooke M, Beet S, Crawford M (eds) Visual representations of speech signals. Wiley, New York, p 27
Margoliash D (1983) Song-specific neurons in the white-crowned sparrow. J Neurosci 3: 1039–1048
Markel JD, Gray AH (1976) Linear prediction of speech. Springer, Berlin Heidelberg New York
Marler P, Pickert R (1984) Species-universal microstructure in the learned song of the swamp sparrow (*Melospiza georgiana*). Anim Behav 32: 673–689
Martindale S (1980) A numerical approach to the analysis of solitary vireo songs. Condor 82: 199–211
May B, Moody DB, Stebbins WC (1988) The significant features of Japanese macaque coo sounds: a psychophysical study. Anim Behav 36: 1432–1444
Nelson DA (1992) Song overproduction and selective attrition lead to song sharing in the field sparrow (*Spizella pusilla*). Behav Ecol Sociobiol 30: 415–424
Nelson DA, Croner LJ (1991) Song categories and their functions in the field sparrow (*Spizella pusilla*). Auk 108: 42–52
Nelson DA, Marler P, Palleroni A (1995) A comparative approach to vocal learning: intraspecific variation in the learning process. Anim Behav 50: 83–97
Nowicki S, Capranica RR (1986) Bilateral syringeal interaction in vocal production of an oscine bird sound. Science 231: 1297–1299
Oppenheim AV, Schafer RW (1989) Discrete-time signal processing. Prentice-Hall, Englewood Cliffs
Owren MJ, Bernacki RH (1988) The acoustic features of vervet monkey alarm calls. J Acoust Soc Am 83: 1927–1935
Randall R (1977) Application of B&K equipment to frequency analysis. Brel & Kjr, Nrum, Denmark
Richard JP (1991) Sound analysis and synthesis using an Amiga micro-computer. Bioacoustics 3: 45–60
Rosen S, Howell P (1991) Signals and systems for speech and hearing. Academic Press, New York
Ryan MJ, Perrill SA, Wilczynski W (1992) Auditory tuning and call frequency predict population-based mating preferences in the cricket frog, *Acris crepitans*. Am Natural 139: 1370–1383
Sabine WS (1900) Reverberation. In: Sabine WS (ed) Collected papers on acoustics. Dover, New York, p 3
Schafer RW (1975) Digital representation of speech signals. Proc IEEE 63:662–677
Sinnott JM (1989) Detection and discrimination of synthetic English vowels by old world monkeys (*Cercopithecus, Macaca*) and humans. J Acoust Soc Am 86: 557–565
Sparling DW, Williams JD (1978) Multivariate analysis of avian vocalizations. J Theor Biol 74: 83–107
Staddon JER, McGeorge LW, Bruce RA, Klein FF (1978) A simple method for the rapid analysis of animal sounds. Z Tierpsychol 48: 306–330
Stephen RO, Hartley JC (1995) Sound production in crickets. J Exp Biol 198:2139–2152
Stoddard PK (1990) Audio computers – theory of operation and guidelines for selection of systems and components. Bioacoustics 2: 217–239
Stremler FG (1977) Introduction to communication systems. Addison-Wesley, Reading, Massachusetts
Suthers RA, Goller F, Hartley RS (1994) Motor dynamics of song production by mimic thrushes. J Neurobiol 25: 917–936
Wells MM, Henry CS (1992) Behavioral responses of green lacewings (Neuroptera: Chrysopidae) to synthetic mating songs. Anim Behav 44: 641–652

Zoloth S, Dooling RJ, Miller R, Peters S (1980) A minicomputer system for the synthesis of animal vocalisations. Z Tierpsychol 54: 151–162

Chapter 4

Application of Filters in Bioacoustics

P. K. Stoddard

1
Introduction

One afternoon in 1981, the screen went blank on the laboratory's real-time digital spectrograph, a recent technological advance at that time. The president and chief engineer of the company selling this spectrograph diagnosed the problem over the phone. "Oh, that's just the filter board. What are you using the analyzer for? Bird song? For the frequency ranges of your signals you don't really need the filters. Go ahead and bypass the filter board." I was worried that an integral component of the machine was not working, but was uncertain about the function of the filter board. If the machine's designer said the filter was unneccesary, then it must be. I removed the offending circuit board, bypassed the connection, and went back to work.

A year later, a colleague brought me a recording he had made of half-masked weaver song and we proceeded to scan the tape on the filterless analyzer. There on the screen appeared one of the most striking examples I had ever seen of the avian *two-voice* phenomenon. Two distinct voices, not harmonically related, appeared almost to mirror each other. A loud rising whistle was accompanied by a softer falling whistle. Sometimes the two voices even crossed each other, making a skewed "X" on the screen.

Six months later, while making publication-quality spectrograms on our antiquated, analog Kay Sonagraph, I found to my dismay and disappointment that the second voice was gone. I thought the upper voice must have been weaker than the lower one and had faded from the tape. Still, there should have been a trace of the second voice, which could in fact still be seen using the real-time spectrograph. I increased the gain on the analog spectrograph, but no second voice appeared. Only then did I remember the missing filter board in the real-time digital machine.

This incident impressed upon me the importance of filters in the daily life of a bioacoustician. Many times since then I have been able to trace peculiar signal artifacts to filters that were either not working or inappropriate to the task at hand. The appropriate use of filters is an essential part of any technical analysis of sound signals. This chapter therefore provides an intuitive explanation of filter theory, filter applications in bioacoustics, and the hazards of misapplying or omitting filtering. Readers may also wish to consult other tutorials on digital filter application that have been written specifically for non-engineers (e.g., Menne 1989; Cook and Miller 1992). For those who are facile in linear algebra and complex arithmetic, the presentation may be too simplistic — the chapter is written specifically for the bioacoustician who wishes to understand filters at a conceptual level, without taking a crash course in electrical engineering. More rigorous treatments can be found in books on signal processing and filter theory (e.g. Haydin 1986; Stearns and David 1988; Hamming 1989; Oppenheim and Schafer 1989).

2
General Uses of Filters and Some Cautions

Filters are frequency-selective resonant devices or algorithms that are used to remove noise from signals, to change the spectral balance or the phase composition of signals, to smooth the analog output of *digital-to-analog* (D/A) converters and as above, to prevent sampling artifacts from contaminating signals during the *analog-to-digital* (A/D) conversion. Filters can also play a critical role in automated signal detection and identification.

Most often, then, filters are used to change signals by damping or excluding certain components while allowing others to pass. However, they also have nonintuitive properties that can pose problems for the unwary. Rather than simply selectively removing or separating the components of a signal, filtering can destroy certain components and irreversibly transform others. For instance, while energy at unwanted frequencies might be attenuated by a filter, the filtering process might be simultaneously distorting the amplitude and phase characteristics of energy in spectral ranges meant to be left unaffected.

3
Anatomy and Performance of a Filter

Figure 1 shows the *gain functions* of four hypothetical filters demonstrating the most common graphical depiction of a filter's characteristics. In these diagrams, the abscissa represents frequency, and can be either discrete (showing individual frequency components) or continuous. The ordinate shows the amplitude of the filtered waveform (the filter output) relative to the unfiltered waveform (the filter input).

As shown in this figure, frequency-selective filters assume two basic attenuation patterns and two composite patterns. In each case, the frequency at which attenuation nominally begins or ends is a *corner frequency*. *Lowpass* filters (upper left panel) have a single corner frequency, allowing the passage of frequencies lower than that point while attenuating higher frequencies. *Highpass* filters (upper right panel) do the opposite, passing frequencies higher than the corner and attenuating lower frequencies. *Bandpass* filters (lower left panel) combine the first two forms to create a frequency band of *unity gain* (where energy passes without attenuation) surrounded by regions of attenuation. A *bandstop* filter (lower right panel) consists of both a highpass and a lowpass filter, and attenuates an intermediate frequency band. Whereas a *graphic equalizer* is an array of bandpass filters with adjacent frequency ranges, a *notch* filter is a sharp, narrow bandstop filter that has corner frequencies and attenuation functions selected to attenuate a narrow frequency band. An application of the latter might be removal of AC power transmission noise, which occurs at either 50 or 60 Hz (depending on the country).

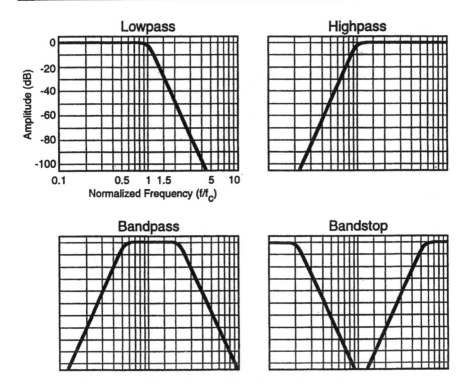

Fig 1. The two basic attenuation patterns, lowpass and highpass, can be combined to pass a particular frequency band (bandpass) or exclude a particular frequency band (bandstop)

Figure 2 illustrates a number of important terms in the context of a lowpass filter. The frequency identified as the corner frequency is the location on the gain function at which signal input is attenuated by 3 dB — the *half-power point*. If, for instance, a lowpass filter with a corner frequency of 10 kHz is applied to a signal that includes energy up to 10 kHz, the highest frequencies will be significantly attenuated. In general, a filter with a corner frequency that is somewhat beyond the frequencies of interest should be used in bioacoustics applications, in order to avoid attenuation of potentially important high-frequency signal components. Note, however, that in some filter designs the transition point has no associated attenuation and is known as the *cut-off frequency* rather than the corner frequency. The frequency range that is not attenuated (or is attenuated by less than 3 dB) is called the *passband*. The frequency range over which maximum attenuation occurs is the *stopband*. A *transition-band* is the frequency range of increasing or decreasing attenuation lying between a corner frequency and the beginning of stopband or passband, respectively.

A filter's *attenuation slope* is specified in decibels per octave (an octave is a doubling or halving of frequency). The greater the number of elements in the filter (its *order* or number of *poles*), the more sharply defined is its gain function. In analog equipment,

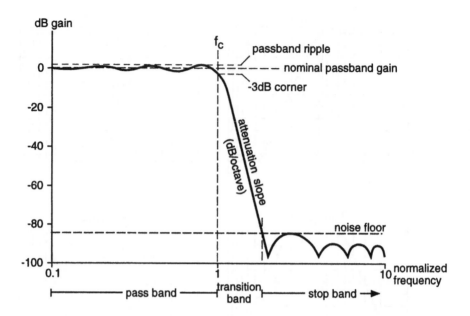

Fig 2. The Anatomy of a frequency-selective filter is illustrated here with a hypothetical lowpass filter. Frequency is normalized so that the corner appears at nominal frequency of 1.0. The attenuation throughout most of the passband is 0 dB (i.e., the gain is 1)

filter order is determined by the number of filter stages in the circuit, with two per stage. In a digital filter, the order is determined by the number of elements in the filter array. Thus, sharp attenuation requires a high-order filter. The optimal filter for attenuation of undesired frequencies is often one with a flat passband and precipitously sloping transition bands. For such purposes, one can obtain sharp filters with attenuation slopes exceeding 100 dB per octave. Engineers and scientists proudly call these "brickwall" filters, a term that was evidently coined to evoke images of powerful filters (and powerful engineers) that crush any undesired frequency components with deep stopbands and vertical transition slopes. In reality, however, even the sharpest filters show finite attenuation slopes and variable attenuation effects over their transition bands.

In addition, sharp attenuation generally comes at a price. For a given filter design, the more acute the attenuation, the greater the occurrence of undesirable artifacts in the passband. One such side-effect is variable gain, called *passband ripple*. The amount of passband ripple is generally included in the performance specifications provided with analog filters and may serve as a criterion for choosing among various designs. Likewise, the maximum amount of tolerable passband ripple can often be specified in the course of designing a digital filter. Increasing the order of a digital filter can reduce the trade-off between ripple and attenuation, although at the cost of more computation.

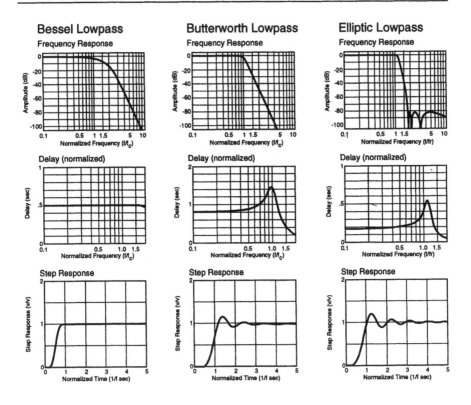

Fig 3. Performance graphs of three common eight-pole, lowpass, analog filter designs. The Bessel filter gives the cleanest step response and least phase distortion but takes an octave to reach its full attenuation slope. The Butterworth filter has the flattest passband and reaches its full attenuation slope over a smaller frequency range. The elliptic function is used when a steep transition band is needed and phase and step response is not important. The elliptic filter's cut-off frequency (or ripple frequency, f_r) has no attenuation. The other two filters attenuate 3 dB at the corner frequency, f_c.

In addition to their amplitude-related effects, filters also delay the output of a signal relative to the input. The *group delay* of a filter is the average time-delay produced in the output as a function of frequency, as shown in the middle row of panels Figure 3. If all frequencies are delayed by a constant interval, phase relations among these components are unchanged. *Constant group delay* is the same as *linear phase delay*, where phase delay is phase divided by frequency, multiplied by -1. Plotted as a function of phase angle and frequency, constant group delay yields a linearly increasing function, because identical delays require increasing proportions of a wavelength as the frequency increases and the period shortens.

Filters with nonunity group delay or nonlinear phase delay in the passband distort the filtered waveform by differentially altering the phases of various spectral components. The examples of filtered waveforms in Figure 4 illustrate linear and non-linear phase delay of different filters. It is clear that the differences are considerable. Scientists working with animal signals should consider the importance of phase information when choosing a filter for their work.

The response of a filter to very brief or *transient* signals is indicated by its *step response*, i.e., the response to a discrete amplitude change called a *square input step* (shown in the bottom panels of Figure 3). The step response reflects the degree of *damping* in a filter. If a filter absorbs a square input step with no overshoot or undershoot in the output, it is said to be critically damped. Critical-damping in a filter is analogous to the action of a shock absorber of a car that completely cushions the vehicle's passengers from any bounces due to bumps in the road. *Underdamping*, or *overshoot*, causes the filter to *ring*, producing ripple in the output. This effect is analogous to oscillation occurring in a heavily laden car when encountering a bump. *Overdamping* causes the filter

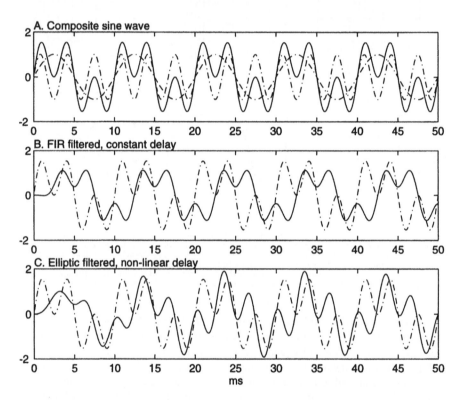

Fig 4. A-C. Examples of linear and nonlinear phase delay. A A composite waveform (solid line) is composed of two sine waves of 100 and 300 Hz (dashed lines) .B-C The composite waveform is lowpass filtered with a symmetric FIR filter and an elliptic IIR filter both with cut-off frequencies set to 330 Hz. The FIR filter (B)has linear phase filter delay; the output waveform (solid line) is delayed relative to the input waveform (dashed line) but the phase relationship between the component sine waves is preserved. C. The elliptic IIR filter has non-linear phase delay; the 300 Hz component is delayed more than the 100-Hz component (solid line). The filter startup delays are evident in the first 2 ms of both filtered outputs. The FIR filter used here also shows some attenuation at the upper edge of the passband as seen in the slight amplitude reduction of the filtered output

to *undershoot*, which is analogous to the rough ride experienced in an unladen truck. Effective filtering of signals with sharp transients requires a good step response. Examples of such signals include the acoustic pulses and clicks of many diurnal insects, as well as the discharges of many electric fishes. A filter with an underdamped step response causes these transient signals to ring, smearing them in time.

4
Properties of Various Analog Filters

An analog filter function can usually be classified as being one of five basic designs: Bessel, Butterworth, elliptic, and Tschebychev (or Chebyshev) types I and II. Figure 3 shows the first three of these types in corresponding order. Digital implementations often emulate these filter-types as well.

Bessel filters produce the least phase distortion, because their phase delays in the passband are linear with frequency. However, a Bessel filter has a gradual attenuation curve and its attenuating effect stretches back farther into the passband as the latter gradually slopes to -3 dB at the corner. Bessel filters require about an octave to reach their full attenuation slope of -6 dB per octave per pole, and its gradual attenuation makes this filter unsuitable to applications for which steep attenuation slopes and flat passbands are essential. Nonetheless, the linear phase delay and damped step response associated with this design are useful in some applications. Digital implementations of the Bessel filter do not have linear phase delay.

Butterworth filters are popular because they show neglible passband ripple. Their phase delay is moderately non-linear in the passband. Each pole contributes about 6 dB attenuation per octave beyond the corner frequency, amounting to a halving of signal magnitude with each doubling or halving of frequency. Thus, an eighth-order (or eight-pole) Butterworth filter produces an attenuation effect of 48 dB per octave.

Elliptic filters have extremely sharp attenuation curves. However, no simple rule relates the order of an elliptic filter to the attenuation slope of the transition band, as in the previous two filter designs. Elliptic filters tend to distort the signal in the outer 20 % of the passband, showing significant passband ripple, group delay, and overshoot. Elliptic notch filters that remove 50- or 60-Hz AC noise are notorious for contaminating the passband with nonlinear phase delay.

Tschebychev type I filters (not shown in the figure) have ripple in the passband, fairly sharp transition bands, and a flat responses in the stopband. Tschebychev type II filters have a flat passband, a less steep attenuation slope, and ripple in the stopband. Tschebychev filters are not used as commonly as the other three filter designs described.

5
Antialiasing and Antiimaging Filters

5.1
A/D Conversion Requires an Analog Lowpass Filter

A signal *alias* is a phantom signal pattern introduced into a waveform (also called a time series) by sampling at too low a rate during A/D conversion. Figure 5 shows how such aliasing occurs. The true signal in this example is a transient that sweeps from 0 to 10 Hz. When the analog waveform is digitized, the A/D converter provides numerical readings of the signal's amplitude at regular points in time (represented by the circled points in Figure 5). When connected, these sample points should approximate the original waveform. If the sampling interval is too long, however, the waveform is undersampled and the sampled frequency actually drops. In the example, the analog waveform is sampled ten times per second. The sampled waveform tracks the frequency of the original from 0 to 5 Hz, but cannot keep up thereafter. The digital version drops back

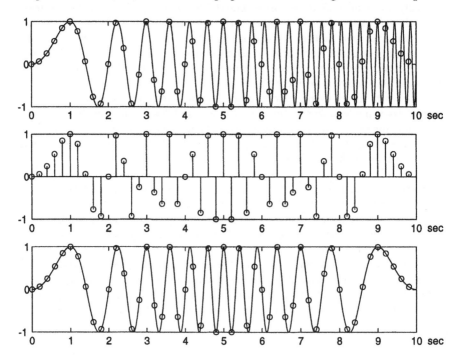

Fig 5. The true signal here (*upper panel*) is a transient that sweeps from 0 to 10 Hz. When the waveform is digitized at 10 samples/sec (*middle panel*), the signal's instantaneous amplitude is measured every tenth of a second, represented by the *circled points*. When connected, these sample points should approximate the original waveform (*lower panel*). However the sampling interval is inadequate and the waveform is undersampled. The waveform reconstructed from sampling actually drops in frequency where the original waveform exceeds half the sample frequency. The false signal that results from undersampling is called an alias

to 0 Hz as the analog signal sweeps upwards to 10 Hz.

According to the *Nyquist theorem*, digital sampling of an analog signal limits the bandwidth of the sample series to a frequency range between 0 Hz and half the sampling rate. Thus, the latter is the highest frequency that can be represented in a digital time series, known as the *Nyquist frequency* (also known as the *folding* frequency). The frequency of an aliased, low-frequency component mirrors that of the higher-frequency counterpart in the original signal, reflected (or folded) downward at the Nyquist frequency. Obviously then, we cannot obtain a digital representation of any signal component with a frequency higher than the Nyquist frequency. Of greater consequence, however, is the risk of obtaining false signals (as I did with my colleague's recordings of half-masked weaver birds).

The first microcomputers to arrive sound-capable from the manufacturer digitized sound at a rate of approximately 22 kHz (actually 22254.5 Hz). When a signal is digitized at 22 kHz, the sampled waveform must be restricted to a passband between 0 and 11 kHz. A Nyquist frequency of 11 kHz should be fine for most frogs and passerine birds, but would not be adequate for rats, bats, dolphins, or other species whose vocalizations contain higher frequencies. These higher frequencies reappear as aliases in the passband, where they become inseparable and often indistinguishable from true signal components. Other, more insidious sources of high-frequency noise can also produce aliases in the passband. In addition to the intended signal, the tape recording may contain a variety of extraneous noises such as wind, footsteps, or the like that will probably contain energy above 11 kHz. Harmonic distortion caused by imperfect electronic processing and especially by occasional overloading during recording and playback can also introduce energy above 11 kHz. When recording with analog equipment, one may not notice the presence of high-frequency energy, particularly high-frequency harmonics.

Alias contamination of digitized signals can be avoided by using an analog, lowpass filter to screen out any energy above the Nyquist frequency before it reaches the A/D converter. This antialiasing filter, if chosen correctly, attenuates frequencies above the Nyquist frequency to insignificant levels while allowing the lower frequencies to pass unaffected. If an inappropriate filter is used, however, the signal may not only be aliased, additional artifacts may be introduced by the filter itself! In practice, it is impossible to obtain a digitized waveform that perfectly represents the original analog signal between 0 Hz and the Nyquist frequency because analog lowpass filters do not have infinite attenuation slopes. If the stopband is set to begin exactly at the Nyquist frequency, some attenuation inevitably occurs in the passband. If the corner frequency is set at the Nyquist frequency, some aliasing will occur. For these practical reasons, sampling rates must be set higher than twice the highest frequency of interest.

5.2
Choosing an Antialiasing Filter

If an A/D converter does not include a built-in antialiasing filter, the user must add one. Similarly, if a digitizing system has a built-in, fixed lowpass filter designed for a sampling rate that is higher than the user would choose (for instance to save storage space), the signals must be prefiltered using a lowpass filter with a lower corner frequency.

Failure to do so will result in aliasing. An adjustable-frequency bandpass filter can be used to prevent aliasing, but often at a cost of several thousand dollars. Less than one hundred dollars will buy a fixed frequency antialiasing filter module that needs only a bipolar power supply and two BNC connectors. Either way, the user needs to select the filter function, the filter order, and the corner frequency.

The amount of attenuation needed at and above the Nyquist frequency depends on the dynamic ranges of both the system and of the signal. A 16-bit digitizing system, for instance, provides a total dynamic range of about 96 dB (each binary bit of quantization increases the dynamic range by 6 dB). A microphone might have a maximum signal-to-noise ratio of 78 dB. A *noise floor* of -78 dB, therefore, is appropriate and sufficient for recordings made under ideal circumstances in which the dynamic range of the microphone is a limiting factor in recording quality. Attempting to achieve a noise floor of -96 dB would be overkill. If recordings are made in the field, however, the signal-to-noise ratio is probably not much better than 34 dB (Wickstrom 1982). That is to say, the signal will be at most 34 dB louder than the average background noise level. A reasonable minimum noise floor in these "real-world" cases is therefore -36 dB.

Let us assume in two examples that one wishes to digitize field recordings of song sparrows (*Melospiza melodia*). The energy in their songs ranges from 1.6 to 10 kHz. Once these songs are in the computer, the plan is to make spectrograms and to conduct playback experiments with birds in the field. In the first example, signals are digitized at 48 kHz and in the second, data storage requirements are cut in half by digitizing at 24 kHz. In both cases, the task is to choose an appropriate analog, antialiasing filter.

Example 1.   48 kHz Sampling Rate. With a sampling rate of 48 kHz, the Nyquist frequency is 24 kHz, 14 kHz above the frequency of highest interest in these examples. The lowpass, antialiasing filter should therefore leave all frequencies below 10 kHz unaffected while reducing energy at frequencies above 24 kHz by at least 36 dB. The corner frequency should be 10 kHz, or perhaps a little higher to accommodate attenuation occurring at the corner.

Figure 3 illustrates the properties of a series of prototypical eight-pole lowpass antialiasing filters modules acquired from a well-known manufacturer. The analog filter functions available include Bessel, Butterworth, and elliptic versions. All the functions are normalized to f / f_c (the ratio of the frequency of interest to the corner frequency), meaning that the corner frequency has a value of 1.

The eight-pole Bessel lowpass filter will be considered first, as it has the best phase and step performance. The corner frequency of the Bessel filter is defined at -3 dB, so it is best to raise it somewhat above 10 kHz. Approximately -1 dB might be deemed acceptable attenuation or ripple in the passband. Inspection of the filter's gain function shows the amplitude is down by 1 dB at $f / f_c = 0.6$, which corresponds to a new corner frequency of about 17 kHz (i.e., 10 kHz / 0.6). The response curve requires all values to be normalized with respect to the corner frequency. The normalized frequency that corresponds to the Nyquist frequency is 1.4 (i.e., $f_N / f_c = 24$ kHz / 17 kHz). The gain at 1.4 is -6 dB, which is not even close to -36 dB. The shallow attenuation slope of the eight-pole Bessel filter makes it unacceptable for this situation.

Like the Bessel design, the corner frequency of the Butterworth filter is defined at -3 dB. Again, the corner frequency should be increased to preserve the integrity of the passband. Attenuation of -1 dB is found in the Butterworth filter gain function at a normalized frequency of 0.9. This more stringent criterion places the corner frequency at 11 kHz (i.e., 10 kHz / 0.9 = 11 kHz). Recalculation of the normalized Nyquist frequency, f_N / f_c, yields a value of 2.2 (i.e., 24 kHz / 11 kHz). The Butterworth filter function shows the gain at a normalized frequency of 2.2 to be about -55 dB, far below the citerion of -36 dB set above. The eight-pole Butterworth lowpass filter with an 11 kHz corner easily prevents aliasing of the digitized field recordings.

Example 2. 24-kHz Sampling Rate. In this example, the data are sampled at only 24 kHz in order to reduce storage requirements. The new Nyquist frequency of 12 kHz is now only 2 kHz above the highest frequency in these signals. This time, then, the task is to find a filter that can provide 36-dB attenuation in the narrow band between 10 kHz and 12 kHz.

The corner is adjusted to 11 kHz for the eight-pole Butterworth filter. At the 12 kHz point, the normalized Nyquist frequency, f / f_c, is 1.1 (i.e., 12 kHz / 11 kHz). Attenuation at this point is only -7.5 dB, far short of the -36 dB that is needed. The Butterworth filter therefore will not work. The eight-pole elliptic filter is well-suited to a 10-kHz cut-off frequency, but also does not provide 36 dB attenuation at 12 kHz. At this point, the normalized Nyquist frequency is 1.2 (i.e., 12 kHz / 10 kHz) and attenuation is only -8.3 dB.

The remaining options are either to sample at a higher rate and reduce the data with *digital decimation*, or to search for a sharper elliptic filter design. The first approach, also known as *downsampling*, produces a lower sampling-rate by discarding a fixed proportion of the digital samples. It is discussed in greater detail below. The second approach takes advantage of the potential for building elliptic filters with extremely sharp attenuation slopes.

It happens that the manufacturer whose eight-pole filters are illustrated in Figure 3 also makes a 7-pole elliptic filter module that provides 20 dB attenuation at a normalized Nyquist frequency of 1.2. This filter produces the desired 36 dB attenuation at f / f_c = 1.3, corresponding to a frequency of 13.4 kHz in this particular case. Using this filter, then, some energy above the Nyquist frequency is aliased, but only between 12 and 13.4 kHz. Frequencies above 13.4 kHz are attenuated by 36 dB and energy in the aliased band is attenuated by at least 20 dB. Energy remaining between 12 and 13.4 kHz folds back across the 12 kHz Nyquist frequency, but the aliases extend no lower than 10.6 kHz (12-(13.4 - 12) = 10.6).

If spectrographic analysis is restricted to the 0 to 10 kHz range, aliases in the signal will not even be visible. In using these songs for playback trials, however, the investigator needs to consider whether subjects will be affected by spurious signals in the 10.6- to 12 kHz range. One important consideration is that the bird's ear has its own "filters". The song sparrow audiogram shows this bird to be approximately 35 dB less sensitive to frequencies above 10 kHz than at its "best" frequency of 4 kHz (Okanoya and Dooling 1987). Song sparrows would therefore probably not notice faint aliases above 10.6 kHz.

The frequency response of the elliptic filter is therefore acceptable, allowing us to consider other performance characteristics. Its normalized group delay curve shows a large amount of nonlinear phase shift well before the cut-off frequency. If phase relations are critical, then, this filter should be rejected. In this particular example, however, signals would be recorded and played back in the *far sound field* (distances over which environmental transmission characteristics affect the signal) and phase relations are therefore probably not critical. Overall, even though this elliptic filter did not meet the initial criteria, the compromises involved in using it turned out to be acceptable.

5.3
D/A Conversion also Requires an Analog Lowpass Filter

When a digital signal is resynthesized using a D/A converter, the output is an analog step function that approximates the smooth analog wave (Figure 6). These tiny steps carry the high-frequency energy typical of square waves. A spectrum of the direct analog output therefore has *signal images*, reflections of the intended output signal that fold upwards accordion-style about multiples of the Nyquist frequency. As a result, a second analog lowpass filter is needed that *follows* D/A conversion to eliminate reconstruction

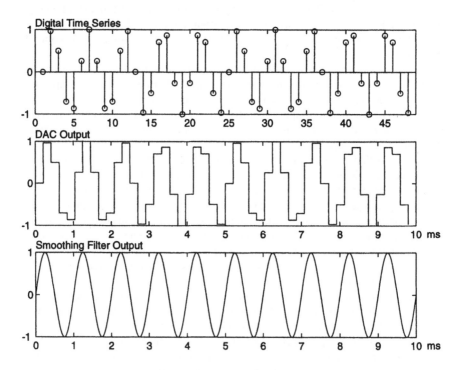

Fig 6. A digital-to-analog converter (DAC) converts a 1 kHz sine wave digitized at 48 kHz (*upper panel*) into an analog step function. Each step is similar to a square wave with energy present at the odd harmonics. An analog lowpass smoothing filter reverts this harmonic energy back to the fundamental (*lower panel*).

images. This *anti-imaging*, or *smoothing* filter recovers the energy in a high-frequency image and puts it back into the smoothed output signal.

If the smoothing filter is absent, inadequate, or incorrectly configured, the output signal sounds peculiar. When played through a speaker, it has an tinny, ringing quality. One can experience the unpleasant sound of imaging by halving the clock rate of the D/A converter without a commensurate halving of filter settings (e.g., playing a signal sampled at 48 kHz at 24 kHz). Many computer audio systems have smoothing filters that are preset in accordance with their highest sample rate. For better or worse, most audio software lets the user play back the signal at half speed with no regard for the smoothing filter. If no smoothing filter is present at all, high-frequency images can eventually damage the high frequency response of the playback speakers.

5.4
Analog Filters: Passive Versus Active Components

Analog filters can be either *passive*, requiring no additional power source, or *active*, requiring some nominal DC source to power their amplifier circuitry. Passive filters typically show some *insertion loss* in the passband, meaning they attenuate the passband by some constant amount and the stopband by a greater amount. The components of a passive filter are all *reactive* (i.e., either *resistive*, *capacitive*, or *inductive*), so the output has high *impedance* (resistance in an AC circuit) and may therefore not be well-suited for driving some kinds of circuitry. A/D converters usually have high input impedance and passive filters can therefore be used effectively for antialiasing. Independence from additional power supplies reduces the overall cost of passive filters and increases their appeal to those with an aversion to soldering. Active filters contain integrated circuit amplifiers and require a clean, bipolar power supply. Active filters have unity gain in the passband and no insertion loss. They are generally built with a low-impedance output stage and can therefore drive just about any kind of equipment except long cables and audio speakers.

6
Analog Versus Digital Filters

Analog filters are constructed from electronic components. An analog filter accepts a continuous voltage signal as input and produces real-time output. Digital filters, by contrast, use a mathematical algorithm to alter the composition of a digital time series such as digitized animal signal. Many frequency-filtering functions can be performed by either analog or digital technology. In these kinds of applications, much of the relevant theory applies to both types. However, digital filters can perform some functions better than analog filters, and can be designed with capabilities that analog filters cannot have.

For instance, digital filters can easily filter a segment of a digitized signal while leaving the rest untouched. With selective digital cutting, filtering, and pasting, a careful person can sometimes separate signals that overlap in time. Digital filters are quiet and do not introduce the noise that is inherent in using a comparable series of analog amplifier circuits. Extremely sharp digital filters can be constructed in a fraction of a second, whereas equivalent analog filters can only be built after testing boxes of identically labeled capacitors for units with particular values. The parameters of digital filter algorithms can be changed with no change in cost, whereas analog counterparts have fixed response functions and one pays a premium for frequency-selectable devices. Digital filters can be made to exhibit virtually any response pattern and can further be designed to detect or isolate particular signals, a feat that is difficult or impossible with analog circuitry. However, digital filters cannot be substituted for analog filters on the A/D input and D/A output because a digital filter cannot prevent aliasing or imaging during the conversion process.

6.1
Designs and Functions of Digital Filters

The most widely used digital filters compute a *weighted moving average* of a digital time series. Energy at frequencies outside the passband is averaged out. The filter characteristics are determined by the weighting function, a combination of a weighting algorithm, and an array of weights (the filter's impulse function). The *finite impulse response* (FIR) filter computes a moving-average of fixed length. Therefore, the contribution of any individual datum to the filtered output is of finite duration. In contrast, the *infinite impulse response* (IIR) filter factors in a proportion of the past moving average to future moving averages. Every datum is thus infinitely present, albeit in ever decreasing proportions, in all future moving averages of the input series.

FIR filters are especially useful because they can be designed to have a perfectly linear phase response in the passband. IIR filters, on the other hand, can be designed to implement a broad array of filter shapes, including close mimics of the traditional analog functions. FIR filters are conceptually and computationally simpler to implement than are IIR filters. IIR filters, however, can often meet a given set of specifications with a much lower-order filter than can a corresponding FIR filter and so may require fewer computations. FIR filters are inherently stable, which means that filters of virtually any steepness of attenuation can be constructed that will yield predictable results. IIR filters, in contrast, are not inherently stable and must be constructed within certain constraints or they may produce unpredictable results.

The key in digital filtering is the design of an appropriate impulse function, an array of weights. This array can be saved and entered in any number of commercial or user-written programs that will perform the filtering of digitized or synthetic signals. Numerous commercial software packages can help users design, test, and implement digital filters. Different filter design routines optimize and trade-off different parameters of the filter function, in particular passband ripple and attenuation slope. FIR filters work by *convolving* an array of weights (the impulse function) against the input waveform (this process is illustrated below in an example problem). In convolution, the weighting array is time-reversed, multiplied point for point against the input waveform, and the resulting product array is summed to produce a single value. These weighted sums are computed as the reversed impulse function is advanced one point at a time across the input waveform. The resulting array of weighted values is the filtered waveform. If the weighting array is symmetrical, the filter will have linear- phase delay, i.e., preserve the signal's phase relations.

FIR filtering is performed efficiently in the frequency domain using the *fast Fourier transform* (FFT) as a shortcut for convolution. The convolution program computes overlapping FFTs of both the input waveform and the filter waveform, then multiplies the Fourier coefficients of each, point for point. An *inverse fast Fourier transform* (IFFT) of the multiplied coefficients results in the filtered (convolved) waveform. In most cases, the shorter segment of the two must be padded with zeros to the length of the longer before the FFT can be computed. For longer signals, such as a birdsong, the digitized wave array can be chopped up into sections and each segment filtered separately by the FFT method (Oppenheim and Schafer 1989).

6.2
Cascaded Filters and Repeated Filtering

Combining filters with different functions is often useful. Two analog filters can be connected in series (*cascaded*) or a digital signal can be filtered successively with different functions. The analog lowpass Bessel filter, for instance, can be improved as an antialiasing filter by being cascaded with an elliptic lowpass filter with a significantly higher corner frequency. Positioning the cut-off frequency of the elliptic filter well beyond the corner of the Bessel filter keeps the phase delay to a minimum but still abbreviates the Bessel filter's gradual transition band.

Filter cascading can be misapplied as well. Suppose a digital filter function improves the signal-to-noise ratio of a digitized signal, but not as much as desired. One can then connect two identical analog filters in series or refilter digitally with the same function and take out more noise each time the signal passes through the filter. Cascading two Butterworth lowpass filters, each with 8 poles (48-dB attenuation per octave) and corner frequencies of 10 kHz, does indeed produce a filter with an attenuation of 96 dB per octave. The result, however, is not a 16-pole Butterworth filter with a 10 kHz corner. As described earlier, the passband is already attenuated by -3 dB at the corner frequency. The cascaded filter doubles this attenuation to -6 dB at 10 kHz corner frequency, effectively shifting the -3 dB corner down to about 9.5 kHz. Repeated filtering is easy and tempting when a software package can filter a waveform with the click of a button. However, the edge of the passband is nibbled away with every filtering.

7
Special Uses of and Considerations Regarding Digital Filters

7.1
Segment Filtering and Elimination of Edge Effects

An especially useful property of digital filters is the ease with which they can be applied to selected segments of waveforms. Digital filtering of individual segments is useful for creating playback stimuli, for instance by removing an extraneous sound from the original recording, or by attenuating a particular harmonic or spectral peak. Because FIR filters compute weighted averages of the waveform, the beginning (or *edge*) of a finite waveform segment does not receive the same treatment as does the rest of the signal. The result is a transient artifact occurring at the onset of the filtered signal. This effect can be eliminated either by filtering the edge with an IIR filter, by filtering the edge with FIR filters of successively shorter length (Menne 1989), or by eliminating the edge altogether. The latter can be achieved by inserting a time-reversed version of the beginning segment at the onset of the waveform, thereby making a new edge in the appended segment. This time-reversal technique usually assures a smooth transition into the original waveform. After filtering, the added segment is removed. However, each of the cutting and pasting operations must occur at transition points where the waveform is at or near zero, to avoid introducing a broadband click.

7.2
Zero-Delay Filtering

Double-filtering provides a way to eliminate the phase delay of a digital FIR or IIR filter. The time-series waveform is filtered in the normal fashion, which produces phase delay in the output. The filtered series is then time-reversed and filtered again with the same function, shifting the delay back to zero. The double-filtered series is again time- reversed and thereby set right. Double-filtering shifts the cut-off frequency as explained in Section 6.2.

7.3
Decimation

In the second of the two anti-alias filter examples considered earlier, the sample rate was halved to reduce data storage requirements. Rather than halving the initial sampling rate, however, one can instead save only every other sample point of a signal that has already been digitized. Simply discarding alternate samples (downsampling by a factor of two) results in an aliased signal identical to that obtained by halving the sampling rate of the A/D converter without changing the corner of the antialiasing filter. Downsampling effectively lowers the Nyquist frequency. In fact, downsampling was used to produce the alias in the undersampled time series of Figure 5. Aliasing can be avoided if, prior to downsampling, the digitized series is refiltered with a sharp digital lowpass filter set to achieve the desired noise floor at the new Nyquist frequency. This process of sample rate reduction, *oversampling* followed by decimation, takes advantage of the ease and low cost of designing a sharp lowpass digital filter.

Decimation is implemented in real-time by commercial digital audio devices. If the device oversamples by a factor of 5 or more, a very clean analog antialiasing filter such as a Bessel design can provide adequate attenuation at the elevated Nyquist frequency while providing linear phase delay and a good step response. The 16-bit "delta-sigma" A/D converter on my current computer oversamples by a factor of 64. Real-time digital filters are also available that accept functions supplied by the user's software. Such a hybrid filter contains an analog antialiasing filter that is applied to the input, an A/D converter to digitize the analog signal, a programmable digital signal processor (DSP chip) that rapidly executes a digital filtering algorithm, a D/A converter that resynthesizes the filtered series as an analog signal, and an analog anti-imaging filter that is applied to the output.

7.4
Simulated Echo

Convolution of digital signals can also be exploited to create echoes and artificial reverberation in either a natural or a synthetic manner. As anything multiplied by one is itself, a digital signal is unchanged by convolution against the unit impulse array {1,0,0,0,0,0,0,0}. If the single unity value appears four points later in the array, i.e., {0,0,0,0,1,0,0,0}, convolution reproduces the original signal while delaying it by four sample points. For example, convolution of a signal against an array of {0.9,0,0,0,0.1,0,0,0} reproduces the signal at nine-tenths the original intensity, followed by an echo occurring four sample points later at one-tenth the original intensity. Digital echo simulation finds use in experimental exploration of echolocation and sound localization (e.g., Keller and Takahashi 1996).

7.5
Recreating Environmental Attenuation and Reverberation

An FIR filter can reproduce the lowpass filtering and reverberation of an environmental sound transmission pathway. For instance, a click might be broadcast and rerecorded some distance away in an environment characterized by multiple irregular, reflective surfaces. Compared to the original, the rerecorded click will show spectral changes and temporal smearing due to attenuation and reverberation effects. Digitizing the rerecorded click then yields an array that can be used to impose similar effects on any digitized signal by convolution (Menne 1989).

7.6
Matched Filters and Signal Detection

The frequency-selective filters described thus far can be used to improve the signal-to-noise ratio of a signal when at least part of the noise band exceeds the signal band. However, frequency-selective filters do not help much when the signal and noise occupy the same frequency band. In some such cases, matched-filter techniques can be used to detect the signal in the noise, provided that consistent differences exist between the two in basic attributes like waveform structure, amplitude envelope, or time-by-frequency characteristics. Matched filters search for particular signal events, providing a quantified measure of the degree to which segments of the sample match the characteristics of a particular target event. A simple technique for matched-filtering is to mathematically compare a relatively noise-free exemplar of the desired signal to a sample consisting of signals that are imbedded in noise. Comparisons based on convolution, for instance, in which a signal is reversed in time and correlated with the original version can be useful in at least partially extracting the signal from a noisy background noise.

This technique depends on *cross-correlation*, in which correlation values for two digital signals are repeatedly computed as one moves in step-wise, point-by-point fashion along the other. The result is a *detection function*, in which the similarity of the two signals is shown by the correlation value calculated at each point in time. If a signal is cross-correlated with a copy of itself (i.e., *autocorrelated*), the detection function reaches as maximum value of 1.0 at the point the two copies are perfectly aligned. Detection functions for non-identical signals vary between 0 and 1.0.

Although cross-correlation is most often applied to time-series waveforms, this technique can be especially useful for detection and comparison of animal vocalizations when applied to spectrographic representations (e.g., Clark et al. 1987). This approach is particularly advantageous when comparing complex acoustic signals. Whereas a time-series waveform has only two dimensions (i.e., time and amplitude), a time spectrogram separates signals into three dimensions (i.e., time, frequency, and amplitude). Adding a third analysis dimension provides additional signal separation, an advantage evident to anyone who has compared a time spectrogram of a birdsong to the time-by-amplitude oscillogram-based waveform display. Cross-correlation of time spectrograms works best when the spectrographic analysis involved represents an optimal combination of frequency versus time resolution (see Beecher 1988, as well as chapters by Clements, Beeman, and Owren and Bernacki in this Volume, for detailed discussion of issues related to spectral and temporal resolution in Fourier-transform-based spectrographic analysis).

8
Reducing Environmental Noise: An Example
Using an FIR Filter and a Simple Matched Filter

A few years ago, I wondered if the interpulse interval within a field cricket's chirps vary with temperature, as do the intervals between the chirps themselves. Crisp autumn weather made trips to the front yard more appealing than a trip to the library. Precise measurement of interpulse interval was easily accomplished by extracting measurements from tape-recorded waveforms digitized directly into the soundport of a Macintosh computer. I stepped outside every couple of hours to record the song of an amorous male field cricket (*Gryllus pennsylvanicus*) as the temperature changed throughout the day. A problem arose in the late afternoon when a male ground cricket (*Allonembius* sp.) moved next to the singing field cricket and dominated my recordings with its persistent trill. Rush-hour traffic had begun and a steady stream of cars whooshed by the yard. The field cricket chirps that were initially so obvious in the waveform became invisible in the noise. I probably could have found field cricket chirps by looking at a spectrogram instead of the waveform, but rejected this option because I wanted better time resolution than a digital spectrogram can provide. Instead of losing the sample, I used digital filtering techniques to extract the signal from the noise. The procedure used is described below.

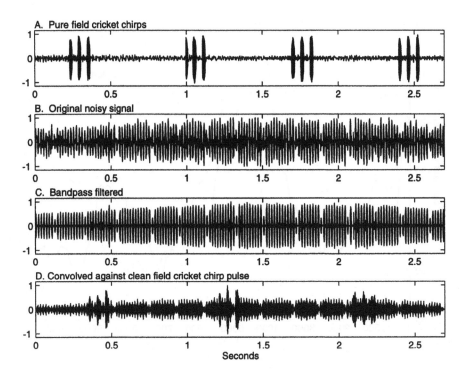

Fig 7 A-D. Cleaning a noisy signal with digital filtering. A. The first waveform is a reasonably pure series of four chirps of a male field cricket. **B.** The second waveform includes three chirps of a field cricket obscured by the trill of a ground cricket and the low frequency noise of passing automobiles. **C.** The noisy signal (B) is bandpass filtered with a 100th order FIR filter. **D.** The result of convolving the bandpass filtered signal C. against a single chirp pulse from signal A. The three chirps are now evident

I began by digitizing two 2.75 s sound segments from the audio tape. The first contained a fairly pure series of four fieldcricket chirps (illustrated in Figure 7A), while the second included a series of 3 fieldcricket chirps contaminated with both the ground cricket's trill and car noise (Figure 7B). The field cricket's chirps were nowhere evident in this noisy segment.

Before attempting to filter out the noise, I needed to better understand the spectral properties both of the target signal and of the noise obscuring it. I therefore plotted power spectrum density estimates of a pure chirp and a section of noise with no chirp (Figure 7A and 7B, respectively), and used Welch's averaged periodogram method to average the squared FFT magnitudes of overlapping 256-point segments. The "pure" chirp turned out to be not so pure – its spectrum is shown as the solid line in Figure 8. As can be seen, the low-frequency noise extended up to about 250 Hz, the chirp's fundamental frequency (basic rate of vibration) was about 4.1 kHz, and its second harmonic was at 8.2 kHz. The noise segment (the dashed line in Figure 8) had a similar, but louder, low-frequency band and was dominated by the trill of the field cricket. This call showed a fundamental of 4.9 kHz and a second harmonic of 9.8 kHz. As these were noisy field

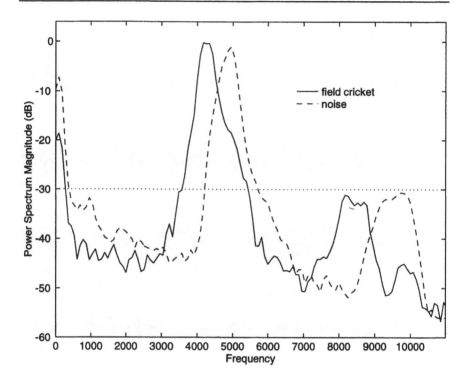

Fig. 8. Normalized spectral density estimates of the clean cricket chirps in Fig. 7A and a section of noise from waveform in Fig. 7B. The field cricket signal (*solid line*) is dominated by the fundamental frequency of its song. The noise spectrum (*dashed line*) is dominated by the loud trill of a ground cricket. The second harmonics are evident an octave above the fundamentals. Low-frequency noise from automobiles is seen at the lower edge of the spectrum

recordings, I decided to set -30 dB as the retention threshold. That is, I did not attempt to preserve components that were lower than -30 dB in relative amplitude (shown by the dotted line in Figure 8).

The low-frequency automobile noise could be removed easily with a digital highpass filter set to a cut-off frequency of 3.5 kHz. Unfortunately, the ground cricket signal overlapped the field cricket signal in time and partially overlapped it in frequency. A digital lowpass filter set to a cut-off of 5.4 kHz could therefore help but would not be sufficient. It was, worth a try. Using a program that designs linear delay FIR filters of arbitrary shape, I requested a 100th-order bandpass filter with the following characteristics:

Frequency (Hz)	0	2500	3500	5400	5800	Nyquist
Relative amplitude	0	1	1	0	0	0

Figure 9 shows the "requested" filter function and the frequency response of the impulse function my program returned. The impulse function itself (shown in Figure 10A) looked nothing like the frequency response curve. Instead its waveform resembles coarsely the waveform structure of the clean field cricket chirp pulse, as is evident in

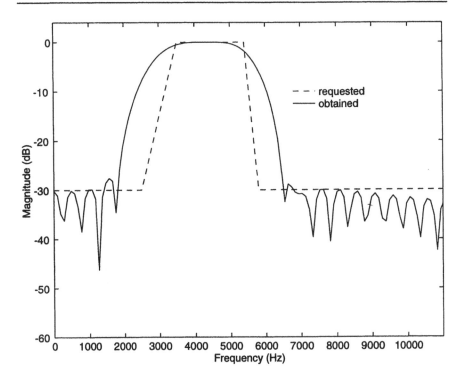

Fig 9. Digital FIR bandpass filter frequency x amplitude response patterns requested (*dashed line*) and obtained (*solid line*) from a digital filter design program

comparing Figures 10A,B. Convolving the FIR impulse function against the noisy signal produced the filtered output shown in Figure 7C. The field-cricket chirps were still obscured by the ground cricket trill – something more was needed.

The repetitive and stereotypical nature of the cricket chirp waveforms lend themselves to matched filtering. On this basis, I felt confident extracting the field cricket chirps from the combined signal by convolving a clean chirp against the combined signal. If an entire chirp were to be used as the model, its interpulse timing would affect the timing of the output. Therefore I selected as my filter function a single pulse from a pure fieldcricket chirp (shown in Figure 10c), processing it using an FIR bandpass filter, as described above, to remove extraneous low-frequency noise. I convolved this filtered pulse (see Figure 10B) against the bandpass-filtered chirp series (illustrated in Figure 7C) and the resulting waveform showed three distinct field-cricket chirps (as in Figure 7D) from which I could easily extract the interpulse intervals.

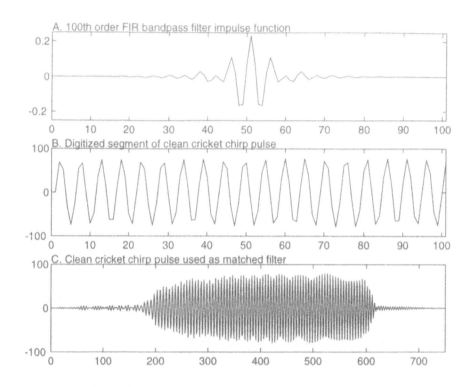

Fig 10A. Impulse function for the digital FIR bandpass filter described in Fig 9. Front-to-back symmetry of the impulse function insures linear phase delay. This impulse function is convolved against a waveform (Fig. 7B) to produce the filtered output (Fig. 7C). B. Section of a clean chirp. Notice the similarity in waveform between the impulse function and this chirp segment. C . The bandpass filtered chirp pulse that is used as a matched filter. This chirp pulse is convolved with the waveform in Fig. 7C to extract the chirp structure seen in Fig. 7D

9
Endnote: Solution to the Puzzle of the Disappearing Half-Masked Weaver Bird Voice

In recording the half-masked weaver song, my colleague had overloaded the analog audio recorder's circuitry, thereby clipping the waveform's amplitude peaks and squaring their tops. The Fourier series of a square wave has energy at all odd-numbered harmonics (h_n, where n = 1, 3, 5...) with the relative power of each being equal to 1 / n. As a result, the original signal, a modulated whistle with a fundamental that swept from about 2 kHz to 3 kHz, was now accompanied by a loud third harmonic that ranged from about 6 kHz to 9 kHz. I had set the real-time analyzer to a frequency range of 0–5 kHz and in so doing apparently had lowered the machine's Nyquist frequency to about 5 kHz. Ordinarily, most of the energy in the third harmonic would have been removed by a lowpass filter, but with no filter board in place, this component had folded downward across the Nyquist frequency. There, it appeared as a downward sweep from about 4 kHz to 1 kHz, crossing the fundamental and bearing no obvious resemblance to a harmonic. Only the antiquated, analog technology of the old Kay Sonagraph had exposed my ignorance and saved me from attempting to publish a description of a recording artifact as an exciting new finding. Had I that day in 1982 access to the digital printout capabilities that are now available, I might have submitted for publication a textbook example of signal aliasing. I wonder how many reviewers would have noticed the error.

Acknowledgments I thank Ben Clopton for introducing me to sampling and filter theory in his "Sound and Math Seminar" at the University of Washington. My exploration of digital filters was facilitated by the DADiSP and DADiSP Filters software packages, provided generously by the DSP Development Corporation for assistance with this chapter. Example signals and figures were prepared with MATLAB and the MATLAB Signal Processing Toolbox (The Mathworks, Inc.) on a Power Macintosh 7100 (Apple Computer) equipped with an Audiomedia II A/D card (Digidesign, Inc.). Support was provided by a summer grant from the Florida International University Foundation.

References

Beecher MD (1988) Spectrographic analysis of animal vocalizations: implications of the "uncertainty principle". Bioacoustics 1:187–208
Clark CW, Marler P, Beeman K (1987) Quantitative analysis of animal vocal phonology: An application to swamp sparrow song. Ethology 76:101–115
Cook EW, Miller GA (1992) Digital filtering: background and tutorial for psychophysiologists. Psychophysiology 29:350–367
Hamming RW (1989) Digital filters. Prentice Hall, Englewood Cliffs, New Jersey
Haydin S (1986) Adaptive filter theory. Prentice Hall, Englewood Cliffs, New Jersey
Keller CH, Takahashi TT (1996) Responses to simulated echoes by neurons in the barn owl's auditory map. J Comp Physiol A 178:499–512
Menne D (1989) Digital filters in auditory physiology. Bioacoustics 2:87–115
Okanoya K, Dooling RJ (1987) Hearing in passerine and psittacine birds: a comparative study of absolute and masked auditory thresholds. J Comp Physiol 101:7–15
Oppenheim AV, Schafer RW (1989) Discrete-time signal processing. Prentice-Hall, Englewood Cliffs,
Stearns S, David R (1988) Signal processing algorithms. Prentice Hall, Englewood Cliffs,
Wickstrom DC (1982) Factors to consider in recording avian sounds. In: Kroodsma DE, Miller EH, Ouellet H (eds) Acoustic communication in birds. vol 1. Academic Press, New York, p 1

Applying Linear Predictive Coding (LPC) to Frequency-spectrum Analysis of Animal Acoustic Signals

M.J. Owren and R.H. Bernacki

1
Introduction

The study of natural acoustic signaling in animals frequently raises questions about the distinctive, or functionally significant, features of the sounds being examined. Such questions have been approached using a variety of methods, but necessarily include acoustic analysis of the signals themselves. Historically, these analyses relied heavily on the analog *sound spectrograph* (also frequently called the *sonagraph* or *sonograph*, after the Kay Elemetrics Corporation's trademark name, "Sonagraph"). Sound spectrography provides a visible record of the frequency components of an acoustic wave-form and had been eagerly anticipated by speech researchers for some time before the requisite technology actually became available shortly after the end of the World War II (see, for example, Koenig et al. 1946; Cooper 1950; Fant 1960; Nearey 1978). This instrument subsequently served as a primary analysis tool for phoneticians and others over a number of years.

However, while the analog sound spectrograph allowed several kinds of measurements to be made simply and with reasonable accuracy, its analysis capabilities were limited. These limitations helped fuel an urgent interest in the development of more sophisticated *digital signal processing* (DSP) algorithms for use in speech analysis, with significant successes in this area occurring as early as the 1960s. While the digitally based spectrogram remains a common format for visual presentation of speech sounds, researchers in this area are unlikely to rely on spectrographic analysis alone in extracting acoustic features of interest. Animal-oriented researchers have also generally made the transition to DSP-based acoustic analyses, but the transition has occurred much more recently. For some, an apparent historical reluctance to acknowledge the weaknesses of classic, analog sound spectrography may now be mirrored by a tendency to make measurements directly from digital spectrograms without significant additional processing.

The purpose of this chapter is to present *linear predictive coding* (LPC) as an addition or potential alternative to more traditional sound spectrography — whether analog or digital. Although LPC is well established in speech research and is applied to the same questions previously approached through analog sound spectrography, it has been virtually ignored in work with nonhuman species. This outcome has been partly due to the general lack of substantive communication between speech researchers and investigators interested in animal acoustic communication. In addition, while bioacousticians examine the signals produced by a variety of species without a unifying model of production processes or common analysis protocols, speech researchers typically face less disparate tasks and are able to base techniques like LPC on generally accepted speech-production models. Finally, both the mathematical sophistication of LPC and, until quite recently, the relative inaccessibility of hardware and software resources to support this level of signal processing have presented major obstacles to the mainstream animal-oriented researcher.

As discussed by Beeman (this Volume), sophisticated DSP packages are now available for relatively inexpensive digital systems and have routinely come into use in bioacoustics laboratories. In this chapter, we hope to encourage bioacousticians to add LPC to their analysis arsenals by describing this approach in simplified, conceptual terms and providing detailed guidelines for its use. This presentation is, however, preceded by a discussion of selected aspects of analog sound spectrography, including some of the limitations and difficulties associated with this technique. Although the analog spectrograph has become obsolete, many aspects of the DSP-based analysis approaches now being used can be traced directly to features of this instrument, or to the measurement techniques that were used with it. Relationships between sound spectrography and LPC are also discussed, along with some of the considerations involved in selecting LPC analysis parameters. Finally, some practical issues surrounding the use of LPC with animal sounds are considered.

2
Sound Spectrography

Following the declassification and subsequent public description of analog sound spectrography in 1946, the spectrograph quickly came into widespread use among speech researchers (Flanagan 1972). Much of its popularity can probably be traced to its conceptual simplicity and ease of operation (Wakita 1976). Both technical (e.g., Bell Telephone Laboratories 1946; Koenig et al. 1946; Potter et al. 1947) and elementary (e.g., Baken 1987; Lieberman and Blumstein 1988) descriptions of spectrograph circuitry and use are readily available.

Briefly, the analog spectrograph produced a visual representation of energy distribution across the frequency spectrum (e.g., up to 8 or 16 kHz) over a short time period (e.g., 2.6 s). The amount of energy present at any given point in the resulting spectrogram was indicated by the darkness of shading on white, heat-sensitive paper. This shading was produced by a current-bearing stylus that burned the paper — the quantity of current passed by the stylus reflected the amount of acoustic energy detected at that moment by the spectrograph. Energy estimation occurred through analog calculation of the average amount of energy occurring within the range of a *bandpass filter* used as the *frequency-analysis window*. The bandwidth of the analysis filter was moderately adjustable, with known resolution trade-off effects in the frequency- and time-domains (see Beecher, 1988; Clements, this Volume). Appropriate interpretation of a spectrogram therefore required that *analysis filter bandwidth* be taken into account, as well as being further complicated by the limited dynamic range of spectrograph paper and the effects of the gain setting, automatic gain control, and *high-frequency shaping* or *preemphasis* (a relative accentuation of high-frequency energy).

In animal research, studies by Thorpe (1954) and Rowell (1962; Rowell and Hinde 1962) marked the beginning of widespread use of spectrograms in the study of avian and nonhuman primate vocal repertoires, respectively. Notes of caution concerning uncritical, "literal" interpretation of spectrograms subsequently appeared. Most notably, Davis (1964) provided an extensive test of spectrograph performance with a variety of known signals and enumerated several difficulties and pitfalls. Later, Gaunt (1983) described the appearance of false harmonics in spectrograms of some bird songs. Unfortunately, the caution advocated by Davis, Gaunt, and others (e.g., Watkins 1967; Greenewalt 1968; Marler 1969; Staddon et al. 1978; Hall-Craggs 1979) did not discernibly influence the use and interpretation of analog sound spectrograms by many animaloriented investigators.

2.1
Analysis Bandwidth and Analog Spectrograph Design

The analysis bandwidths typically available on analog spectrographs (as well as some later digital versions) were originally selected so as to allow speech researchers to measure both the frequency of the periodic laryngeal vibrations associated with *voiced* vowel segments and to characterize the varying resonance characteristics of the supralaryngeal vocal tract (the latter affecting both voiced and *unvoiced* sounds). Such analysis continues to be of great interest in the study of acoustic phonetics, and reflects a tactic of treating the speech signal as being a linear combination of two general components — the underlying *source energy* and a subsequent *vocal tract filter*. This technique is consistent with the *acoustic theory of speech production*, often referred to as the *source-filter model* (Chiba and Kajiyama 1941; Fant 1960), and is discussed in detail by Rubin and Vatikiotis-Bateson (this Volume; see also Lieberman and Blumstein 1988; Kent and Read 1992; Titze 1994 for recent nontechnical reviews). In general, this production-based, source-filter approach allows the investigator to characterize salient features of the speech signal very economically. The source-filter perspective has come to be increasingly applied to nonhuman animal vocal production as well, for instance in nonhuman primates (for recent reviews see Fitch and Hauser 1995; Owren and Linker 1995; Owren et al., 1997) and songbirds (reviewed by Gaunt and Nowicki, this Volume).

In accordance with the limitations and production expense associated with analog electronic circuitry, the traditional spectrograph came equipped with only a small number of analysis-filter settings. Typically, the settings that were available were ones meant to be used with the voices of adult human males. In adult-male speech, voiced segments characteristically exhibit a vocal fold vibration rate (known as the *fundamental frequency*, or F_0) of 100 to 125 Hz (Baken 1987). For such a signal, moving a 45-Hz-wide bandpass filter (the analog spectrograph's typical *narrowband* setting) across the frequency spectrum reveals individual energy peaks corresponding to the fundamental frequency of vibration and associated harmonic partials, as shown in Figure 1A. The approximately 22-ms temporal window corresponding to this analysis bandwidth is too

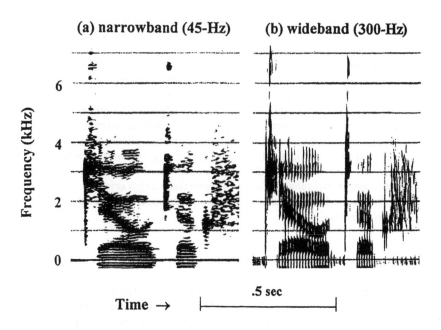

(a) narrowband (45-Hz)   (b) wideband (300-Hz)

Fig. 1 A . A narrowband (45-Hz filter) analog spectrogram of the words "Joe took" spoken by author M.J. Owren. B. A wideband (300-Hz filter) analog spectrogram of the same utterance. (From Owren and Linker 1995, used with permission.)

long to capture individual cycles of glottal opening and closing (see Beecher 1988), but the F_0 is revealed in the regular spacing of energy peaks in the frequency domain at integer multiples of the basic vibration rate. The F_0 can also be extracted using a wider analysis filter, which decreases frequency-domain resolution but increases time-domain resolution. The results of applying an analysis bandwidth of 300 Hz are shown in Figure 1B. This typical *wideband* setting (a 450-Hz setting was also often available on the analog sound spectrograph) provides a time-domain resolution of approximately 3 ms. In this case, individual glottal energy pulses become visible as vertical striations separated by periods of glottal closure.

In speech analysis, however, wideband analog spectrograms were more typically used to reveal the resonance characteristics of the supralaryngeal vocal tract than to measure F_0. The influence of such resonances in shaping the sound is also evident in Figure 1. In the absence of vocal tract filtering effects, one would expect the amplitude of successive harmonic partials of a quasi-periodic glottal source vibration to decrease monotonically at a rate of approximately 12 dB per octave (or frequency doubling). Taking into account a gain of approximately 6 dB per octave due to radiation characteristics associated with the mouth opening, a net loss of 6 dB per octave is expected (see, for instance, Titze 1994). The amplitude pattern in the utterance shown in Figure 1a does not decrease across the frequency range in this orderly fashion. Instead, supralaryngeal vocal tract *resonances* (that amplify particular frequency ranges) and *anti-*

resonances (that attenuate particular frequency ranges) differentially shape the frequency spectrum of the source waveform during speech production. Patterns of resonances and anti-resonances that are associated with various articulatory positions of the vocal tract are traditionally thought to give rise to distinctive spectral patterns underlying various speech sounds (e.g., Joos 1948). However, it is the resonances (given the special name *formants*) that have been particularly emphasized in understanding the vocal tract *transfer function*, or input-output relation.

As an example of the utility of wideband spectrographic analysis in revealing a distinctive spectral pattern, consider an F_0 of 100 Hz. Using a 300-Hz analysis window, one can expect that multiple harmonics will always be simultaneously detected by the filter as it traverses the frequency spectrum. Areas in which vocal tract resonances produce high overall energy levels appear as dark bands on the resulting spectrogram, reflecting the high average amplitude of harmonics in these regions. Conversely, loss of energy in excess of the expected frequency-dependent *roll-off* in the source waveform spectrum indicates either a lack of vocal tract resonance in that frequency range or the presence of an anti-resonance. Thus, wideband analog spectrograph filters were designed to span at least two harmonics in the speech of a typical adult male while encountering no more than one formant region at a time. The spectrograph's high-frequency shaping capability was added to accentuate harmonics in the upper spectrum that might reveal the presence of formants that would not be apparent due to the inherently low underlying energy levels in the source waveform at higher frequencies. High-frequency shaping was used to boost energy by 6 dB per octave, thereby overriding the 6-dB roll-off discussed above.

The two analysis filter bandwidths implemented on analog sound spectrographs were often fortuitously rather well matched to the frequency and temporal properties of birdsong (Beecher 1988). Unfortunately, the match between these filter bandwidths and other bioacoustic signals was often not very close. As a result, the settings were often used interchangeably, without evident recognition of their inherent design characteristics. One common problem was the use of the wideband setting to analyze signal with F_0 values higher than one-half the analysis filter bandwidth (an error also committed by author M.J. Owren). In other words, instead of characterizing the average amplitude of two or more adjacent harmonics, the filter simply smeared the energy from individual harmonics across a range corresponding to its own bandwidth. This effect is shown in Figure 2A, using digital spectrograms. Here, a narrowband representation reveals the harmonic structure of a highly tonal Japanese macaque (*Macaca fuscata*) *coo* call while wideband analysis distorts the energy of each harmonic without revealing formant locations. The dark bands produced in this fashion have in fact sometimes been mistaken for the broad spectral peaks that do reflect vocal tract formants.

Overall, then, unless a sound is appropriately low-pitched, narrowband analysis typically provides the most defensible means of using sound spectrography to examine acoustic energy distribution across the frequency range. For the majority of nonhuman animal signals, the wideband setting should probably be reserved for maximizing time-domain resolution in order to examine the sound's temporal properties. As discussed in detail by Beecher (1988), an analysis filter used in digital sound spectrography can and should be explicitly adjusted to closely match the particular characteristics of the signal of interest.

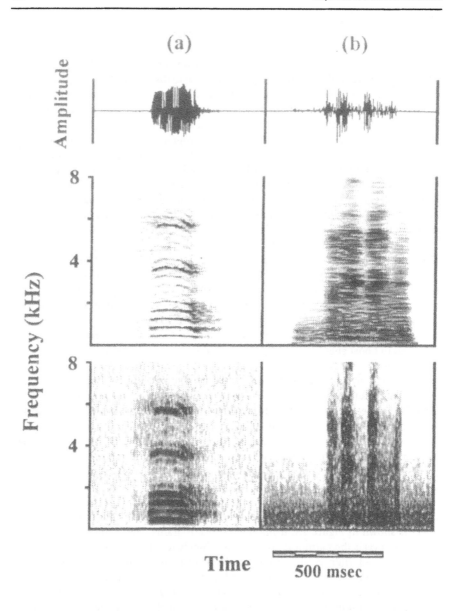

Fig. 2A. *Top* The waveform of *coo call* produced by an adult female Japanese macaque (20-kHz sampling rate). *Middle* and *bottom* narrowband (45-Hz filter) and wideband (300-Hz filter) digital spectrograms of this call, respectively. **B.** *Top* The waveform of a *snake alarm call* produced by an adult female vervet monkey (20-kHz sampling rate). *Middle and bottom* narrowband (45-Hz filter) and wideband (300-Hz filter) digital spectrograms of this sound, respectively. This call was recorded by Robert Seyfarth and Dorothy Cheney

2.2
Amplitude Representation in Analog Spectrograms

Using an analog sound spectrograph to measure amplitude characteristics was made difficult by the limited dynamic range of the thermal-sensitive paper used both in this machine and some subsequent hybrid analog-digital versions. The actual range of shading of such paper encompassed only 12 dB. As an additional complication, most analog spectrographs included automatic gain control (AGC) circuitry that could automatically vary the intensity of the input sound at any given point. Its function was specifically to compress frequency-dependent amplitude differences in the signal during analysis, thereby showing "more" energy within the limited representational range of the thermal paper. Of course, an unfortunate side effect (sometimes unknowingly experienced by bioacousticians) was that relative amplitude information across the frequency range was no longer presented accurately. Although an *amplitude contour* feature was included on many models to create a separate and more accurate representation of relative amplitude across the frequency range, this capability was very rarely used.

The problems associated with the limited amplitude range and resolution of analog spectrography were particularly acute in analysis of noisy sounds. Applied to either completely or partially aperiodic signals (i.e., with either no fundamental frequency or showing an unstable source vibration), such analysis typically did not reveal source energy characteristics, vocal tract filtering characteristics, or the net distribution of spectral energy — whether narrowband or wideband filters were used. Exactly the same difficulties apply to detecting spectral patterning through unaided visual inspection of digital spectrograms, as illustrated in Figure 2B, which shows narrowband and wideband representations of a noisy *snake alarm call* produced by a vervet monkey (*Cercopithecus aethiops*).

Historically, the complications associated with amplitude representation were generally ignored. This tradition continues in the use of *bandwidth* (the range of visible energy across the frequency spectrum) as a measurement variable characterizing animal acoustic signals shown in spectrographic form. As the relative intensities of various frequency components can be significantly influenced by analysis parameters such as relative gain setting, AGC, and high-frequency shaping, spectrographic analysis alone is a poor means of determining the spectral range of salient energy present in a given signal. Bandwidth is also necessarily and complexly influenced by other factors, including production amplitude, environmental transmission characteristics, and signal amplitude at the microphone, each of which can dramatically affect the absolute amplitude of high-frequency components. Acquisition of virtually any given set of animal acoustic signals involves substantial variation in the associated audio-recording conditions. For example, both for a particular individual and among individuals, sound samples will have been recorded from a variety of distances and angles of incidence. While the bandwidth variable appears necessarily to be vulnerable to such variation, one can expect the most biologically important acoustic features of communication sounds to be those that are robust to analogous, minor differences in the circumstances of production and reception. Overall, there is ample reason to question the inherent validity and reliability of the bandwidth measure.

2.3
Conclusions

While taking full advantage of the visual "road map" that sound spectrography provides, animal-oriented researchers have tended to ignore the inherent limitations of this technique as an analysis tool. The particular design features of the analog sound spectrograph played a significant role in shaping both the forms of the analyses that were undertaken and the interpretation of resulting measurement data. While the visual representation associated with this instrument justifiably continues to enjoy particular popularity, there is every reason to take full advantage of DSP algorithms in further investigating the importance of potential acoustic features that are identified, but not necessarily analyzed, from spectrographic images. The spectrogram may be most useful in helping an investigator to form hypotheses about distinctive acoustic features. It is not particularly well-suited for detailed analysis, specifically including the case of characterizing the spectral content and patterning of a signal.

3
Linear Predictive Coding

In the 1960s and 1970s, a number of independently developed approaches to analysis and synthesis of human speech emerged, gained prominence, and became unified under the collective heading of linear predictive coding, or LPC (Wakita 1976). One strength of LPC lies in providing objective estimates of formant characteristics and assisting in F_0 extraction, the same parameters previously sought using sound spectrography. The former is illustrated in Figure 3A, showing the waveform, wideband digital spectrogram, and comparable LPC-based spectrogram of a second instance of the utterance "Joe took" by author M.J. Owren. In Figure 3B, single spectral slices are shown for the short vowel segment marked by vertical lines in Figure 3A. The top panel shows a Fourier-transform-based power spectrum (see Beeman and Clements, this Volume, for detailed descriptions of this spectral analysis technique). This frequency-by-amplitude characterization is fine-grained and reveals the harmonic structure of the vocalic segment (as also shown in the narrowband spectrogram of Figure 1). An LPC-based smooth spectrum is superimposed on this power spectrum in the middle panel. These two characterizations are computed independently, but are closely matched. The LPC model provides a mathematical characterization from which the vocal tract resonances underlying this sound's overall spectral energy patterning can be derived. The LPC-based spectrogram of Figure 3A represents a series of spectral slices like the one illustrated in Figure 3B, in which only spectral energy peaks and their amplitudes are shown. Spectral envelopes that differ in overall slope appear in Figure 3B, a comparison that is discussed in Section 4.2.3.

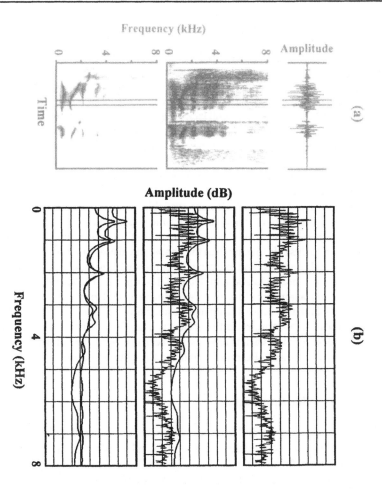

Fig. 3. A *Top* The waveform of a second rendition of the words "Joe took" by author M.J. Owren (16-kHz sampling rate). *Middle* and *bottom* wideband and LPC-based digital spectrograms of this utterance, respectively. B *Top* A 512-point Fourier transform of the segment enclosed by vertical lines on the waveform and spectrogram panels; *Middle* a corresponding 20-coefficient, LPC-based envelope superimposed on this power spectrum. The smooth spectrum was computed with a preemphasis factor of 1.0 and is shown in the bottom panel with a second envelope that was identically derived, except with preemphasis set to 0.0

Figure 4 shows smooth spectral envelopes superimposed in comparable fashion on Fourier-transform-based power spectra for the calls illustrated in Figure 2. For the *coo*, the slice is drawn from the midpoint of the call. For the *alarm call*, the slice represents the entire second pulse of this noisy sound, which is the one that is highest in amplitude. The corresponding waveform segment is shown at the top in each case, along with four envelopes representing different degrees of "fit" between a smooth spectrum and the "raw" spectrum. For these sounds, LPC functions can be used to successfully characterize both general and more detailed aspects of spectral patterning (see Section 3.4.2 for further discussion).

Various LPC modeling approaches have given rise to a number of associated DSP processing methods, and a range of applications are therefore possible. This chapter focuses primarily on using LPC to produce a smooth spectral envelope that either simply characterizes major energy peaks or can be interpreted as a representation of formant characteristics. Due to a general lack of information concerning the nature of the source energy in nonhuman animal source-filter sound-production systems, the latter application — attempting to characterize the transfer function — must be done with caution. However, as this information becomes available for various species and is incorporated into LPC-based implementations, detailed models of the filter component of production systems may be routinely sought in nonhuman animal studies as well.

A number of descriptions of LPC are available, but generally consist of technical mathematical treatments rather than more practical, application-oriented presentations. Markel and Gray's (1976) classic primer on LPC in speech research is somewhat of an exception to this rule inasmuch as the nontechnical user can fruitfully read around the mathematical equations and examine the numerous figures. Wakita (1976) is highly recommended as a nontechnical and detailed description of LPC and other common speech processing algorithms that provides intuitive insight into both the mathematical methods used and the acoustics-related assumptions of each modeling approach. O'Shaugnessy's (1987) more recent overview is a readable, albeit technical survey of important procedures in a variety of areas of speech research. The following description of the conceptual bases of LPC relies heavily on these three sources.

3.1
Three LPC Modeling Approaches

The terms *linear prediction* and linear predictive coding are used to designate any of a number of approaches to modeling the speech signal (Makhoul 1975; Markel and Gray 1976). Most of these formulations produce spectral envelopes in which only frequency peaks are included — spectral *dips* are specifically disregarded. Three theoretically distinct LPC approaches, the *linear-regression, inverse-filter,* and *partial-correlation* models, are discussed here.

3.1.1
Linear-Regression

The linear-regression model is based on linear least-squares estimation techniques developed in the late 18th century (Markel and Gray 1976). The mathematical roots of LPC were thus roughly contemporaneous with the development of the Fourier transform. This approach operates in the time domain and its strategy is to approximate successive points of the digitally sampled signal as linear combinations of past values. The resulting linear-regression representation can be used to produce a smooth version of the waveform, with corresponding simplified frequency characteristics.

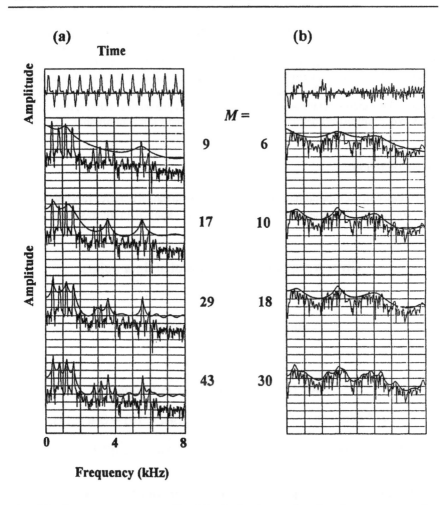

Fig. 4. Waveform segments, corresponding 512-point Fourier-transform-based power spectra, and autocorrelation-based LPC spectral envelopes are shown for A the Japanese macaque *coo call* and B the vervet monkey *snake alarm call* shown in Fig. 2. The waveform segments are from the midpoint of the *coo* and over the highest amplitude pulse of the *alarm call*, respectively. The smooth spectral envelopes for each segment were produced by varying *M*, the number of autocorrelation coefficients used

The linear-regression conception is illustrated in Figure 5. Given, for instance, a sequence of 100 sampled data points, a set of simultaneous linear regression equations is defined. Each equation is based on a number of coefficients, *M*. In Figure 5, *M* is equal to 10. In this example, then, the eleventh point of the sequence is the first that can be predicted, which occurs by weighting each of the preceding ten sample points by a unique coefficient (a_1 to a_{10}). The twelfth point is predicted as the sum of weighted points 2 through 11, etc. Predicting each sample point in the sequence in this manner produces a set of simultaneous linear equations that essentially constitute a problem in multiple regression. The solution to this problem is a single regression equation that predicts successive sample points of the time segment on the basis of the previous M points while minimizing overall prediction error (in the least-squares sense). Any of several methods can be used to compute the M *regression coefficients* that constitute the solution to this problem, including *autocorrelation-* and *covariance-based* techniques. These methods take advantage of redundancy properties in the regression equations to determine the coefficients in a computationally efficient manner.

3.1.2
Inverse Filtering

As it is based in the time domain, the linear-regression model requires very few assumptions concerning the nature of the signals to which it is applied. In contrast, inverse-filter

> ## Given a sequence of 100 sampled data points,
>
> ## $S(1) ... S(100)$
>
> ## find a set of 10 coefficients,
>
> ## $a_1, a_2, ... a_{10}$
>
> ## such that a simple linear equation,
>
> ## $a_1 * S(n) + a_2 * S(n + 1) + ... a_{10} * S(n + 10) = S(n + 11)$
>
> ## is best satisfied in the least-squares sense over all 100 points.
>
> ## This problem corresponds to a set of 100 simultaneous equations,
>
> ## $a_1 * S(1) + a_2 * S(2) + ... a_{10} * S(10) = S(11)$
>
> ## $a_1 * S(2) + a_2 * S(3) + ... a_{10} * S(11) = S(12)$
>
> ## $a_1 * S(3) + a_2 * S(4) + ... a_{10} * S(12) = S(13)$
>
> ## $\vdots$
>
> ## and can be solved through multiple regression.

Fig. 5. The conceptual basis of linear prediction as the least-squares solution to a set of simultaneous equations

modeling is a form of LPC that is specifically based on the source-filter conception of sound production. In this case, it is assumed that the acoustic signal represents a linear combination of a sound source and a subsequent transfer function (for instance, glottal vibration and vocal tract resonance effects in voiced speech). The source is further as-sumed to be either a quasi-periodic impulse or white noise, both of which are charac-terized by a *flat* or *white* spectrum. For periodic speech sounds then, a source-energy function with a characteristic 12-dB per octave energy roll-off is not sought. Instead, the 6-dB gain associated with lip radiation in humans is supplemented by the 6-dB preem-phasis applied during analysis and the combination is presumed to compensate for the original frequency-dependent energy loss. As illustrated in Figure 6, the inverse filter that best recovers the input (the hypothesized source function) from the output (the sound itself) can therefore be obtained. This filter is represented by a polynomial char-acterized by a series of coefficient values referred to as the *filter coefficients* (the *a para-meters*). The reciprocal of this filter approximates the transfer function of the vocal tract that produced the sound.

3.1.3
Partial Correlation

The partial-correlation model of LPC (also called the *lattice-filter method*) is also ex-plicitly tied to the source-filter framework. However, the theoretical assumptions made in this case are even more detailed than in inverse-filter-based analysis. In partial cor-relation, the researcher attempts to characterize energy reflections occurring in the supralaryngeal vocal tract as the source energy passes between smaller and larger di-ameter sections of this tube. In speech production, one such reflection occurs at the lips, where the waveform originating at the glottis encounters the boundary between the large outside air mass and the much smaller vocal tract. The *impedance* (resistance) mismatch that results causes partial reflection of the waveform back towards the glottis. Other reflections occur within the vocal tract where changes in cross-sectional area also produce impedance boundaries (Rabiner and Schafer 1978).

In this model, then, vocal tract resonances result from dynamic interactions between the glottal wave and its reflections, which cause frequency-dependent reinforcement (and cancellation). These effects are determined by relationships among vocal tract length and shape, the speed of sound conduction, and the wavelength associated with each frequency component of the source waveform. In partial correlation, the vocal tract is represented as a contiguous series of discrete tube segments or cylinders of equal length but unequal cross-sectional area, as illustrated in Figure 7. Each resulting bound-ary causes partial reflection of the glottal waveform.

Interactions between the original wave front and its reflections are characterized using a series of *partial-correlation* or *reflection coefficients* (the *k parameters*). Each coefficient parcels out the correlated quantity between the waveform and the expected reflection-based physical effects at a particular location in the vocal tract. This corre-lated quantity can be subtracted from the original waveform, leaving a residual from which the next correlated quantity can be removed. Note that there is not a direct con-

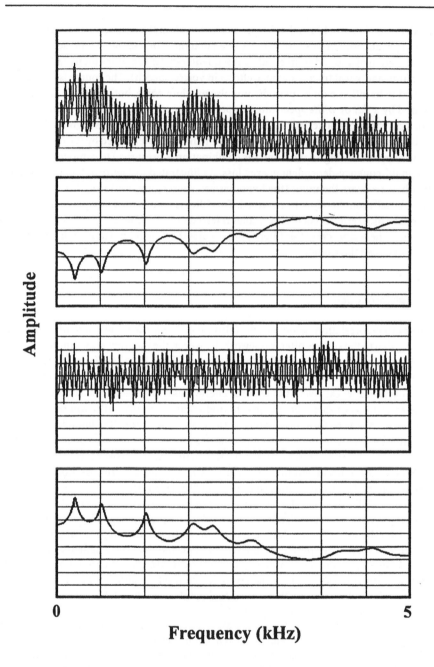

Frequency (kHz)

Fig. 6. The inverse-filter conception of LPC. The power spectrum shown in the *top panel* represents the output of a production system in which quasi-periodic source energy passes through a filter. The spectrum of the source energy is assumed to be flat, as shown in the *third panel*, which allows derivation of a filter function that "whitens" the system output. As shown in the second panel. The inverse of this filter, shown in the *bottom panel*, models the transfer function associated with the underlying production system

nection between the coefficients and frequency peaks observed in the spectral envelope that is produced. However, both spectral peaks and area ratios for the tube segments can be recovered.

3.2
Filters Derived from Regression Coefficients

The connection between the linear-regression, inverse-filter, and partial-correlation models lies in the fact that the coefficients of the inverse-filter model are identical to the regression coefficients derived through linear prediction, and are indirectly related to the reflection coefficients. In other words, each of these models produces results from which equivalent smooth spectral envelopes can be derived (see Rabiner and Schafer 1978 for for a detailed technical review). In spite of operating from different theoretical assumptions in requisitely diverse representational domains, then, the three models produce results that, where they overlap, are identical or equivalent. Filter coefficients and regression coefficients can be derived from the reflection coefficients, and the *residual* or *error signal* produced in the linear-regression model corresponds to the source waveform as recovered through inverse-filtering or partial correlation. This intersection among the three models allows some flexibility in the implementation of LPC-related computation subroutines. When the researcher specifies the linear-regression or inverse-filter models, for instance, it is often possible to obtain the requisite regression or filter coefficients more rapidly by first deriving reflection coefficients (and performing transformation operations) than by computing the desired coefficients directly. In fact, the most powerful speech analysis implementations calculate each coefficient type so as to provide both the smooth spectral envelope and the area ratios that are charac-

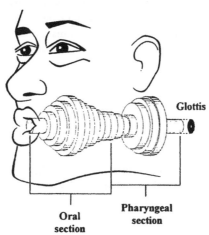

Glottis

Oral section

Pharyngeal section

Fig. 7. A male speaker of American English producing the vowel /u/. The vocal tract is modeled as a tube consisting of a series of 20 segments of equal length but variable cross-sectional diameter. Using the partial-correlation model, the vocal tract transfer function is computed based on the reflection effects associated with impedance mismatches occurring at the boundaries of adjacent segments. (From Titze ©1994, all rights reserved. Adapted by permission from Allyn and Bacon)

teristic of the vocal tract configuration used to produce a given sound.

3.3
Quantifying the Frequency Spectrum

A major strength of LPC analysis lies in its ability to provide quantitative, replicable characterizations of the frequency spectrum. Each of the three models described above can be used to produces a smooth spectrum — either indirectly, through a transformation from the time domain to the frequency domain, or directly, as the inverse of the vocal tract filter model. Given the mathematical equivalence that exists among the computations that follow from each model, one can conceive of LPC as producing a generalized "polynomial". The coefficients and variables found in the polynomial vary, depending on whether the model produces a time-domain, frequency-domain, or some other form of representation. In each case, however, the coefficients of the various polynomials are identical or equivalent, and critical features of the corresponding smooth spectrum described by the polynomial, such as the frequencies, amplitudes, and bandwidths of spectral peaks, can be obtained.

One approach to obtaining these values is to solve the polynomial, finding its roots. Available algorithms include the *Newton-Raphson approximation* and similar "cut-and-try" techniques described in standard, advanced algebra texts. Approximations along these lines are faster but less accurate in estimating frequency characteristics than are more sophisticated and exact numerical analysis methods. However, the latter require intensive computation and sometimes fail to provide an outcome when there is no physically realistic root (e.g., due to misassumptions in the underlying model). A second approach is to first produce the smooth spectrum by computing a Fourier transform of the polynomial coefficients and then to derive peak frequencies and amplitudes from the envelope itself. Bandwidths of the spectral peaks are estimated through parabolic mathematical curve-fitting at these locations. Direct measurements from the smooth spectral envelope can be made very quickly, but are less accurate than are root-solving procedures.

3.4
Analysis Parameters

Using LPC requires the investigator to select values for several important analysis parameters and the particular values that are specified play a critical role in shaping the resulting spectral envelope. The parameters discussed here are the size and shape of the analysis window, the number of coefficients used in the regression calculation, and the preemphasis factor that is applied. The terminology used is associated with the linear-regression technique, specifically as described by Markel and Gray (1976).

3.4.1
Analysis Window Size and Shape

The smooth spectral envelope produced using LPC requires simultaneous consideration of a series of sample points, as illustrated in Figure 5. Each such series constitutes a time-domain analysis window that, metaphorically at least, is positioned over a segment of the waveform. The size of the analysis window is specified in terms of the number of sample points, N, that it spans. Complete analysis of a sound includes traversing the entire signal in stepwise fashion, creating a number of segments whose spectral characteristics can be presented either in spectrogram format (as in Figure 3), or in a *waterfall display*. In the latter, the frequency-by-amplitude spectral slices are presented one behind the other and offset along a diagonal representing the temporal dimension.

The fundamental consideration in selecting window size is the rate at which the signal changes. Time- and hence frequency-domain characteristics of the segment being analyzed should be *stationary* (i.e., remaining constant) throughout the length of the window. This requirement can be restrictive, for instance in the case of fast-moving frequency modulations, but typically it is not. For example, given a digital sampling rate of 20 kHz, sample points occur at intervals of 0.05 ms. A window size of 128 points corresponds to a rather short 6.4-ms segment and even shorter windows can be used effectively. In LPC, unlike sound spectrography, resolution in the time domain (determined by analysis window width and step size) does not necessarily exhibit a direct inverse relationship with resolution in the frequency domain (accuracy of spectral measurement). The latter is primarily related to the number of coefficients in the polynomial, although window size and the computation technique used in producing the smooth spectrum can also be important (each factor is discussed in Section 3.4.2).

Selecting larger N values does improve the accuracy of the spectral envelope when frequency characteristics are subject to non-systematic variation. In other words, the spectral representation of a stationary noisy signal (or a stationary tonal signal that has noisy components) will usually improve with increasing window size. However, while selecting a larger window does not guarantee more accurate analysis, it does decrease the likelihood that the windowed segment will be stationary and can therefore result in smearing of frequency modulations. Larger window sizes also entail longer computation times.

Window length can be adjusted to better fit the characteristics of a nonstationary signal by specifying overlapping placement of adjacent analysis windows as the waveform is traversed. Selecting a high degree of overlap produces a smoothing effect in which critical attributes of spectral peaks are more likely to show continuous rather than abrupt change over time. However, inaccuracies introduced in this fashion may be more than offset by the improvement that results from the requisite increase in the rate at which the waveform is being "sampled". Ultimately, experience with the particular signal of interest is likely to be the best guide in selecting an optimal combination of window length and degree of overlap, while taking into account both the rate of change in critical signal characteristics and overall analysis goals.

The shape of the analysis window must also be specified, as discussed by both Beeman and Clements (this Volume) in the context of Fourier analysis. Window shape determines the relative weighting or importance assigned to each sampled point of calculations in which multiple sampled values are involved. A *rectangular* window, for instance, weights all sample points equally. A tapered or *weighting* window (e.g., *Hanning*, *Hamming*, or *Blackman*) is used to attenuate frequency-domain distortions resulting from the amplitude discontinuities that routinely occur at the endpoints of a rectangular window, by weighting samples in the middle more heavily.

3.4.2
Number of Coefficients

M, the number of coefficients used in the polynomial, lies at the heart of LPC analysis. As a rule of thumb, one can expect the number of peaks found by LPC algorithms to be $(M / 2) -1$ (Markel and Gray 1976). Speech processing routines typically allow M to vary from 3 to 30 or more, and can go significantly higher. For voiced speech sounds, Markel and Gray (1976) suggest that M be set to the sampling rate in kilohertz, plus 4. A 10-kHz sampling rate is customary in speech research and a value of 14 is therefore used to characterize the four to five formant peaks that often occur within the resulting 5-kHz-wide spectrum in the speech of adult human males. Noisy speech sounds typically exhibit only a few important spectral peaks and a value of 4 to 6 is suggested for this frequency range. Conceptually, the value selected for M is inversely related to analysis filter bandwidth in sound spectrography. Lower values of M correspond to wider analysis filters, while higher values correspond to narrower bandwidths. A small M value is therefore appropriate for characterizing formant locations, while a high value may or may not be useful, depending on the nature of the sound.

In partial correlation, M corresponds strictly to the number of segments used in the vocal tract model and is therefore selected in conjunction with vocal tract length and shape. In the linear-regression and inverse-filter models, however, there is some leeway is setting this parameter. A primary consideration (applicable to all three cases) is to adjust M in accordance with the digital sampling rate used. Higher sampling rates require that relatively higher M values be used. In addition, the investigator should specifically select a value that results in an appropriate fit between the LPC spectrum and a corresponding fast Fourier transform (FFT) representation of the waveform segment. The optimal closeness of this fit depends explicitly on the purposes of the analysis at hand.

The LPC envelopes shown in Figure 4 illustrate the effect of increasing M values. For both calls shown, the M value represented in the top panel is too low for the envelopes to provide useful characterizations of the spectrum. However, spectra shown in the panel immediately below are useful. For the *coo* call, the envelope successfully captures the apparent four-peak character of this spectrum. Here, the first peak represents the fundamental and second harmonic, harmonics 3 and 4 form a second peak, harmonics 7 to 12 make up a third peak, and harmonics 13 and 14 constitute a fourth peak. As noted earlier, no final conclusions can be drawn concerning how the amplitude pattern shown across the harmonics exactly reflects the effects of source energy and vocal tract resonance characteristics. However, given that resonance effects are inevitable in any naturally occurring sound-production system, it is not risky to claim that the tube-like vocal tract involved in producing this sound is showing the resonance effects that characterize tubes (e.g., Negus 1949; Fitch and Hauser 1995). Further, it is likely that the peaks in the envelope directly reflect the presence of formants.

Nonetheless, in the absence of specific knowledge of the source-energy spectrum, some uncertainty remains concerning the relative contributions of the source and filter components. Given the general similarity of the vocal folds in many mammalian species, it is likely that the amplitude of harmonics in the source energy of a *coo* call decreases with frequency (see Titze 1994 for a detailed review of vocal fold action in humans, nonhumans, and mechanical models). But it is also clear that detailed aspects of the glottal wave can affect the relative amplitudes of specific harmonics and that differences among mammalian species in vocal fold anatomy (including their size, shape, and tissue characteristics) is sufficient to produce significant variation in the details of that wave (see Schön Ybarra 1995). Thus, the low relative amplitude of the second harmonic may be a reflection of some source-related factor, or, as indicated in the four-peak smooth spectrum, may result from being located roughly between two adjacent formants. The reappearance of strong harmonics well above these two lower peaks specifically suggests the presence of formants in higher frequency regions. Overall then, the occurrence and circumstances of prominent spectral energy peaks can provide strong, preliminary evidence concerning the vocal tract transfer function. Although not leading to unequivocal conclusions concerning the relationship of the sound spectrum to underlying production mechanisms, this evidence allows the researcher to formulate specific, testable hypotheses about those mechanisms.

The *snake alarm call*, being noisy, shows no harmonic structure. Nonetheless, a discernible pattern of energy distribution is well-characterized by the LPC-based envelope. Again, this smooth spectrum is likely to reflect characteristics of both the source energy and subsequent vocal tract transfer function involved. In examining vervet monkey calls, Owren and Bernacki (1988) argued that these peaks (and similar, but distinctive peaks in *eagle alarm calls*) probably represent vocal tract resonance effects. However, this proposal has not been specifically evaluated in relation to spectral peak locations in other vervet calls, vocal tract length, and articulatory maneuvers in this species, or the possibility that the vervet's vocal folds are capable of rough, but stable vibration modes due to anatomical specializations (cf. Schön Ybarra 1995).

Smooth-spectrum characterizations are most informative when found to be consistent across biologically significant segments of the signal in question. In analyzing a noisy sound, selecting an M value that is too high is likely to produce peaks whose characteristics vary randomly across adjacent spectral slices (in accordance with the random nature of noisy signals). Thus, the spectral peaks shown for the snake call in the two lowest panels show little stability over time (if multiple adjacent envelopes are viewed). Therefore, rather than increasing M to capture each nuance of a given Fourier transform, the lowest M value that can both reveal peaks of possible interest and consistently "find" those peaks across adjacent call segments should be used. The situation is different for a harmonically structured sound, for which high values of M can be used to produce good characterizations of sounds with very simple spectra, as well as capturing the locations and amplitudes of individual components in a harmonically rich spectrum.

3.4.3
Preemphasis (High-Frequency Shaping)

As noted earlier, high-frequency shaping circuitry was included on the analog sound spectrograph to compensate for the observed net 6-dB per octave roll-off associated with pre- and post-vocal-tract filtering mechanisms in human speech production. In DSP approaches to speech analysis, preemphasis algorithms are used for the same purpose but can usually be set to a range of values. The results of varying the preemphasis factor are shown in Figure 3B. In the middle panel, full preemphasis has been applied in deriving the smooth spectrum. In the bottom panel, this outcome is contrasted with an envelope calculated in identical fashion, but without preemphasis. From this illustration, it is apparent that preemphasis decreases overall spectral tilt and can allow clearer differentiation of peaks occurring at higher frequencies. As noted earlier, the use of a high preemphasis setting is particularly useful in conjunction with the inverse-filter model in "whitening" the spectrum of the hypothesized source.

Preemphasis can be implemented in LPC by preselecting a value for the first reflection coefficient associated with the partial-correlation model. In this case, the user specifies a preemphasis factor that might range from 0.0 (no preemphasis) to 1.0 (full preemphasis). The user-selected value is then automatically transformed to an appropriate coefficient, whose effect is to influence overall spectral slope without affecting the locations of resulting frequency peaks. In a typical polynomial, then, the first coefficient in the series can be used either to influence overall spectral slope (preemphasis is applied) or as a measure of spectral slope (preemphasis is not applied — the first reflection coefficient is free to vary).

3.5
Computational Methods Used in LPC Analysis

The key to a clear understanding of LPC is to keep in mind that while the theoretical models involved are distinct, the results that are ultimately obtained are inherently equivalent in each case. A variety of computational methods are used in LPC, reflecting the diverse underpinnings of these various models. Three *autoregressive* (AR) computation methods used to compute LPC polynomials are autocorrelation, covariance, and partial correlation (PARCOR). Due to the equivalence of the smooth spectrum ultimately obtained with each method, any of the three can be used to analyze the spectral characteristics of a sound, regardless of the characteristics of the theoretical model that originally gave rise to the computational technique. The critical factor is to exercise care in interpreting the relationship between the smooth spectrum and underlying production processes. In addition, however, each computation method has specific advantages and disadvantages. It is therefore also important to select the technique that is best suited to the characteristics of the particular waveform being analyzed.

3.5.1
Autocorrelation

In the linear-regression model, each point in the analysis window of length N is predicted using the M preceding points. In associated autocorrelation-based computations, a weighting window is used to taper each end of the selected time segment to zero amplitude, thereby fulfilling the necessary assumption of independence between the target segment and the rest of the waveform. Points preceding and following the window are assumed to be zero.

As in sound spectrography, frequency resolution in autocorrelation-based computations varies inversely with window size. As a result, selection of window size must balance the considerations of achieving acceptable frequency resolution with the danger of smearing possible frequency modulations. In addition, the use of a (tapered) weighting window necessarily means that the time-domain data are being somewhat distorted. Together, these two factors act to set a lower limit on analysis window size. In practice, the window must be wide enough to include at least 2.5 cycles of a periodic sound in order to produce acceptable spectral representations. For nonperiodic sounds, the size of the window should reflect the necessary compromise between attaining acceptable frequency resolution while maintaining accurate representation of frequency changes.

The greatest advantage of the autocorrelation method is that it can be used with practically any sound. Its analyses are robust to variation in waveform characteristics and can therefore be applied to a variety of vibrational patterns, pulsatile productions, or trills. In addition, autocorrelation coefficients can be computed efficiently and *stable* functions always result. As noted earlier, some LPC models can produce polynomial representations whose solutions are physically impossible. Phrased in the terminology of filters, an *unstable* function of this kind is one that produces constant or exponentially increasing oscillation in response to an input. While instability is avoided, a disadvantage of autocorrelation is that the minimum analysis window size required is larger than if either covariance or PARCOR methods are used. Furthermore, spectral envelopes based on autocorrelation are less precise than those provided by these other techniques. This loss in precision goes hand-in-hand with the gain in breadth of applicability.

3.5.2
Covariance

Least-squares calculations based on the covariance method are made directly from the sampled values without modification, which is equivalent to using a rectangular analysis window. These calculations include the M points preceding the windowed time segment. Avoiding the data distortion caused by using a weighting window (as required by autocorrelation) produces precise spectral envelopes while allowing the use of shorter analysis segments. In theory, this lower limit approaches $N = M$. In practice, $N = 2M$ is a more realistic minimum value. Larger windows can also be freely used. If a very short analysis window is selected, however, it is necessary either that the signal be continuous and nonmodulated, or that the window be aligned with and matched in length to individual pitch periods in the source function (*pitch-synchronous analysis*). Covariance computations with short windows therefore require the use of pitch-tracking algorithms, which may not always perform accurately. For windows of all sizes, more computation time is required than with autocorrelation and the function that is produced can be unstable. Such instability can cause the occurrence of high-amplitude peaks in the 0-Hz and half-sampling-rate regions.

3.5.3
Partial Correlation

The advantages of PARCOR computation are essentially those of the covariance method — spectral precision and robustness to window-size variation. In addition, the PARCOR method always produces a stable function. Disadvantages of PARCOR, like those of covariance, are that the computations are relatively time-consuming and that pitch-synchronous analysis may be required when a small analysis window is used.

4
Applying LPC to Animal Signals

As an analysis tool, LPC has a number of strengths. Some of these are simply related to the flexibility and precision of DSP-based processing compared, for example, to analog sound spectrography. However, the automated nature of this approach provides a particularly compelling argument for using LPC. Digitally based spectrographic representations, in contrast, are frequently specifically designed to be used in conjunction with visually guided, cursor-based measurements (see Beeman, this Volume). While the digital nature of the underlying processing represents a significant improvement in numerical accuracy over similar data derived from an analog spectrogram, the researcher is nonetheless required to make a range of subjective measurement decisions. These decisions are likely to be very difficult to describe and communicate to others. The decision-making process involved may therefore significantly compromise the objectivity and consistency of selecting and measuring important spectral peaks (particularly in the case of noisy sounds). LPC, in contrast, provides replicable, quantified estimates of major frequency peak locations, amplitudes, and bandwidths. For a given signal, measurement results need depend only on parameter selection and analysis window placement. Thus, each important aspect of the decision-making process can be more easily described and communicated.

An additional advantageous feature of LPC is that the information present in the original waveform is significantly reduced. In both analog and (FFT-based) digital spectrography, the time-domain waveform is transformed to the frequency domain without significantly decreasing the amount of data represented. The analog spectrograph was designed to alter signal information through averaging, but did not reduce it per se. Fourier analysis transforms the time-domain waveform into frequency, amplitude, and phase characteristics, but again without reduction — the original signal can be fully recovered through an inverse transform.

LPC can be thought of as specifically modifying the characteristics of the time-domain waveform and effectively discarding much of its variability. This point is illustrated in Figures 3 and 4. The Fourier transforms shown in these figures were computed using 512-point windows, producing power spectra that each contain 256 separate frequency values. The LPC analyses that quantify the major spectral peaks in each sound do so using only 10 to 20 polynomial coefficients. Depending on interpretation, this procedure can be seen as producing either a simplified, smooth spectral envelope or as characterizing a signal's most prominent, and possibly most salient, acoustic features. Because of these data-reduction properties, a more manageable, quantitative characterization of the frequency spectrum is provided. Statistical analysis can then proceed based either on the frequency-, amplitude-, and bandwidth-values of the major spectral peaks, or more generically, on the coefficient set itself. In addition, a given signal can be synthesized from its LPC-derived characteristics, allowing the investigator to test empirically whether communicatively significant features of the acoustic signal were captured by the analysis.

4.1
Applicability

As noted in Section 3.1, the theoretical underpinnings of the linear-regression, inverse-filter, and partial-correlation models are quite different. Each model has some assumptions in common with the other two, but there are critical discrepancies as well. Surprisingly then, when the various computational methods inspired by the three models are applied to a given signal in corresponding fashion, the outcomes are identical. This remarkable coincidence provides an opportunity for researchers interested in acoustic signals produced by nonhuman animals to take advantage of the very powerful analysis capabilities of LPC, even when the underlying production processes are quite unlike those of human speech.

To use LPC effectively, the researcher first selects a computational method whose particular strengths and weaknesses provide a good match to the characteristics of the signal of interest and the goals of the analysis. These strengths and weaknesses were outlined in Section 3.5. Second, analysis outcomes are interpreted in accordance with the assumptions of each theoretical model and their respective connections to underlying sound-production processes.

One assumption that is shared by all three models is that the signal segment falling within the analysis window is stationary. A second common assumption is that the signal can be characterized in terms of frequency peaks alone. LPC does not capture possible dips in the spectrum, where, for instance, energy present in certain portions of the frequency spectrum of the source waveform may have been specifically attenuated by vocal-tract filtering. Such anti-resonances almost certainly exist in a variety of animal sound-production systems, but have rarely been investigated (although see Haimoff 1983).

As it operates in the time domain and is not tied to production processes at all, linear regression is the most broadly applicable of the three models. It has been used effectively with a range of both tonal and noisy speech sounds as well as in time-series analyses of nonspeech signals like seismic waves (Robinson 1967). As with many other statistical procedures, linear regression requires the rather standard assumption that the residual or error signal will be a random, Gaussian process. Although this model would therefore seem to be best suited to noise-based sounds, it nonetheless operates well with a variety of signals and is relatively robust to violations of the Gaussian assumption.

The inverse-filter model does not require that the signal be Gaussian in nature, but does assume an underlying linear production system involving a quasi-periodic or aperiodic source waveform that subsequently passes through an independent filter. From a theoretical perspective, inverse-filter-based analysis is more clearly applicable to both periodic and aperiodic signals than is linear prediction. Strictly speaking, however, the production model itself may only match the sound-production systems of a relatively small number of animals. The partial correlation model is even more specific in being based on a tube model of sound production, in this case one in which a tube-like vocal tract varies in cross-sectional area.

Overall, there is little danger of going astray in using LPC to derive a smooth spectrum and thereby characterize the major frequency peaks of an animal acoustic signal — so long as no conclusions not drawn concerning the underlying sound-production processes. If independent evidence suggests that the sound was produced using a linear source-filter system where the source waveform can be estimated or measured, the resulting envelope can be interpreted as a filter function. If, in addition, the filter can be modeled as a series of coupled tube-segments of varying area and the overall length of the tube is known, the cross-sectional areas of each can be derived.

It should be noted that some of the assumptions of LPC are routinely disregarded in the acoustic analysis of speech. The practical, experience-based approach to LPC provided by Markel and Gray (1976) reflects the fact that the characteristics of any naturally produced speech violates at least some of the assumptions associated with these models. First, for instance, it is known that human speech production is not a linear process. Instead, some *coupling* occurs between source vibrations and the subsequent filter — in other words, particular vocal tract positions can influence the characteristics of the glottal waveform during voiced segments of the speech waveform. Second, cavities below the glottis that are not included in the source-filter model also affect frequency characteristics of speech signals. Finally, anti-resonances due to side-cavities coupled to the vocal tract (i.e., energy shunts such as the nasal cavity) act to attenuate the source waveform in a frequency-dependent manner. Successful synthesis of highly intelligible speech signals from LPC analysis data, whether periodic or aperiodic, demonstrates their robustness to violations of theoretical assumptions. LPC synthesis produces both well-formed affricates (e.g., /dʒ /) and fricatives (e.g., /s/) in spite of the fact that these noisy sounds are based on a turbulent sound source placed well forward in the vocal tract and are particularly unlikely to conform to the assumptions of the inverse-filter model.

4.2
Analysis Parameters

4.2.1
Analysis Window Size and Shape

Fourier transforms should be used in conjunction with LPC whenever possible to visually verify the fit between the smooth spectral envelope and an unreduced spectral representation, as illustrated in Figures 3 and 4. A reasonable tactic is to use the same
window size for both analyses. Commonly used FFT algorithms require that N be selected as a power of 2, and it is therefore convenient to select LPC analysis window sizes
in the same manner. Note, however, that the discrete Fourier transform (DFT) and
padded FFT are not restricted in this way (see for instance Markel and Gray 1976; Beeman, this Volume; Clements, this Volume). As mentioned, computations based on
autocorrelation require that the window include at least 2.5 pulses of a periodic sound.
Window sizes can vary more freely when the covariance and PARCOR methods are
used, but require pulse alignment when N is small.

4.2.2
Number of Coefficients

Detailed spectral attributes of simple tonal sounds can be automatically and reliably
quantified by specifying a large number of coefficients. Figure 8 shows a song produced
by a grey warbler (*Gergygone igata*), whose single-component spectrum has been characterized using 20-coefficient, autocorrelation-based LPC functions. This technique is
particularly applicable in cases where there is a high fundamental frequency and few or
no additional harmonics. Under these circumstances, LPC will show each harmonic as
a separate peak and the spectral envelope should not be interpreted as a transfer function.
    A smaller value is selected for M to obtain a smooth, reduced envelope if the inves-
tigator either wishes to simplify the spectrum of a complex signal (whether noisy or
tonal) or can assume a linear source-filter production mechanism and is attempting to
recover the filter's resonance characteristics. In these cases, each resulting LPC peak
encompasses a spectral region consisting of a set of adjacent harmonics or a broad peak
in a noisy spectrum. If an interpretation of vocal tract filtering based on cross-sectional
area ratios and PARCOR analysis is warranted, the number of coefficients is selected
so as to match the number of vocal tract impedance boundaries characteristic of that
species.

4.2.3
Preemphasis

Strictly speaking, there are probably few occasions when preemphasis is theoretically
appropriate for animal acoustic signals. Production characteristics are unknown for

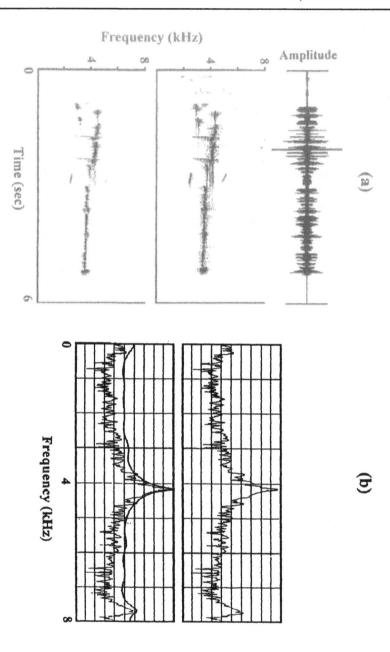

Fig. 8A *Top* The waveform of a grey warbler song (16-kHz sampling rate); *Middle and bottom panel* show narrowband and LPC-based digital spectrograms of this rapidly changing, frequency-modulated sound. **B** A 512-point Fourier transform of the segment marked by a vertical line on the waveform. A corresponding 20-coefficient, LPC-based function superimposed on this power spectrum. This smooth spectrum envelope was computed without preemphasis

most signals and the most conservative approach is therefore to forgo preemphasis. Again, however, the most important consideration is that the investigator makes an informed decision and takes the consequences of that decision into account in subsequent data interpretation.

On the one hand, there is no direct relationship between signal preemphasis settings and production processes in nonhuman animals, as there is with respect to source energy and lip radiation characteristics in humans. In nonhuman sound-production systems involving vocal tracts, for instance, data concerning the spectrum of the source energy are rarely available. Further, the 6-dB per octave gain associated with the typical head size and mouth features of adult male humans cannot be generalized to other, anatomically dissimilar species. A bird's beak, for example, provides better coupling between the internal and external sound transmission environments with less overall effect on the frequency spectrum of the radiated signal. Thus, for nonhuman vocalizers, relative peak amplitude and overall spectral tilt values in the LPC-based envelope cannot be viewed as being reliable characterizations of the corresponding properties of the underlying vocal tract transfer function, whether or not preemphasis is applied. On the other hand, a researcher using LPC simply as a means of providing a tractable characterization of major spectral features might employ preemphasis when there are high-frequency peaks of interest whose amplitudes are well below those of the low-frequency peaks. In this case, a preemphasis setting of 0.8 or above will accentuate these peaks. In such cases, it should be borne in mind that in applying preemphasis, a somewhat distorted depiction of the energy is being created.

As noted above, spectral slope can be reliably quantified in the normal course of LPC analysis if signal preemphasis is not applied. Owren and Bernacki (1988) used this strategy to measure overall spectral slope ($TILT$) in vervet monkey *alarm calls* as 100 times the first reflection coefficient ($TILT = RC[1] \times 100$). A generally flat spectral shape results in a $TILT$ value close to zero, while rising and falling envelopes are associated with positive (0 to 100) and negative (0 to -100) $TILT$ values, respectively. A measure such as $TILT$ provides some clear advantages when compared to the bandwidth measure that was strongly criticized in Section 3. Specifically, while bandwidth results depend on the absolute quantity of high-frequency energy that is detected, $TILT$ measures the relative shape of the overall spectrum and is therefore less likely to be unduly influenced by variability within and among individuals, or by various sound-production and audio-recording factors. Overall slope can also be measured either from an LPC-derived smooth spectrum or an FFT-based power spectrum by exporting the individual values making up the spectrum to a spreadsheet program, and fitting a straight line to the series using linear regression.

4.3
Some Previous Applications

Applications of LPC analysis have been relatively rare among animal acoustics researchers. Nonetheless, a number of examples are available and several particularly instructive ones will be described. In a seminal and rigorous application of LPC to animal sounds, Carterette et al. (1979) analyzed tonal *isolation calls* produced by domestic kittens less than ten weeks of age. Waterfall displays based on 9 or 18 coefficients (preemphasis was not discussed) and an analysis window of 8.33 ms (call length could exceed 1 s) were compared with wideband analog spectrograms. Using 9 coefficients produced the most satisfactory results, revealing three stable energy peaks that were interpreted as reflecting the kitten's vocal tract resonance characteristics. The resulting center frequency, amplitude, and bandwidth measurements of these three peaks were consistent with those predicted by modeling the kitten's vocal tract as a simple, single-tube, source-filter system.

Owren and Bernacki (1988) applied LPC to *snake* and *eagle alarm calls* recorded from free-ranging vervet monkeys. Both sound types consist primarily of repeated, atonal energy pulses. Parameter selection was based on preliminary visual comparison of LPC envelopes with FFT power spectra computed at intervals through the calls. Setting preemphasis to 0.0, an *M* value of 6 was found to provide the most consistent fit to a 5-kHz frequency range, both when the analysis window spanned an entire pulse or some shorter segment. Peak-related measurements were obtained through root-solving the LPC polynomials and using first reflection coefficients to characterize spectral tilt, as noted above. Statistical comparisons that included values obtained for peak frequencies, amplitudes, and bandwidths revealed a variety of significantly different spectral features in these two calls, including the two to three spectral peaks whose interpretation was briefly discussed in Section 3.4.2. Synthetic calls based on LPC data were subsequently presented to vervets trained to classify *snake* and *eagle alarm* sounds in an appetitive operant-conditioning-based task (Owren 1990). This study showed both that the synthesized calls captured perceptually salient features of the natural sounds and that LPC-based filter functions could be used to create novel, synthetic versions "from scratch".

Seyfarth et al. (1994) measured the frequency characteristics of *double grunts* produced by wild mountain gorillas (*Gorilla gorilla beringei*) using a ten-coefficient, autocorrelation-based polynomial. The resulting spectra were interpreted in a source-filter manner with formant frequencies, amplitudes, and bandwidths being derived from the spectral envelope. The number of coefficients was selected on the basis of comparing LPC and FFT characterizations of corresponding 25-ms segments. The report does not specifically note how measurements were extracted from the smooth spectral envelopes. However, the results indicated that *double grunts* include at least two subtypes with distinct second-formant frequency values.

Rendall (1996) used 12 and 18 coefficient polynomials (with full preemphasis) to characterize the spectra of tonal *coos*, noisy, pulsed *grunts*, and very noisy *screams* produced by adult female rhesus macaques (*Macaca mulatta*). Comparisons were made both within call-types and among the three sounds. For the latter, a set of 8 peaks was eventually used, but "missing" peaks were allowed to occur (producing greater consistency in the face of variation in the number of identifiable peaks for a particular call or animal). A notable feature of this research was that the coefficients themselves were used in statistical analyses that classified the vocalizations by individual caller. This approach was very successful, demonstrating that the various sounds differed in the degree to which spectral features were individually distinctive by caller. Analysis results were corroborated in playback experiments involving two of the call-types, *coos* (see also Rendall et al. 1996) and *screams*. Rendall (1996) interpreted LPC-based spectral peaks as reflecting formants and discussed possible individual distinctiveness and kinship-based similarities in the apparent vocal tract filtering effects evident in the calls. This work is a particularly good source for informative discussion of practical issues that can arise in analyzing LPC-based measurements.

5
LPC Synthesis

Although worth much more extensive treatment, LPC-based synthesis techniques can only be briefly discussed here. In speech research, it was the successful application of LPC to synthesis that particularly spurred the rapid acceptance of this general approach (Wakita 1976). Two areas in which LPC techniques have great potential for bioacoustics-related applications are in using LPC functions as spectral shaping tools, and in modeling and testing hypothesized components of source-filter production processes in animals.

In bioacoustics, the intellectual predecessor of these sorts of applications was Capranica's (1966) use of analog electronics to create bullfrog mating calls. His technique was to produce a mating call facsimile using periodic pulse trains or white noise as input to four resonant circuits arranged in parallel. As discussed by Beeman (this Volume), DSP algorithms can be applied in exactly this fashion to synthesize stimuli with particular spectral characteristics. For instance, noise or sinusoids can be used as input components that are combined through simple arithmetic operations or can be explicitly filtered so as to take on particular spectral characteristics.

Digital filtering, of course, requires a function that will produce the desired spectral characteristics (discussed by Stoddard, this Volume). In bioacoustics-related applications, these features are typically ones found in the natural sounds themselves and can only sometimes be well-modeled using predefined, mathematical functions. In many cases, then, LPC analysis probably provides the better spectral shaping tool, in the form of the function that is derived. The coefficients of this function can entered into a filter-creation subroutine without modification, and then used with a requisitely selected source waveform (e.g., Owren 1990). Regardless of the original underlying production process, such functions can be used to create both naturalistic, replica sounds, and stimuli in which one or more aspects of source energy and filter characteristics have been manipulated so as to modify some aspect of interest in the synthesized signal. For a case in which the sound production processes are well-approximated by a source-filter system whose characteristics are generally known, LPC-based synthesis (and, for similar reasons, formant-based systems such as Klatt's 1980, cascade-parallel synthesizer) can be used to create experimental stimuli entirely from scratch (see also Rubin and Vatikiotis-Bateson, this Volume).

6
Conclusions

Seeking functionally significant features in sound signals is frequently a problematical undertaking, perhaps due to the complexity of both contributing disciplines, communication and acoustics. Compared with researchers that are primarily interested in human speech, animal-oriented researchers face additional difficulties. Examples include extreme variation in both the signal forms and production processes that characterize different animal species. Such variation impedes the development of common analysis strategies among bioacousticians and encourages a division of resources and effort by species and signal-type. In addition, there are many fewer investigators and a smaller resource-base in bioacoustics than in speech research. It seems inevitable, then, that a significant proportion of the innovative techniques and instruments used by animal researchers in the future will be borrowed from speech research, as they have been in the past.

LPC is one such technique, and it has great potential value for the endeavor of understanding animal acoustic signals. This tool is a broadly applicable method for characterizing the frequency spectrum that is complementary to but independent of Fourier-transform-based approaches. The value of LPC has already been shown in literally thousands of applications to speech sounds. LPC is not, of course, a panacea for the difficulties of spectral analysis in bioacoustics. Each of its models and computation methods are subject to caveats and limitations — no single analysis approach will ever be appropriate or sufficient for every animal signal of interest. Application of LPC also demands some sophistication on the part of the user, requiring a conceptual understanding of the assumptions associated with the various theoretical models and computation methods. However, the potential benefits amply justify the effort needed to understand its proper use.

Unlike sound spectrography, for instance, LPC is well suited to analyzing complex frequency spectra. This general task is a common one in bioacoustics, where many signals are either noisy or exhibit closely spaced harmonic structures. Unfortunately, researchers have most often settled for using measurement techniques and variables that have fallen far short of characterizing the spectral richness of such sounds. LPC can also be applied to tracking individual harmonics in sounds with simpler frequency content. In each case, considerations of quantification, flexibility, and reliability all suggest that this technique could routinely be extremely useful in studying animal sounds.

7
Summary

The analog sound spectrograph long served as the primary tool for frequency-spectrum analysis of animal sounds due to its ease of operation and apparent conceptual simplicity. However, difficulties and limitations inherent in the spectrograph were often overlooked, including the rationale underlying wideband analysis, the effect of high-frequency shaping or preemphasis, and limitations in amplitude representation. Many of the features of analog sound spectrography have been carried over to digital signal processing environments. In speech research, the spectrographic approach to frequency analysis has been largely replaced by theoretical models and computational methods that are collectively known as linear predictive coding (LPC). Three of these models are based on linear prediction, inverse filtering, and partial correlation. These three approaches differ in their theoretical assumptions and associated computational techniques, but provide identical results when applied to a given signal in equivalent fashion. LPC characterizes the frequency spectrum as a series of peaks with quantified frequencies, amplitudes, bandwidths, and overall spectral tilt.

The linear-prediction model is based in the time domain and requires few assumptions about the waveform. Most importantly, successive waveform segments encompassed by a sliding analysis window are assumed to be stationary. The inverse-filter approach is rooted in a source-filter sound-production system in which source energy passes through a resonant chamber whose properties can be characterized by a transfer or filter function. The net output of the system is a linear combination of the original waveform and subsequent filtering effects. The characteristics of the transfer function are sought as the reciprocal of the inverse filter that best recovers the source energy, which is assumed to be either quasi-periodic or aperiodic. The partial-correlation model is based on a source-filter production model system involving a tube-like vocal tract of known length. The vocal tract is modeled as a contiguous series of segments of equal length but variable area, in which partial reflection of the waveform can occur at each segment boundary. The reflective effects of each segment are then parceled out.

The computational methods used in LPC analysis were developed separately, in accordance with the varying rationales of the emerging theoretical models. The techniques underlying these approaches include calculations based on autocorrelation, covariance, and partial correlation (PARCOR), all of which are known as autoregressive (AR) methods. Autocorrelation is typically the most robust of the three and can be applied to a wide variety of signals, regardless of their origins. It produces stable functions based on regression coefficients and is relatively immune to source energy fluctuations. However, a relatively long and weighted (tapered) analysis window that spans at least 2.5 pulses of a periodic sound is required, and autocorrelation generally produces less precise spectral envelopes than do either the covariance and PARCOR methods. The latter methods produce functions based on filter and reflection coefficients, respectively, and both use an unweighted (rectangular) analysis window whose length can be greatly varied. However, using a very short, unweighted window requires either that the signal of interest is continuous or that the window be pulse-aligned. These covariance and PARCOR methods require longer computation times than does autocorrelation. While the covariance technique can produce filters that are unstable, PARCOR always derives stable filters.

Due to the equivalence of the outcomes associated with the various computational methods, LPC analysis can readily be applied to sounds produced by nonhuman animals. In so doing, the computation technique should be selected so as to match the characteristics of the signal of interest. Due to the varying theoretical assumptions of the models that gave rise to the different computational methods, however, some care is required in interpreting LPC-based results vis à vis underlying sound-production

systems. The linear-prediction model allows the researcher to use LPC-related computational methods freely with virtually any time-domain waveform so long as the function that is produced is only interpreted as a smooth spectrum based on energy peaks.
If the sound is produced by a linear, source-filter system, and characteristics of the
source energy are known, the inverse-filter model allows interpretation of the smooth
spectrum as a transfer function. In addition, if this system can be characterized as a
series of contiguous segments that form a tube of known length, the partial correlation
model allows computation of the cross-sectional area of each segment. In many cases,
the results of LPC analyses are robust to violations of the assumptions of these underlying models.

The most important parameters to specify in LPC analysis are analysis window size
(N) and shape, the number of regression, filter, or reflection coefficients (M), and
preemphasis (a factor used to manipulate overall spectral tilt). Selection of analysis
window size and shape varies according to computational method, but should always
take into account the rate of change in spectral characteristics over time. The number
of coefficients is specified so as to produce a good fit between the LPC-based smooth
spectrum and a Fourier transformation of a corresponding waveform segment, or to
match the length and number of segments in the vocal tract under study. Preemphasis
can be used to bring out potentially important high-frequency harmonics and spectral
peaks by fixing the value of the first reflection coefficient. However, allowing this coefficient to vary provides a quantified measure of overall spectral tilt.

LPC has great potential value for studies of animal acoustic signals that may involve
either simple and or complex frequency spectra. Applied to a signal with relatively few
frequency components, LPC can be used to automatically provide a detailed, quantitative characterization of its frequency content. Used with a signals showing more complex frequency patterningcontent, LPC produces a replicable smooth spectral envelope.
Depending on the assumptions that can be made concerning the production system
that gave rise to the sound, this envelope can be interpreted either as revealing of prominent spectral peaks in the sound or as a possible vocal tract transfer filter function. In
some cases, LPC can be used to model more detailed aspects of vocal tract configuration
during the production process. To date, little of the potential of LPC has been realized
by bioacoustics researchers.

Acknowledgments Helpful comments on previous drafts of this chapter were provided by Edward Carterette, Tecumseh Fitch, Carey Yeager, and two anonymous reviewers. Thanks are extended to David Pisoni for use of the spectrographic facilities in
the Speech Research Laboratory at Indiana University (supported in part by NIH grant
DC-0111-17) and to Robert Seyfarth and Dorothy Cheney for the vervet monkey snake
alarm call shown in Figures 2 and 4. Acquisition of hardware and software used in
acoustic analysis and figure preparation was partially supported by awards to Michael
Owren from Faculty Development and Seed Money funds at the University of Colorado
at Denver and from the Dean's Development Fund and Department of Psychology at
Reed College, Portland, Oregon.

References

Baken RJ (1987) Clinical measurement of speech and voice. College-Hill, Boston

Beecher MD (1988) Spectrographic analysis of animal vocalizations: implications of the "uncertainty principle". Bioacoustics 1: 187-208

Bell Telephone Laboratories (1946) Technical aspects of visible speech. J Acoust Soc Am 17: 1-89

Capranica RR (1966) Vocal response of the bullfrog to natural and synthetic mating calls. J Acoust Soc Am 40: 1131-1139

Carterette EC, Shipley C, Buchwald JS (1979) Linear prediction theory of vocalization in cat and kitten. In: Lindblom B, Öhman S (eds) Frontiers of speech communication research. Academic Press, New York, p 245

Chiba T, Kajiyama J (1941) The vowel: its nature and structure. Tokyo-Kaiseikan, Tokyo

Cooper FS (1950) Spectrum analysis. J Acoust Soc Am 22: 761-762

Davis LI (1964) Biological acoustics and the use of the sound spectrograph. Southwest Naturalist 9: 118-145

Fant G (1960) Acoustic theory of speech production. Mouton, The Hague

Fitch WT, Hauser MD (1995) Vocal production in nonhuman primates: acoustics, physiology, and functional constraints on 'honest' advertisement. Am J Primatol 37: 191-219

Flanagan JL (1972) Speech analysis, synthesis, and perception, 2nd edn. Springer, Berlin Heidelberg New York

Gaunt AS (1983) On sonograms, harmonics, and assumptions. Condor 85: 259-261

Greenewalt CH (1968) Bird song: acoustics and physiology. Smithsonian, Washington DC

Haimoff EH (1983) Occurrence of anti-resonance in the song of the siamang (*Hylobates syndactylus*). Am J Primatol 5: 249-256

Hall-Craggs J (1979) Sound spectrographic analysis: suggestions for facilitating auditory imagery. Condor 81: 185-192

Joos M (1948) Acoustic phonetics. Language 24 (Suppl): 1-136

Kent RD, Read C (1992) The acoustic analysis of speech. Singular, San Diego

Klatt DH (1980) Software for a cascade/parallel formant synthesizer. J Acoust Soc Am 67: 971-995

Koenig W, Dunn HK, Lacy LY (1946) The sound spectrograph. J Acoust Soc Am 18: 19-49

Lieberman P, Blumstein SE (1988) Speech physiology, speech perception, and acoustic phonetics. Cambridge Univ, New York

Makhoul J (1975) Linear prediction: a tutorial review. Proc IEEE 63: 561-580

Markel JD, Gray AH (1976) Linear prediction of speech. Springer Berlin Heidelberg New York

Marler P (1969) Tonal quality of bird sounds. In: Hinde RA (ed), Bird vocalizations. Cambridge Univ Press, New York, p 5

Nearey TM (1978) Phonetic feature systems for vowels. Indiana Univ Linguistics Club, Bloomington

O'Shaugnessy D (1987) Speech communication. Addison-Wesley Reading

Owren MJ (1990) Classification of alarm calls by vervet monkeys, (*Cercopithecus aethiops*). II. Synthetic calls. J Comp Psychol 104: 29-41

Owren MJ, Bernacki RH (1988) The acoustic features of vervet monkey (*Cercopithecus aethiops*) alarm calls. J Acoust Soc Am 83: 1927-1935

Owren MJ, Linker CD (1995) Some analysis methods that may be useful to acoustic primatologists. In: Zimmermann E, Newman JD, Jürgens U (eds) Current topics in primate vocal communication. Plenum, New York, p 1

Owren MJ, Seyfarth RM, Cheney DL (1997) The acoustic features of vowel-like *grunt* calls in chacma baboons (*Papio cynocephalus ursinus*): implications for production processes and functions. Acoust Soc Am, 101: 2951-2963

Potter RK, Kopp GA, Green HC (1947) Visible speech. Van Nostrand, New York

Rabiner LR, Schafer RW (1978) Digital processing of speech signals. Prentice-Hall, Englewood-Cliffs

Rendall D (1996) Social communication and vocal recognition in free-ranging rhesus monkeys (*Macaca mulatta*). Doctoral Diss, University of California, Davis

Rendall D, Rodman PS, Emond RE (1996) Vocal recognition of individuals and kin in free-ranging rhesus monkeys. Anim Behav 51: 1007-1015

Robinson EA (1967) Predictive decomposition of time series with application to seismic exploration. Geophysics 32: 418-484

Rowell TE (1962) Agonistic noises of the rhesus monkey (*Macaca mulatta*). Symp Zool Soc London 8: 91-96

Rowell TE, Hinde RA (1962) Vocal communication by the rhesus monkey (*Macaca mulatta*). Proc Zool Soc Lond 138: 279-294

Schön Ybarra M (1995) A comparative approach to the nonhuman primate vocal tract: implications for sound production. In: Zimmermann E, Newman JD, Jürgens U (eds) Current topics in primate vocal communication. Plenum, New York, p 185

Seyfarth RM, Cheney DL (1984) The acoustic features of vervet monkey grunts. J Acoust Soc Am 75: 1623-1628

Seyfarth RM, Cheney DL, Harcourt AH, Stewart K (1994) The acoustic features of double-grunts by mountain gorillas and their relation to behavior. Am J Primatol 33: 31-50

Staddon JER, McGeorge LW, Bruce RA, Klein FF (1978) A simple method for the rapid analysis of animal sounds. Z Tierpsychol 48: 306-330

Thorpe WH (1954) The process of song learning in the chaffinch as studied by means of the sound spectrograph. Nature 173: 465-469

Titze IR (1994) Principles of voice production. Prentice Hall, Englewood Cliffs

Wakita H (1976) Instrumentation for the study of speech acoustics. In: Lass NJ (ed) Contemporary issues in experimental phonetics. Academic, New York, p 3

Watkins WA (1967) The harmonic interval: fact or artifact in spectral analysis of pulse trains. In: Tavolga WN (ed) Marine bioacoustics, vol 2. Pergamon, New York, p 15

SECTION II

SOUND PRODUCTION AND TRANSMISSION

CHAPTER 6

Acoustic Communication Under the Sea

P. L. TYACK

1
Introduction

When one dives under the sea, it usually sounds relatively quiet. This is because the human ear is adapted to hear airborne sound. Our middle ear is designed to operate with air on both sides of the eardrum and it does not transmit sound very well when the outer ear canal is flooded with water. However, if one listens with a *hydrophone*, or underwater microphone, it is immediately apparent that the sea is full of sound. Some of this stems from physical processes, such as earthquakes or wind and waves. Humans and other animals also take advantage of unusual properties of underwater sound for communication and gathering information about their environment.

Of all the ways to transmit information over long distances through the sea, sound is the best. Light has the lowest attenuation through seawater of any frequency of electromagnetic radiation, yet even sunlight is reduced to about 1 % of its surface intensity at a depth of 100 m in clear water (Clarke and Denton 1962). The signals of most bioluminescent animals are unlikely to be visible beyond tens of meters. On the other hand, sound propagates so well underwater that a depth charge exploded off Australia can be heard in Bermuda (Shockley et al. 1982). These long ranges for underwater acoustic propagation differ strongly from most terrestrial environments in which one typically can see farther than one can hear or yell. There are also costs associated with acoustic displays for marine animals. Since underwater sound travels rapidly over great distances, acoustic signals can be intercepted by predators or animals other than the intended recipient. It is often advantageous to direct a display to one particular recipient in order to avoid this problem of interception. Limiting the audience of a display is often more difficult for acoustic displays than for visual or tactile ones.

Because animal sounds tend to carry farther underwater than one can see, marine biologists usually find it easier to hear a sound-producing animal than to see it. There are still loud sounds heard frequently in the ocean for which the source is completely unknown. One has the picturesque name of Wenz's boing. A Navy expert on ambient ocean noise named Wenz described a curious "boing" sound heard by sonar crews on submarines and from bottom mounted hydrophones in the North Pacific (Wenz 1964). Thompson and Friedl (1982) analyzed the seasonal occurrence of these "boings" over 2 years, and found that the source was heard in Hawaiian waters during the wintertime. Even though the boing has been recorded for over three decades, no one knows what kind of animal makes it. Besides the boing, the range of animal acoustic communication under the sea is rich. Biologists have reported that three groups of marine animals produce sounds underwater: crustaceans, fish, and marine mammals.

1.1
Invertebrates

Marine crustaceans can produce surprisingly loud impulse sounds. Spiny lobsters, *Panulirus* sp., produce a sound by moving a hard part of the body against a serrated area of their exoskeleton (Moulton 1957). This kind of sound production is called *stridulation*, and is similar to making sounds by moving one's fingernail along the teeth of a comb. Snapping shrimp of the genera *Crangon, Alpheus,* and *Synalpheus* make sounds by snapping the finger of a specially modified claw. Snapping shrimp often hold the finger cocked open in a retracted position, and the snapping sound is correlated with the rapid closure of the finger on the claw. The finger has a hard knob which fits into a socket in the claw. When the knob hits the socket, a jet of water flows out of a groove near the socket as a sharp snapping noise is produced. It is remarkable to think that an important source of ocean noise in coastal tropical and subtropical waters stems from millions of snapping sounds of these small crustaceans.

1.2
Fish

The common names of many fish, such as grunts, croakers, and drumfish, indicate that people have known for centuries that fish make sounds. However, fish sounds were only recorded with fidelity from the animals' natural environment after World War II, when hydrophones and portable recording devices became available to biologists. The growth of snorkeling and diving also provides opportunities for hearing the sounds of fishes. For example, if you have snorkeled in a coral reef, you may have heard the loud sounds produced by parrot fish as they bite into the coral. This sound is a necessary consequence of parrotfish feeding movements. It is not necessarily produced by the fish in order to communicate, but other fish nearby may attend to this reliable signal of feeding behavior. Many fish also have hard teeth in the pharynx and can produce stridulatory sounds using either these hard tissues or other hard tissues such as bone or spines of the dorsal or pectoral fins. Many different groups of fish also use an air-filled *swim bladder* to produce sounds. The swim bladder is located in the abdomen and initially evolved for buoyancy control in fish. Many species have evolved secondary uses of the bladder as an acoustic resonator. Swim bladder sounds are produced when contractions of specialized "sonic" muscles cause changes in the pressure and volume of the gas in the swim bladder. The vibrating surface of the resonating swim bladder acts as an efficient low frequency underwater loudspeaker (Tavolga 1964, pp. 195-211). The importance of the swim bladder has been demonstrated in experiments showing greatly reduced sound output in fish with damaged, deflated, or water-filled swim bladders (e.g., Tavolga 1962; Winn and Marshall 1963).

The lore of people who fish often includes caution about making noises that might disturb fish, but in the last century there has been considerable controversy among biologists about whether fish can hear. Along with his pioneering work on the dance language of honeybees, the German ethologist Karl von Frisch (1938) also performed an influential demonstration of fish hearing. Many earlier tests had shown that fish did not respond strongly to artificial noises. Frisch reasoned that the sounds simply might have had little significance to the fish. He trained a fish by whistling just before feeding it, and the fish soon would swim over at the sound of the whistle. By testing hearing abilities in fish after removal of various parts of the inner ear, von Frisch was able to determine which organs are responsible for hearing. Frisch had a genius for simple elegant experiments, and his work stands as an object lesson for students of biology that the more carefully thought out an experiment is, the simpler it often can be. Readers interested in a recent review on the role of underwater sound in fish behavior, including both sound production and hearing, can refer to Hawkins (1993). An elementary discussion of the hearing of fish and other animals is presented in Stebbins (1983). More detailed treatment of fish hearing can be found in Popper and Platt (1993). Extensive ethological study over the past several decades has shown that fish use sound for a broad variety of functions including reproductive advertisement displays, territorial defense, fighting assessment, and alarm (Myrberg 1981).

1.3
Marine Mammals

The sounds of marine mammals are such a part of our culture now that the reader may be surprised to learn that the first recordings identified from a marine mammal species were made in 1949. These were recordings of beluga whales, *Delphinapterus leucas*, reported by Schevill and Lawrence (1949, 1950) in the lower Saguenay River near the St. Lawrence estuary. Shortly afterwards, there was rapid growth in studies of how dolphins echolocate using high frequency click sounds and of field studies of low frequency sounds made by baleen whales. In the 1950s and 1960s, marine mammal bioacoustics was concerned primarily with identifying which species produced which sounds heard underwater. Much of this research were funded by naval research organizations because biological sources of noise can interfere with human use of sound in the sea. Readers interested in a historical perspective on marine bioacoustics during this time can refer to Schevill et al. (1962) and Tavolga (1964). Studies of echolocation in captive dolphins flourished particularly because of the obvious naval interest in exploring how dolphins achieve their remarkable echolocation abilities.

Controlled experiments on hearing and echolocation in dolphins have been critical for our understanding of auditory processes in cetaceans. For example, research on dolphins emphasizes that they can perform very rapid auditory processing for echolocation signals (Au 1993). Mammalian auditory systems appear to integrate energy over short intervals of time. This can be measured in experiments that vary either the duration of a signal or the interval of time between two clicks. When the interval is less than the integration time, the animal perceives as louder either the longer signal or the two-click signal. Experiments changing the duration of pure tones suggest dolphins have time constants on the order of 10 ms, close to that of humans, for frequencies from 500 to 10000 Hz. Bottlenose dolphins have also had the integration time tested using pairs of high frequency clicks and the interval measured is near 265µs. The 40-fold difference in integration time for these two kinds of signals suggests either that dolphins have a frequency-specific difference in the temporal characteristics of their peripheral hearing or that they have two different kinds of central auditory processing. Bullock and Ridgway (1972) found that different areas of the dolphin brain process these two kinds of signals. When they presented click sounds, there was a rapid response of an area of the midbrain called the *inferior colliculus*. Broad areas of the cortex of the brain showed slow responses to frequency modulated tonal sounds. This evidence supports the interpretation that differences in central processing may affect the integration time for these different stimuli.

There have been few studies relating acoustic communication to social behavior in marine animals compared with terrestrial ones. This stems in part from the agendas of funding agencies and in part because of difficulties observing marine animals in the wild. During the 1950s and 1960s, the communicative functions of vocalizations were only investigated in a few marine mammal species that could be studied in captivity, especially the bottlenose dolphin (*Tursiops truncatus*). More recent reviews of acoustic communication in marine animals (e.g., Myrberg 1981; Hawkins and Myrberg 1983) develop more of a communicative perspective, particularly for smaller animals that can easily be studied in relatively naturalistic captive settings. New methods have more recently opened up windows on the social behavior and communication of marine mammals (Costa 1993). For example, the United States Navy has recently made networks of bottom-mounted hydrophone arrays available to biologists, even though these were closely guarded military secrets during the Cold War. These arrays allow biologists to track vocalizing whales hundreds of kilometers away. The increased power and reduced size of electronics have also allowed the development of new devices that can be attached directly to wild animals. For example, new tagging technology has allowed biologists to record the acoustic stimuli an animal hears, as well as its vocal and physiological responses, even if the animal is 1000 m deep and hundreds of kilometers away from where it was tagged. Bothlong-range acoustic detections of whale vocalizations and the ability to tag highly mobile animals have radically expanded the effective ranges of our ability to study marine mammal communication. This chapter will focus on some of these new techniques and results. Since the author's own expertise covers marine mammals, and since several earlier chapters on marine bioacoustics emphasize fish (e.g., Hawkins and Myrberg 1983), this chapter will emphasize examples from marine mammals.

1.4
Organization of This Chapter

The basic goal of this chapter is to present acoustic communication by marine organisms in the context of ocean acoustics. The author first discusses some basic features of ocean acoustics and propagation of sound in the sea, and uses this information to discuss long-range communication by whales. The equation used to estimate the operating range of a sonar is applied to dolphin sonar. Hearing and the use of underwater sounds in communication are then discussed for invertebrates, fish, and marine mammals, with special emphasis on agonistic signals, reproductive advertisement displays, and recognition signals. Several issues of particular importance to marine mammals are then detailed, including vocal learning among marine mammals, how vocal learning affects communication, and promising new bioacoustic techniques. The first paragraph of Section 1.3 mentions that one early impetus for studying the sounds of marine organisms came from potential interference with human activities. As we learn how much marine animals have come to rely upon sound for their own critical activities, and as endangered marine mammal species have received increased protection, there has been increased concern about the impact of anthropogenic noise on marine animals (Green et al. 1994). The chapter closes with a discussion of this issue.

2
Elementary Acoustics

When a loudspeaker is turned on, the cone moves in and out, and it causes molecules in the surrounding air to move. As the nearby molecules move out, they cause an increase in pressure that induces an outward motion of neighboring particles. This *particle motion* induced by a moving or vibrating source can propagate through any compressible medium including gases such as air or liquids such as seawater. Sound induces motion in any elastic objects in the medium. In order to be sensed by a nervous system or electronic gear, this mechanical energy must be converted into electrical energy. In animals, this is typically generated when motion of the external medium bends hair-like *cilia* on specialized sensory cells. This bending of the cilia modifies the voltage difference between the inside and the outside of the sensory cell. This voltage difference in turn can modify the rate of action potentials produced by the cell. Engineers, on the other hand, typically convert mechanical into electrical energy using crystals or ceramics that have the *piezoelectric* property, i.e., they generate a voltage difference when subjected to pressure forces. Much of the advancement in marine bioacoustics of the past few decades stems from the ready availability of small efficient underwater microphones, or hydrophones, using these piezoelectric ceramics. These ceramic sensors have a high impedance, so if one needs a cable more than 10 m or so long, one must buffer the signal to a lower impedance with a preamplifier. The preamplifier and ceramic sensor are typically potted in a waterproof compound with the same density and speed of sound as water. The entire assembly can be on the order of 1 cm in diameter, small enough to attach to a large animal or to install in a small aquarium. If one needs to quantify actual pressure levels of sound, one can either purchase calibrated hydrophones, or calibrate them oneself with respect to a calibrated reference hydrophone.

2.1
Frequency and Wavelength

A sound that we perceive as a pure tone has a sinusoidal pattern of pressure fluctuations. The *frequency* of these pressure fluctuations is measured in cycles per second. The modern name for this unit of frequency is the Hertz (Hz), and just as 1000 meters are called a kilometer, 1000 Hertz are called a kiloHertz (kHz). The *wavelength* of this tonal sound is the distance from one measurement of the maximum pressure to the next maximum. Sound passes through a homogeneous medium with a constant speed, c. The speed of sound in water is approximately 1500 m/s, roughly five times the value in air, 340 m/s. The speed of sound c relates the frequency f to the wavelength λ by the following formula: $c = \lambda f$. Thus the wavelength of an underwater sound of frequency f has about five times the wavelength of the same frequency in air. Imagine the significance of this difference for the sound localization abilities described by Pye and Langbauer (this Volume). An animal that locates a sound of a particular frequency by measuring the delays in arrival time of a signal at both ears would need five times the separation between ears underwater to achieve the same effective size as in air. Not all sounds are tonal. Sounds that have energy in a range of frequencies, say in the frequency range between 200 and 300 Hz, would be described as having a *bandwidth* of 100 Hz.

2.2
Sound Pressure Level and the Decibel

It is seldom possible to follow acoustic measurements without understanding the unusual measurement scale adopted by acousticians. This scale, called the *decibel* (dB), may be facetiously thought of as a way to limit acoustic calculations to adding and subtracting integers from 1 to 200 or so. Physicists are used to complex mathematical manipulations of variables in a small set of basic units, even if the numbers vary from 10^{-lots} to $10^{+even\ more}$. By comparison, an acoustic engineer creates a complex scale to keep later calculations simple. Most acoustic calculations involve multiplication, so the acoustic engineer performs a logarithmic transformation to convert the multiplication operation to addition — remember that the log (A × B) = log (A) + log (B). This logarithmic transformation is called the *bel* scale, in honor of Alexander Graham Bell. However, if one wants to round off the numbers to integers, this logarithmic transformation limits precision too much. To get around this, the acoustic engineer multiplies the logarithm by 10 to convert the bel to the decibel or dB.

Sound transmits energy via the propagating pressure fluctuations. Sound *intensity* is the amount of energy per unit time (power) flowing through a unit of area. The intensity of a sound equals the acoustic pressure squared divided by a proportionality factor which is specific for each medium. This factor is called the specific acoustic resistance of the medium, and equals the density of the medium, ρ, times the speed of sound, c.

$$Intensity = \frac{Pressure^2}{\rho c} \qquad (1)$$

Seawater has a higher acoustic resistance than air. The ratio of the acoustic resistance of seawater divided by that of air is about 3571 or in decibel terms = 10 log (3571) = 35.5 dB. For a given sound pressure, the intensity of this sound in water would be 35.5 dB less than the intensity for the same pressure level in air. If I and I_{ref} are two intensities, their difference in dB is calculated as follows:

$$Intensity = 10 \, log\frac{I}{I_{ref}} \, dB \qquad (2)$$

In order that intensity levels and pressure levels be comparable in dB, the sound pressure level (SPL) is defined as follows:

$$SPL = 20 \, log\frac{P}{P_{ref}} \, dB \qquad (3)$$

This maintains the appropriate proportionality of intensity and pressure (if $I \propto P^2$ then log $I \propto 2$ log P) for sounds in the same medium. As an example, take a sound measured to be ten times the pressure reference. This would be 20 dB by the definition of SPL in Eq.(3). Since intensity is proportional to pressure squared, the intensity of this sound would be 10^2 or 100 times the intensity of the reference. This would still be 20 dB re the reference intensity, by the definition of intensity in Eq.(2).

The primary definition of the decibel is as a ratio of intensities. The decibel always compares a pressure or intensity to a reference unit. Both intensities and pressures are referred to a unit of pressure, 1 μPa, in a form of shorthand. When an intensity is referred to pressure of 1 μPa, it really means referred to the intensity of a continuous sound of pressure equal to 1 μPa. Acousticians seldom measure intensity directly though, because microphones and hydrophones measure pressure. There have been a plethora of references used over the years (see Urick 1983, p. 14-15). One standard for airborne acoustics was the lowest pressure level humans could nominally hear, 0.000204 dyne/cm^2. The current standard reference is less anthropocentric and more closely related to standard units. In SI units, pressure is measured in newtons per meter squared, or Pascals (Pa). The standard reference pressure for underwater sound is 10^{-6} Pa or 1 μPa. This equals 10^{-5} dyne/cm^2 and is -26 dB with respect to the airborne sound standard. In order to convert a sound level referred to the airborne standard of 0.000204 dyne/cm^2 to the 1 μPa standard, add 26 dB. Another earlier reference pressure is the μbar, which equals 1 dyne/cm^2. If one μPa equals 10^{-5} dyne/cm^2 then it is 20 log (10^{-5}) = -100 with respect to 1 μbar. Thus, if there is a reference to a sound pressure level with respect to 1 μbar, just add 100 to get the equivalent level with respect to 1 μPa.

3
Propagation of Sound in the Sea

The propagation of sound energy is described mathematically by the *wave equation*. Two solutions are typically used to model the propagation of underwater sound: *normal mode theory* and *ray theory*. In this section, the author will fall back upon more qualitative descriptions of familiar ray phenomena such as refraction. The reader interested in a more quantitative introductory treatment of underwater sound should refer to Clay and Medwin (1977). The three major factors affecting the speed of sound in seawater are temperature, salinity, and pressure. Water is warmed near the surface of the ocean in most locations and seasons. This leads to an increase in the speed of sound as one approaches the sea surface. Below this warmed layer there is less change of temperature with depth, and sound speed is dominated by increasing pressure. At greater depths, the sound speed increases with increasing pressure (Ewing and Worzel 1948).

Like light passing through a lens, sound rays are refracted when they pass from a medium with one sound speed c_1 to a medium with a different sound speed c_2. The angles of these sound rays are measured between the rays and a line normal to the interface between the two media (Figure 1). The change in angle of such a ray at a sound speed boundary is defined by *Snell's law*:

$$\frac{sin(\theta_1)}{c_1} = \frac{sin(\theta_2)}{c_2} \tag{4}$$

If one visualizes a sound front emanating from a source at a depth corresponding to the minimum in sound speed, one can see that a horizontal ray would continue horizontally as long as it remained in a zone of constant c. However, rays with vertical components would start refracting back towards the axis of minimum speed as they traveled up or down into areas of faster c. A simple numerical example may illustrate this. Suppose a ray launches in water with a sound speed of c_1 with an angle of $45°$ with respect to vertical in the upward direction and encounters a transition to a horizontal

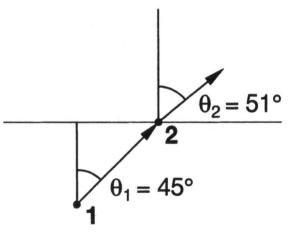

Fig. 1. Refraction using Snell's law. A sound ray starts at point 1 in a medium with a speed of sound c_1 and with an angle of θ_1. At point 2, it encounters a horizontal layer above which the sound speed, c_2, is 1.1 times faster. This change in the speed of sound from c_1 to c_2 causes the sound to refract to an angle of θ_2. Both angles are measured with respect to the vertical

layer of water in which the sound speed, c_2, is 10 % faster (Figure 1). In this case, where c_2=1.1 x c_1, the new ray angle θ_2=arcsin [1.1 (sin45°)]=51°, an angle closer to horizontal. The upward traveling ray thus deflects downward as it travels into water where the speed of sound is faster. A ray traveling downward from the depth of the minimum sound speed would similarly refract back towards that depth. Refraction in the deep ocean thus creates a channel in which sound energy concentrates at depth.

3.1
Transmission Loss

Imagine a source producing a sound impulse in a uniform unbounded space. At the moment the sound is produced, all of the sound energy will be centered at the source. After time t when the sound front has traveled out to range r, the energy will be distributed over a sphere of area = $4\pi r^2$. Just as for gravity, the intensity of the signal at any one point decreases as the square of the range. This is called the *inverse-square law* for gravity and *spherical spreading* for underwater acoustics. This loss of intensity as one moves farther from the source is called *transmission loss* (TL). In the dB nomenclature of acousticians, this transmission loss is calculated as follows:

$$\text{Spherical spreading:} \quad TL = 10\,log\frac{I}{I_{ref}} = 10log\frac{4\pi r^2}{4\pi r_{ref}^2} = 20\,log\frac{r}{r_{ref}} \qquad (5)$$

The typical reference range is 1 m yielding TL=20 log r. Sound propagates by spherical spreading when the sound energy is free to spread in all directions. This occurs until the sound encounters a boundary with different acoustic properties such as the sea surface or ocean floor. As mentioned in Section 3, sound may also encounter areas of water with different acoustic properties within the ocean which can cause refraction. When one is far from a sound source compared with the ocean depth, the sound energy may be concentrated by refraction in the deep ocean sound channel. This sound can be thought of as spreading in a plane, to a first approximation. In this case, sound energy at time t would be distributed over a circle of area = $2\pi r$. Since the sound energy is not really restricted to a plane, but more like to a short wide cylinder, it is known as cylindrical spreading and follows a TL = 10 log r dB.

$$\text{Cylindrical spreading:} \quad TL = 10\,log\frac{I}{I_{ref}} = 10log\frac{2\pi r}{2\pi r_{ref}} = 10\,log\frac{r}{r_{ref}} \qquad (6)$$

If sound interacts with the sea surface or ocean bottom, then there is some loss of acoustic energy to these interfaces. In these intermediate conditions and transmission in coastal waters, the transmission loss is often somewhere between 10 and 20 log r.

3.2
Absorption

Sound spreading is a dilution factor and is not a true loss of sound energy. *Absorption* on the other hand is conversion of acoustic energy to heat. The attenuation of sound due to absorption is a constant number of dB per unit distance, but this constant is dependent upon signal frequency, as the following approximate figures in Table 1 indicate.

Table 1. Absorption coefficient for 14°C seawater at sea level (approximated from Figure 3.3.1 in Clay Medwin 1977, pp. 100-101)

Frequency	Absorption coefficient (dB/m)
100 Hz	10^{-6}
1 kHz	10^{-4}
10 kHz	10^{-3}
40 kHz	10^{-2}
300 kHz	10^{-1}
2 MHz	1

While absorption yields trivial effects at frequencies below 100 Hz, it can significantly limit the range of higher frequencies, particularly above 40 kHz or so, where the loss is more than 1 dB/100 m.

3.3
Sonar Equation

During World War II, acoustic engineers worked out equations to predict the performance of different sonar systems. These *sonar equations* are still used to predict the detection range of an underwater sound (Urick 1983). These equations are convenient for back-of-the-envelope-type calculations, but predict only in an average sense. A simple form of the *passive sonar equation* is:

$$RL(dB) = SL(dB) - TL(dB)$$

(7)

We have already seen the last term on the right side of the equation, *transmission loss* (Sect. 3.1). Source level is the sound intensity measured in decibels with respect to a reference distance, usually 1 m. Transmission loss is the difference between the source level at the reference distance and the intensity predicted for the range of the receiver. RL is the *received level* of sound at the receiving hydrophone, measured in dB. The likelihood of detecting a signal does not just depend upon the received level but also depends upon the external environmental noise as well as any internal noise in the receiver. It is the ratio of signal to noise that is used to calculate the probability of detection. Using the logarithmic terms of the dB scale, this *signal to noise ratio* (SNR) is expressed as a subtraction:

$$SNR(dB) = RL(dB) - NL(dB)$$

(8)

where NL is the *noise level.*

While the signal to noise ratio is a simple concept, there are some complications that may improve performance. If a directional receiver is able to listen only in the direction of an incoming signal, then it will boost the strength of both signal and noise in this direction at the expense of reception from other directions. If the noise comes from all directions and the signal only from one direction, then the directional receiver increases the strength of the signal compared with the noise. Receivers can also be designed to match the frequency and time characteristics of the sonar signal. A receiver with a band-width well tuned to the signal bandwidth buys the equivalent of extra gain from reduction in noise outside the frequency range of the signal. Noise is often described in terms of a *spectrum level*, which is the noise level in a frequency band that is 1 Hz wide. The intensity of sound across a broader band of frequencies is called the *band level*. If the noise has the same energy at different frequencies across a frequency band, then the noise level within a frequency band W Hz wide will be:

$$Band \ level = spectrum \ level + 10 \ log \ W \tag{9}$$

If the receiver integrates signals over some period of time, then it is also beneficial for the integration time to be well matched to the duration of the sonar signal. If a receiver has an integration time, t_{int}, that is longer than a short pulse, t_{pulse}, then the effective source level of the sound, SL', will be reduced by the following amount:

$$SL'(dB) = SL(dB) + 10log\frac{t_{pulse}}{t_{int}} \tag{10}$$

3.3.1
Long-Range Sound Transmission by Whales

We now have enough information to understand how some whale sounds are adapted for long-distance communication. Finback whales, *Balaenoptera physalus*, produce pulses with energy in a range roughly between 15 and 30 Hz (Watkins et al. 1987), near the lowest frequencies that humans can hear. Absorption is negligible at these frequencies. These finback pulses qualify as infrasonic signals as discussed by Pye and Lang-bauer (this Volume). Each pulse lasts on the order of 1 s and contains 20 cycles. The source level of the pulses ranges from about 170 to 180 dB re 1 µPa (Patterson and Hamilton 1964). Particularly during the breeding season, finbacks produce series of pulses in a regularly repeating pattern in bouts that may last more than 1 day (Watkins et al. 1987). These sounds were first detected by geologists who were monitoring hydrophones installed on the sea floor in order to study submarine earthquakes (Patterson and Hamilton 1964). Figure 2 is a record from a seismographic drum recorder made on 18 November 1978 which shows a regular series of these pulses.

Figure 3 from Spiesberger and Fristrup (1990) shows the ray paths predicted for a sound pulse produced by a finback whale at 35 m depth, reaching a receiver at 500 m depth, 400 km away. These paths are predicted for a sound speed profile typical of temperate or tropical seas. The computer model used to predict the propagation of this sound suggests that a 186 dB finback signal produced at 35 m depth in the deep ocean

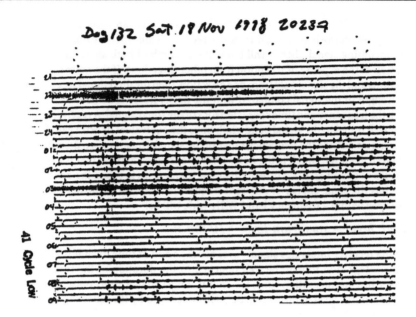

Fig. 2. Seismographic drum recording from a hydrophone off Bermuda. The hydrophone signal was filtered to cut out energy above 41 Hz. This filtered signal then drives a pen which records the waveform on the paper. Regular marks every minute mark timing of the signal, with each line taking 20 min. In order to expand the record, the *right-hand side* has been cut off, so only 7 out of each 20 min are shown. The signal starts at 202307, or 20:23 h. Around 22:00 h a ship passes by, marked by steadily increasing and then decreasing energy. Between 23:00 h and 24:00 h, a finback whale starts producing its regular pulses, which continue through the passage of another ship near 03:00 h. The whale finally stops sometime near 04:00 h

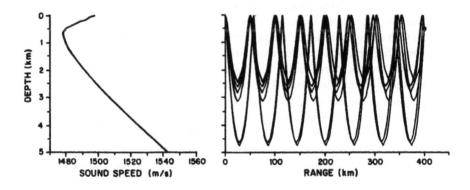

Fig.3. *Left* Typical profile of sound speed vs. depth for temperate or tropical seas. This particular profile was taken in the northeast Pacific near 31°N and 157°W. *Right* calculated ray paths for a 20-Hz finback pulse produced at range 0 and a depth of 35 m. Rays detected at a receiver 500 m deep and 400 km away are shown. The ray paths illustrate the general patterns of propagation of this kind of signal in deep temperate or tropical seas. Reprinted with permission from Spiesberger and Fristrup (1990) © The University of Chicago Press

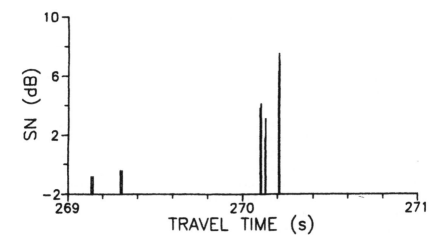

Fig. 4. Travel times and signal to noise ratios for the different acoustic ray paths shown in Fig. 3. Signal-to-noise ratios are calculated for a receiver which correlates a stored version of the outgoing pulse against the incoming signals. Reprinted with permission from Spiesberger and Fristrup (1990) © The University of Chicago Press

would be easily detectable out to a range of 400 km for a receiver at a depth of 500 m. The separate arrivals of the different rays arriving at this receiver 400 km away are illustrated in Figure 4, also from Spiesberger and Fristrup (1990). Many of the ray arrivals are well above (by 4 — 6 dB) typical ambient noise levels. As can be seen, the single pulse made by the whale is greatly modified by the channel. The separate rays from one pulse will arrive at different sound levels over a period of approximately 1 s. Since the pulse itself lasts 1 s, these different arrivals will superimpose, yielding a complex pattern. The simple acoustic structure of the finback calls may simplify recognition of these signals even when different arrivals are superimposed. The high rate of repetition and redundancy of these simple pulses in a well-defined sequence is also useful for communication in a noisy channel.

Refraction in the deep ocean causes most acoustic rays to center around the depth of minimum sound speed, c, which is called the axis of the sound channel. This channel is around 1 km deep in tropical or temperate oceans. If sound energy concentrates at depths near 1000 meters, then this would be a good depth for receiving the signals. However, finback whales are not known to dive deeper than a few hundred meters. The acoustic rays converge near the surface every 52 km, in what is known as a convergence zone (Figure 3). While baleen whales are not known to dive near the axis of the deep sound channel, a whale swimming through a convergence zone might experience an increase in sound level of 10-20 dB, perhaps even within a few tens of meters (Urick 1983). Other environments have very different propagation conditions. Arctic propagation under ice cover is dominated by upward refraction and downward reflection off the ice. This causes sound energy to concentrate in shallower waters and favors the frequency band from 15 to 30 Hz. Even away from ice cover, polar waters are subject to less surface warming, and this causes the sound velocity minimum to occur much closer to the surface in polar seas. Baleen whales in polar waters could dive nearer the axis of sound spreading than in temperate or tropical deep ocean.

A correlation of these 20 Hz series with the finback reproductive season suggests that they may function as reproductive advertisement displays (Watkins et al. 1987). Finback whales disperse into tropical oceans during the mating season, unlike other species such as humpback and gray whales which are thought to congregate in well-defined breeding grounds. The functional importance of a signal adapted to long-range communication is obvious for animals that disperse over ocean basins for breeding. However, even though biologists can hear finback pulses at ranges of hundreds of kilometers by listening at appropriate depths, the effective range of these signals for finbacks themselves is not known. Until very recently, marine bioacousticians have had few methods which would allow us to detect whether a whale was responding to the calls of a whale more than several kilometers away. The US Navy hydrophone arrays mentioned in Section 1.3 can provide such data, but first let us make a rough estimate of the range of detection for finback pulses.

3.3.2
Using the Sonar Equation to Estimate the Range of Detection for Finback Pulses

The threshold of detection for a signal is often considered to occur when the signal level equals the background level of masking noise. In order to compare these two, it is necessary to establish the appropriate frequency band for comparing signal and noise. Many finback pulses have maximum energy in a frequency range from about 18 to 23 Hz. This 5 Hz bandwidth comprises about one-third of the octave from 15 to 30 Hz. Research on hearing in mammals suggests that for noise to mask a tonal sound effectively, it must be within approximately one-third octave of the frequency of the signal (for a review of this issue applied to marine mammals, see Richardson et al. 1995). Thus, both the frequency range of the finback pulse and the bandwidth of mammalian hearing suggest measuring both signal and noise in the 18 – 23 Hz one-third octave interval in the 15 – 30 Hz octave band. The passive sonar equation allows us to make a rough estimate of the average intensity of a signal as a function of range. In the case of the finback pulses we need not concern ourselves with absorption loss, since even at 100 Hz the loss over 4×10^5 m is <1 dB. On the other hand if we were analyzing a 10 kHz signal, absorption would be 4×10^5 m $\times$ 10^{-3} dB/m = 400 dB, well in excess of the spreading loss. Suppose we assume spherical spreading from the reference distance of 1 m to the depth of 5000 m and cylindrical spreading from 5 km to 400 km, then the calculation of transmission loss is as shown in Table 2.

Spiesberger and Fristrup (1990) used a source level of 186 dB for their finback pulse.

	Δ Range	TL (dB)
Table 2. Calculation of transmission loss (TL) using sherical spreading (20 log *r*) to 5000 m and cylindrical spreading (10 log *r*) from 5 to 400 km	20 log 5000/1m	74
	10 log 400/5km	19
	Attenuation	<<1
	Total	93

SL	186
- TL	-93
= RL	93
-NL (18-23 Hz)	-77
=SNR	16

Table 3. Calculations in decibels of signal to noise ratio (SNR) from modification of the passive sonar-equation [Eq. (7)]: RL=SL - TL, and, [Eq. (8)]: SNR = RL - NL

If we use the transmission loss estimated in Table 2, then the received level would be RL = SL-TL = 186 - 93 = 93 dB (Table 3). Now let us estimate the signal to noise ratio for the finback signal. Spiesberger and Fristrup (1990) estimate the noise level in the 18—23 Hz third octave band to be approximately 77 dB in the area they selected. This would indicate that the signal to noise ratio would be SNR = RL - NL = 93 - 77 = 16 dB louder than the ambient noise level (Table 3).

This result is 8 dB or more higher than the signal to noise ratio shown in Figure 4, which is predicted from ray theory. This kind of difference is not exceptional. The sonar equations give results that are highly averaged over time, range, and depth. They are best used for initial general tests of potential operating range. More detailed analytical methods are available, particularly when one knows more about a particular application. Ray theory is often used for deep water in the open ocean. Mode theory is more appropriate for propagation in water that is shallow with respect to the wavelengths of sound being used. There are also computer programs which can simplify solutions of equations for wave propagation. The reader interested in reviews of these more complex analyses can refer to Clay and Medwin (1977) for ray and mode theories and Tappert (1977) for parabolic equation solutions to the wave equation.

3.3.3
Echolocating Dolphins

Dolphins have highly developed abilities for echolocation. They can detect distant objects acoustically by producing loud clicks and then listening for echoes. These abilities are truly remarkable. For example, trained bottlenose dolphins can detect the presence of a 2.54-cm solid steel sphere at 72 m, nearly a football field away (Murchison 1980). They can discriminate targets that are identical in terms of shape and differing only in composition (e.g., Kamminga and van der Ree 1976). Most of these echolocation sounds are pulses of short duration and high frequency. By minimizing duration, dolphins are able to minimize the chance that the echo from the start of the signal will arrive before the signal has ended. Depending upon how dolphins process echoes, this can also improve the precision of estimating the distance to the target and facilitates discrimination of multiple echoes.

The optimal frequency of a sound used for echolocation depends upon the expected target size. Absorption imposes a significant penalty for higher frequencies, but small targets can best be detected by short wavelength λ or high frequency signals. In the last century, Lord Rayleigh solved the frequency dependence of sound scattering from small rigid spherical targets of radius r; this is called *Rayleigh scattering* (Rayleigh 1945). A spherical target reflects maximum energy when the wavelength of the sound impinging on it equals the circumference of the sphere, or $\lambda=2\pi r$. There is a sharp drop off of echo strength from signals with wavelength λ>2πr. If one refers to Section 2.1 one sees that λ=c/f. If we equate the two λ terms we get c/f = 2πr. We can rearrange these to derive the relationship f = c/2πr to calculate the optimal frequency for reflecting sound energy off a spherical target of radius r. Higher frequencies than this would still be effective sonar signals, but frequencies below f would show a sharp decrease in effectiveness with decreasing frequency.

Except for the dolphins trained to tell acousticians how they echolocate, most dolphins are presumably less interested in rigid spheres as targets than in fish, squid, sharks, conspecifics, and large obstacles. Living organisms have properties that differ from rigid spheres (Stanton et al. 1995), but let us use the formula for rigid spheres for first approximations. If a dolphin needs to use echolocation to detect targets about 1 cm in size or with a "radius" of 0.5 cm, it would do well to use a frequency f = c/2πr = 1500/(2π × 0.005) = near 50 kHz. This is in fact on the low end of the typical frequency range of dolphin echolocation clicks (Au 1993). The echolocation signals of dolphins include energy up to about 150 kHz, which would be well suited to detecting spherical targets with radii as small as 1.5 mm. For larger targets, these higher frequencies could improve the resolving power of dolphin echolocation for features about 1 mm in size. The hearing of toothed whales is also tuned to be most sensitive at these high frequencies. Figure 5 shows audiograms for two of the toothed whale species which have been kept in captivity under

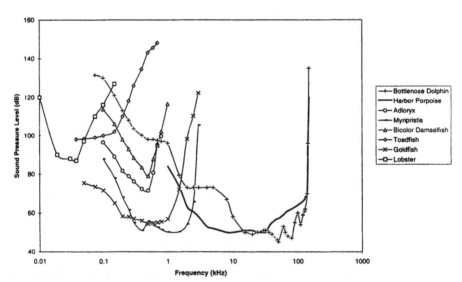

Fig. 5. Audiograms of selected marine animal species. The lines on the right indicate the audiogram of the bottlenose dolphin, *Tursiops truncatus*, indicated by "+"*Symbols*, (Johnson 1966) and of the harbor porpoise, *Phocoena phocoena*, indicated by the *thick black line* (Andersen 1970). The lines centering in the 100-1000 Hz range are of fish — *Adioryx* indicated by the circle symbols and *Myripristis* indicated by the *small filled black rectangles* (Coombs and Popper 1979); bicolor damselfish, *Pomacentrus partitus*, indicated by *the triangle symbols* (Myrberg and Spires 1980); toadfish, *Opsanus tau*, indicated by the diamond symbols (Fish and Offut 1972); and Goldfish, *Carassius auratus*, indicated by the "x" *symbols* (Jacobs and Tavolga 1967). The line far left with the *square symbols* indicates audiogram of the American lobster, *Homarus americanus*, (Offut 1970). Sound pressure levels are in decibels re 1μPa

conditions where they could be trained in order to test their hearing sensitivity — the bottlenose dolphin (*Tursiops truncatus*) and harbor porpoise (*Phocoena phocoena*).

Echolocation has only been studied in a small minority of dolphins and toothed whales, a taxonomic group called the odontocete cetaceans. There is considerable variety in clicks recorded from these animals. Figure 6 shows two examples of odontocete echolocation pulses: one from the bottlenose dolphin (*Tursiops truncatus*) and one from the harbor porpoise (*Phocoena phocoena*). The *Tursiops* pulse has a very sharp rise time on the order of tens of microseconds, and a very high maximum peak to peak sound pressure level, >220 dB re 1 μPa SPL at 1 m. The pulses of bottlenose dolphins from open waters are typically reported to have energy in a rather broad range of frequencies from about 100 to 130 kHz (Au 1993). The one illustrated in Figure 6 has a peak frequency of 117 kHz and a bandwidth of 37 kHz. The harbor porpoise has a high frequency component to its echolocation pulse tuned around a center frequency of 120–150 kHz with sound pressure levels of around 150–160 dB (Møhl and Anderson 1973; Kamminga and Wiersma 1981, Goodson et al. 1995). These have a longer duration (hundreds vs. tens of microseconds) and narrower bandwidth (10–15 kHz vs. 30–60 kHz) than typical clicks from bottlenose dolphins (Au 1993). Odontocetes have some ability to vary their echolocation pulses. For example, when the echolocation clicks of a beluga whale, *Delphinapterus leucas*, were tested in San Diego, the clicks had peak energy in the frequency band 40–60 kHz (Au et al. 1985). When the beluga was moved to Kaneohe Bay, Hawaii, where there is a high level of noise in these frequencies, it shifted its clicks to have peak energy mostly above 100 kHz.

If a sound transmitter can direct its output in the direction of the target, it buys the

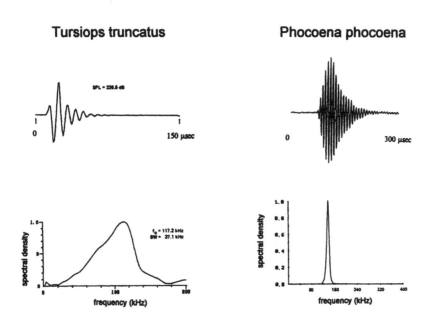

Fig.6. Waveform (*top*) and spectra (*bottom*) of clicks from bottlenose dolphins, *Tursiops truncatus*, (*left*) and harbor porpoise, *Phocoena phocoena* (*right*). The bottlenose dolphin diagrams show an average from an entire click train adapted from Au (1980). The harbor porpoise diagrams, adapted from Kamminga (1988), show a single click from a young animal

equivalent of a higher source level. In general, for a sound source of a given size, the higher the transmitting frequency, the more directional one can make the signal. This benefit of directional signals may be another reason why dolphins have evolved such high frequency echolocation signals. The clicks of bottlenose dolphins are highly directional, losing half their energy (3 dB) when just 10° off the main axis. Figure 7 shows the vertical beam pattern of *Tursiops* echolocation clicks (Au et al. 1978). The source levels indicated for *Tursiops* clicks were recorded in the center of the beam, so they take this gain into account. The transmit beam pattern of odontocete clicks is not necessarily fixed. The clicks of a false killer whale, *Pseudorca crassidens*, were recorded by Au et al. (1995) while the whale discriminated artificial targets. The whale produced four different kinds of click signals, each one with a different beam pattern. There was also quite a strong relationship between the source level of a click and its center frequency, with louder clicks having higher center frequencies. While the acoustic structure of odontocete clicks has been well documented as animals echolocate on artificial targets, we know very little about how clicks may vary when these animals are working on different echolocation tasks in the wild. Future research may find even more variability in the clicks used by dolphins to solve the problems in the natural environment for which echolocation evolved.

As was discussed when SNR was introduced, the ability to hear better along the target axis can improve the performance of a sonar system by reducing signals and noise from other directions. Dolphins have directional hearing that is well correlated with the beam pattern of their echolocation clicks (Au and Moore 1984). If NL is the noise level measured by an omnidirectional hydrophone, then the receiving directivity (DI) increases the signal to noise ratio (SNR) by DI if the noise is omnidirectional and the signal is

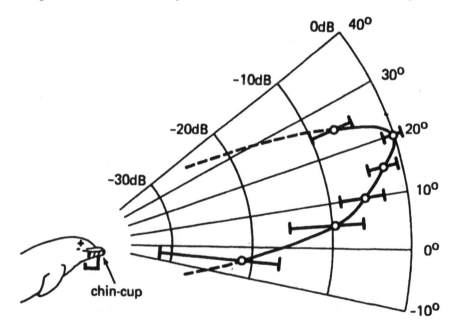

Fig. 7. Broadband beam patterns for the echolocation signals of bottlenose dolphins, *Tursiops truncatus*, in the vertical plane (Reprinted with permission from Au et al. 1978 ©Acoustical Society of America)

unidirectional. Sonar engineers call this the receiving directivity index (Urick 1983), and they define it as

$$DI = 10 \log \frac{SNR \; directional \; receiver}{SNR \; omnidirectional \; receiver} \tag{11}$$

Another way to put this is that the noise masking level for a directional receiver equals the NL (omnidirectional) - DI.

3.3.4
Using the Noise-Limited Active Sonar Equation to Estimate the Range of Dolphin Echolocation

A sonar with a sensitive receiver is typically limited by ambient noise rather than the inherent sensitivity of the receiver. A sonar that detects echoes from its own sounds is called an active sonar, and the performance of a sensitive active sonar is calculated from the noise limited *active sonar equation*. For simplicity, the following equation assumes that the signal is detected when the level of the echo equals or exceeds that of the noise:

Echo level $\geq$ *noise masking level*

$$SL - 2TL + TS \geq NL - DI \; . \tag{12}$$

The difference between the active and passive equations is that the active equation must account for the transmission loss from the sound to the target, the ratio of echo intensity to sound intensity incident on the target *[target strength* = TS = 10 log (intensity of reflected sound at 1 m)/(incident intensity)] and the transmission loss on the return of the echo to the receiver.

We now know enough to estimate the effective range of bottlenose dolphin sonar. For this example we will assume that a bottlenose dolphin in Kaneohe Bay, Hawaii emits a 220 dB echolocation click in order to detect a 2.54 cm diameter ball bearing 72 m away (remember that this is the demonstrated range at which this target could be detected). Let us start with the left half of Eq. (12). First we must estimate the transmission loss (Table 4). Au (1993) states that spherical spreading is a good approximation for transmission loss in this area. The equation for spherical spreading [Eq.(5)] is 20 log (range/reference range of 1 m). For a range of 72 m, this is 20 log (72) = 20 × 1.86 = 37 dB. Dolphins in Kaneohe Bay tend to use relatively high frequency clicks, so let us assume that this click had peak energy near 120 kHz. If the peak energy of the click was at 120 kHz, the absorption would be somewhere between 10^{-1} and 10^{-2} dB/m from Table 1. If we take an intermediate value of 3×10^{-2} dB/m, then the estimated loss for a 72 m path would be 2 dB.

We must be careful to ensure that the source level of the dolphin click is corrected to estimate the energy the dolphin actually hears from the echo. We will focus on the domain of time, and will need to take into account the physical processes of sound transmission and reflection from the target as well as the timing of the dolphin's auditory system. The most important of these factors is the short duration of the echoloca-

Δ Range	TL
	(dB)
20 log 72/1 m	37
Absorption	2
Total	39

Table 4. Calculation of transmission loss (TL) using spherical spreading (20 log r) to 72 m

tion clicks of dolphins compared with the integration time of dolphin hearing. For example, the peak to peak sound pressure level of 226 dB shown in Figure 6 occurred over just one wave of the click. At 120 kHz, this would correspond to a duration of 1/120 000 s = 8.3 μs. Au (1993) reviews evidence that dolphins integrate click energy over an interval of 265 μs. How does this compare with the changes in the click as it propagates and reflects off the target? As this click propagates, slight inhomogeneities in the water may create several ray paths between the dolphin and the target. As we saw for the finback pulse (Sect. 3.3.2), this may spread the energy of the pulse over a longer time period. As the click reflects from various parts of the target, this will also spread the energy of the pulse. The time it would take from when the pulse hits the front of the target to when it hits the rear of the target would be 0.0254 m / 1500 m/s = 16.9 μs. However, if dolphins integrate sound energy over an interval of 265 μs, then the timing of dolphin hearing will have the greatest impact on the perceived source level because the dolphin will integrate the click energy over a period more than an order of magnitude greater than the click duration. The ideal way to calculate this energy would be to integrate the energy of one click over 265 μs. We can approximate this by assuming that most of the click energy occurs over a duration of 8.3 μs and averaging this over 265 μs. By Eq.(10) this correction would be SL' = SL + 10 log (8.3/265) = SL - 15 dB. The target strength of the 2.54 cm diameter ball bearing used as the target has been measured at -42 dB (Au 1993; p 147).

The simplified form of the active sonar equation [Eq.(12)] that we are using assumes that the dolphin will detect the click when the energy from the echo equals the noise level. We will now use the right half of Eq.(12) to estimate whether the dolphin could detect the target at a range of 100 m (Table 6). The noise level present in Kaneohe Bay at 120 kHz is near 54 dB re 1 μPa²/Hz (Au 1993). This level is given as a *spectrum* level which is the level for a 1 Hz band. This must be corrected for the actual band over which the dolphin receives the click. These clicks tend to have a bandwidth of about 40 kHz (e.g., Figure 6). This bandwidth is quite close to the bandwidth over which bottlenose dolphins integrate sound energy at this frequency (Au and Moore 1990). Thus, both the bandwidth of dolphin hearing and of the click would suggest that we estimate the noise

SL	220
Correction for integration time	-15
-2 TL	-78
+ TS	-42
= RL	85

Table 5. Calculation in decibels of received level (RL) from the left-hand side of the active sonar equation [Eq.(12)]: RL = SL -2TL + TS

Spectrum noise level at 120kHz	54
Correction for bandwith of click	+46
Correction for directivity of dolphin hearing	-20
Noise level	80

Table 6. Calculation in decibels of noise level from the right-hand side of the active sonar equation [Eq. (12)]

level for a band of about 40 kHz around 120 kHz. If we assume that the noise level at 120 kHz is representative, then we can use Eq. (9) to correct from the 1 Hz spectrum level to a 40-kHz band level as follows: Band level = spectrum level + 10 log (40 000/1) = 54+46 dB. The highly directional hearing of dolphins at 120 kHz gives a receiving directivity index of 20 dB at 120 kHz (Au 1993). This reduces the effective noise level by 20 dB.

These rough calculations would suggest that the dolphin would hear an 85 dB echo from a 2.54 cm ball at a range of 72 m, compared with a noise level of about 80 dB. This would indicate that the dolphin detects the echo at a signal to noise ratio of about 5 dB. This estimate is close to the theoretical detection threshold where the energy of the echo equals the ambient noise (0 dB SNR), and is closer than the comparison between the sonar equation and ray theory for the finback pulses. This may stem from the way the numbers in the dolphin active sonar equation were measured for a very specific application. However, Murchison (1980) points out that the dolphin click may reflect off the sea floor, sea surface, and any other inhomogeneities in its beam. Reverberation from these sources may raise the noise floor above the ambient sound from sources other than the dolphin, and this may mean that the dolphin's detection occurs at a lower signal to noise ratio than would be estimated ignoring reverberation.

Now it is the rare dolphin that spends its days echolocating on ball bearings. The kinds of targets of biological relevance for odontocetes include prey, predators, obstacles, and the seafloor and sea surface. Targets of biological importance may have echoes that differ from those of artificial targets. There has been considerable research by bioacousticians analyzing echo information from manmade sonars in order to classify biological targets (for a recent review see Stanton et al. 1995). Echolocating dolphins use subtle details of the returning echo to classify artificial targets (Helweg et al. 1996), and they may select or classify targets in their natural environment the same way. However, we know little about how dolphins use their sonar to detect or select prey or predators. Some animals have gas-filled cavities such as the lungs of mammals or the swim bladders of fish. The gas—water interface reflects sound energy and these gas-filled cavities provide good sonar targets. Swim bladders also have a resonance much lower in frequency than would be predicted by the $f = c/2\pi r$ calculation for rigid spheres. For example, a small anchovy at 1 atm of pressure had a resonant frequency of 1275 Hz (Batzler and Pickwell 1970). Little is currently known about whether odontocetes may modify their sonar signals in order to take advantage of these features of biologically important sonar targets.

Most animals have a density similar to water, and animals such as fish without swim bladders, squid, copepods, or krill will have lower target strengths than air-filled targets such as a swim bladder fish. In spite of the potentially low target strength of squid, many biologists have hypothesized that sperm whales may use their clicks, which are lower in frequency than those of dolphins, to echolocate on their prey, squid, in the deep ocean. Goold and Jones (1995) present a rough estimate based upon the sonar equation, suggesting that echoes from sperm whale clicks reflecting off of squid might be detectable to ranges of up to hundreds of meters. This estimate rests upon some untested assumptions and nothing is known about whether or how sperm whales echolocate, so confirmation of these results will have to rely upon direct empirical studies. It will be a significant challenge to attempt to study the sensory mechanisms sperm whales use to forage as they dive hundreds of meters deep for up to an hour or more.

4
Hearing Under the Sea: How Do Marine Animals Receive Acoustic Signals?

In order to detect sound, animals require a receptor that can transduce the forces of particle motion or pressure changes into neural signals. Most of these *mechanoreceptors* in species as diverse as invertebrates, fish, and mammals involve cells with hair-like *cilia* on their surfaces. Movement of these cilia modify the voltage between the inside and the outside of the receptor cells, and this voltage difference can modify the rate of action potentials that communicate the signal to other parts of the nervous system. If the cilia are on the surface of the body, they can detect the flow of water or hydrodynamic forces along the animal's body. Many invertebrate and vertebrate species also have internal sensory organs containing sensory cells with cilia. The stimulus that excites the sensory cells depends upon the structure of the sensory organ. For example, the semicircular canals detect fluid motion within the canal induced by rotation of the head. There are other sensory organs in which the cilia are coupled to a dense mass. When the body of an animal accelerates, this mass will tend to remain stationary. The inertial forces on this mass will bend the cilia, generating a sensory signal proportional to the acceleration of the animal. If this organ can detect the acceleration forces of gravity, it may function for maintenance of equilibrium. In invertebrates, the fluid-filled sensory organ which contains a dense mass is called the *statocyst*. The statocyst is known to be involved in orientation and equilibrium in some invertebrate species. While we know from behavioral conditioning experiments that marine invertebrates like the lobster *Homarus americanus* can hear, there have been few studies to demonstrate which organs are responsible.

The inner ear of fishes contains sensory organs in which heavy calcified masses called the *otoliths* are in close contact with the cilia of sensory cells. Hawkins and Myrberg (1983) reviewed experiments showing that these otolith organs are sensitive to sound. Most marine organisms are about the same density as seawater, which makes them acoustically transparent. This means that when a marine organism is exposed to sound, the particle displacement in the surrounding medium may cause the whole body of the animal to move slightly back and forth. The dense mass moves less, generating a force that stimulates the coupled sensory cells. This is a mechanism for detecting particle motion, which is a vector quantity with both a magnitude and an associated direction.

It has been suggested that directional hearing in fish may stem from orientation-specific sensory cells detecting particular directions of particle motion (Popper and Platt 1993). For a review of directional hearing in fish and other nonmammalian vertebrates, consult Fay and Feng (1987).

There is great variability in the hearing ranges of fish as is indicated in the middle section of Figure 5, which shows audiograms of a variety of fish species. This figure also includes an invertebrate and two cetaceans. Many fish appear to rely primarily on the detection of particle motion that was just described in the preceding paragraph. They have relatively low sensitivity which declines rapidly above 100-200 Hz. The toadfish indicated in Figure 5 by the line with the diamond symbols, is an example of this kind of audiogram. Some other species of fish have evolved specializations for hearing involving a coupling between a gas-filled cavity such as the swim bladder and the ear. Gas is highly compressible and is much less dense than water, so when a gas-filled space is ensonified, it will reverberate with the pressure fluctuations created by sound. The motion of the swim bladder may then be conducted to the ear, amplifying an acoustic signal for the ear to detect. This allows fishes with these specializations to detect the pressure fluctuations induced by sound. Fishes with these specializations have more sensitive hearing and better high frequency hearing. The importance of the coupling between the swim bladder and the inner ear can be seen by comparing the hearing of two genera of squirrelfish. The fish with the most sensitive hearing indicated in Figure 5 is a squirrelfish of the genus *Myripristis*, which is indicated by a line with small filled retangles. There is a good coupling between the anterior swim bladder and the inner ear in this genus. By comparison, even though the genus *Adioryx* is a member of the same family of fish, it has less sensitive hearing, particularly at higher frequencies, as is indicated by the line with circle symbols. The swim bladder in *Adioryx* is not as well coupled with the inner ear as in *Myripristis*, indicating how significant an effect this coupling can have on hearing. Other groups of fishes have independently evolved other forms of coupling between the swim bladder and the inner ear. For example, the goldfish has a series of bones coupling the swim bladder and the ear. Even though the mechanism for the coupling differs from *Myripristis*, the goldfish also has better hearing, particularly at high frequencies, than unspecialized fish such as the toadfish, *Opsanus*.

The terrestrial mammalian ear is typically divided into three sections. The *outer ear* is a canal which ends at the *eardrum* or *tympanic membrane*. The eardrum moves due to the pressure fluctuations of sound in air. The *middle ear* contains a series of bones which couples this movement of the eardrum to the inner ear. The *inner ear* of mammals contains the *cochlea*, which is the organ in which sound energy is converted into neural signals. Sound enters the cochlea via the *oval window* and causes a membrane, called the *basilar membrane*, to vibrate. This membrane is mechanically tuned to vibrate at different frequencies. Near the oval window, the basilar membrane is stiff and narrow, causing it to vibrate when excited with high frequencies. Farther into the cochlea, the basilar membrane becomes wider and floppier, making it more sensitive to lower frequencies. The mammalian ear thus appears to encode sound differently from the fish ear. Sound pressure is the primary stimulus and the basilar membrane acts as series of filters, each of which only respond to a small band of frequencies. Sensory cells at dif ferent positions along the basilar membrane are excited by different frequencies, and their rate of firing is proportional to the amount of sound energy in the frequency band to which they are sensitive.

Marine mammals share basic patterns of mammalian hearing, but cetacean ears have adaptations for marine life. The eardrum and middle ear in terrestrial mammals functions to efficiently transmit airborne sound to the inner ear where the sound is detected in a fluid. Such matching is not required for an animal living in the water, and cetaceans do not have an air-filled external ear canal. The problem for cetaceans is isolating the ears acoustically, and the inner ear is surrounded by an extremely dense bone which is isolated from the skull. Norris (1968) proposed that sound enters the dolphin head through a thin section of bone in the lower jaw and is conducted to the inner ear via fatty tissue which acts as a waveguide. This has recently been verified by Aroyan (1996) who developed a technique to map acoustic parameters using data from a computerized tomography (CT) scan of a dolphin's head provided by Cranford (1992).

It is obvious that most of the energy of an acoustic signal must occur within the hearing range of an animal to be an effective stimulus, and there is a correlation between the frequencies at which most fish species can hear best and the frequencies typical of their vocalizations (e.g., Stabentheimer 1988). Given the small sizes of fish, it is striking how low in frequency most of their calls are, with most energy typically below 1 kHz. This clearly matches the auditory sensitivities illustrated in Figure 5. As Figures 5 and 6 show, there is also a close correspondence between the frequencies of best hearing for bottlenose dolphins and harbor porpoises and the dominant frequencies of their echolocation clicks. The frequency range of hearing has never been tested in baleen whales. Hearing is usually tested by training an animal, and baleen whales are so big that only a few have been kept for short periods in captivity. However, both their low frequency vocalizations and the frequency tuning of their cochlea suggests they are specialized for low frequency hearing (Ketten 1994).

The hearing abilities of animals have not evolved exclusively for detecting sounds of conspecifics. For example, the goldfish, *Carassius auratus*, has good hearing for a fish (Figure 5), but has never been heard to produce a sound (Popper and Platt 1993). Fish must also use passive listening to learn about their biological and physical environment. For example, it has been suggested that the ability to detect high frequencies may have evolved in fish to allow them to detect the loud echolocation clicks of their odontocete predators (Offut 1968). Some recent research suggests that fish such as alewives, herring, and cod are able to detect intense sounds of frequencies much higher than is typical of their vocalizations. For example, Astrup and Møhl (1993) provide evidence that cod, *Gadus morhua*, detect short 38 kHz pulses at 194 dB re 1 μPa. This is a very high frequency for fish to detect, and the only known natural sources of this intensity and frequency are the clicks of echolocating toothed whales. Cod may be able to detect the loudest clicks of echolocating toothed whales at ranges of 10–30 m. Both alewives, *Alosa pseudoharengus*, and blueback herring, *Alosa aestivalis*, formed tighter schools and moved away from playback of sounds from 110 to 140 kHz at levels above 160 dB re 1 μPa (alewives; Dunning et al. 1992) to 180 dB re 1 μPa (herring; Nestler et al. 1992). While these results are suggestive, it remains to be seen how fish respond to the echolocation sounds of their actual predators. Given the extensive evidence of elaborate strategies and countermeasures between echolocating bats and their insect prey (e.g., Surlykke 1988), it is amazing how little comparable evidence is available for marine echolocators.

Perhaps more interesting than the obvious frequency match between signal and receptor is the comparison of time vs. frequency cues in fish vs. mammals. Many fish calls are either amplitude modulated or are composed of pulses. Hawkins and Myrberg (1983) review studies suggesting that fish are remarkably sensitive to slight variation in the timing and duration of their calls, while frequencies can be changed considerably with little change in response. Fish do not appear to be very sensitive to changes in frequency, and they do not have a peripheral frequency analyzer as highly developed as the mammalian cochlea. Most species of fish whose auditory frequency discrimination has been tested are capable of discriminating tones that differ by 5–10 % (Hawkins 1981). Instead of relying so heavily on a spatially organized array of frequency tuned sensory cells as in mammals, fish apparently rely more upon synchronizing neural responses with the temporal characteristics of the acoustic signal. By contrast, mammalian hearing is highly sensitive to frequency. Among mammals, dolphins have extraordinarily good abilities of discriminating different frequencies, and can detect a change of as little as 0.2 % in frequency (Thompson and Herman 1975). The number of sensory cells in their inner ear is higher than that for the human ear (Ketten 1994). The frequency range in which dolphins can discriminate frequencies the best is near 10 kHz. This matches the frequency range of dolphin whistles, which are frequency modulated tonal sounds. Most researchers who have studied whistles have concluded that frequency cues are critical elements of these communication signals.

5
Communicative Significance of Marine Animal Sounds

Most of the research in the previous section stems from studies in which captive animals are trained to respond to artificial stimuli, such as the pure tones used to measure hearing sensitivity. Marine mammals are well known for being easily trainable, but both invertebrates (Offut 1970) and fish (e.g., Tavolga et al. 1981) can also be trained to respond to sound for studies of hearing. Experimental studies of captive animals working on artificial problems have also been crucial for demonstrating the echolocation abilities of toothed whales. Schusterman (1980) reviews many of the conditioning techniques used to study echolocation in marine mammals. A very different research method has been used for most studies of animal communication. The ethological tradition of studying animal communication emphasizes studying animals in naturalistic settings. Ethologists start with a catalog of behaviors or vocalizations typical of their study species. As mentioned earlier (Sect. 1.3), the early phase of marine bioacoustics consisted primarily of cataloging the vocalizations of different marine species. Ethologists use this kind of catalog of species-specific displays as a starting point. Once the displays are well defined, ethologists observe animals to define the contexts in which an animal makes a particular display, and to define the responses of animals that receive the signal. Descriptive studies often find that a response may frequently follow a particular signal, but these studies cannot test whether a signal actually causes the response. These associations are tested in experiments in which the investigator presents a replicate of a signal and observes how an animal responds. There is a long tradition of testing responses of fish to artificial visual (Tinbergen 1951) or acoustic (Moulton 1956) stimuli in order to tease apart the pattern of signal and response that make up a communication system. The acoustic version of these experiments involves playbacks of recorded sound to animals in the environment in which the communication system usually functions (Nelson and Marler 1990). Sound playback experiments of this sort are critical for identifying which features elicit which response, and for defining sound categories in terms of their function in natural settings.

The logic of underwater playback experiments is the same as the airborne version (see Hopp and Morton, this Volume). Since vision is so limited underwater, it is often easier to set up realistic experiments underwater where the subject cannot see as far as the range of the speaker and has no reason not to expect the likely presence of the animal whose sounds are being played. Playback of natural vocalizations allows one to study the function of animal vocalizations. Synthetic stimuli are also often very useful for sound playbacks. This requires the ability to extract the acoustic features one thinks are relevant, modify one or more features, and resynthesize the modified sound (see discussion by Beeman, this Volume). The technical problems of accurate reproduction of playback stimuli are alleviated by a service of the US Navy. The Navy runs an Underwater Sound Reference Detachment (USRD, Orlando Florida) which maintains a stock of calibrated underwater sound projectors and reference hydrophones (USRD 1982). With careful selection of recorder and power amplifier, these projectors allow highly accurate reproduction of most marine biological acoustic signals. However, one must always monitor the playback stimuli even more carefully than in airborne applications, because underwater sounds cannot be heard very well without electronic monitoring. Researchers playing back sounds in air are more likely to detect technical problems with the unaided ear.

Several examples of communication in marine invertebrates, fish, and marine mammals will now be discussed.

5.1
Invertebrates

Little is known about the communicative significance of acoustic displays in marine invertebrates. Interesting work has been conducted with snapping shrimp of the genera *Crangon* and *Synalpheus*. These shrimp produce loud snapping sounds which are reported to have primary energy between 2 and 15 kHz (Everest et al. 1948). MacGinitie and MacGinitie (1949) suggested that snapping shrimp of the genus *Crangon* produce a sound intense enough to stun their prey, and that these snaps function in feeding. Later work emphasizes the role of snapping as an aggressive display, particularly in defense of a shelter, and suggests that the water jet may be as important a stimulus as the acoustic attributes of the display (Hazlett and Winn 1962; Nolan and Salmon 1970; Schein 1977). In the laboratory, when two snapping shrimp of the genus *Alpheus* are put in a jar with one shelter, they will usually go through a series of threats, including cocking the claw and snapping. A shrimp may damage the leg or antenna of an adversary if it makes contact as it snaps its claw. Most fights appear to be resolved by an exchange of ritualized threats without injury, however. Larger animals usually win over smaller ones. In contests between individuals belonging to different species but having the same body size, the species with the larger claw tends to win. This suggests that shrimp pay particular attention to the claw when assessing the fighting ability of an adversary. It appears that the visual appearance, acoustic properties of the snap, and tactile properties of the water jet are used for this fighting assessment. Less is known about perception of underwater sound in marine invertebrates than about production, and use of this sensory channel for communication remains underexplored in marine invertebrates compared with vision and the chemical sense.

5.2
Fish

Fish were early subjects of ethological research. Even after von Frisch (1938) demonstrated hearing in fish, most early ethological study of communication in fish emphasized visual stimuli such as coloration and markings on the body (e.g., Baerends and Baerends-van Roon 1950). However, acoustic signals may play an important communicative role even in fish with pronounced coloration displays. For example, Myrberg (1972) showed that bicolor damselfish (*Pomacentrus partitus*) produce different sounds and coloration patterns when they are courting, fighting, or feeding. These fish were studied both in a large aquarium in the laboratory, where every detail of behavior could be observed, and on a coral reef in the sea, using an underwater video system. Adult males of this species are very territorial, and their stable residence allowed them to be identified in the wild over many months. Males compete to attract females that are ready to spawn, and they have specific coloration, motion patterns, and chirp or grunt sounds associated with different phases of their courtship displays. When bicolor damselfish are engaged in a hostile or agonistic interaction, they have a very different coloration pattern, produce a *pop* sound, and engage in a different set of behaviors. When feeding, these fish have yet another coloration pattern and produce a faint stridulation sound different from either the courtship or agonistic sounds. Males will often attack a fish approaching their territory, and presumably a male produces courtship displays to a female to signal that he will not attack her, to signal that he is ready to mate, and to influence the female's choice of a mate. If a male detects that courtship is going on nearby, he may start producing competitive courtship displays as well (Kenyon 1994).

In order to test whether these damselfish sounds had a signal function, Myrberg (1972) played recordings back to males both in the field and in the laboratory. Playback of the courtship sounds had a pronounced effect on these males. Within 30–60 s of the onset of courtship sounds, males tended to switch from the normal coloration pattern to one of the two courtship patterns (Myrberg 1972). These playbacks of courtship sounds also elicited high rates of courtship behaviors and sounds from these males. Playback of the agonistic pops to males which were engaged in some courtship behavior, by contrast, inhibited both courtship acts and sounds. Spanier (1979) played back sounds of four different species of damselfishes, and found that each species responded more strongly to sounds of conspecifics. In keeping with the proposed importance of temporal cues for acoustic communication in fish, Spanier (1979) found that the most important acoustic features for this discrimination were the pulse interval and number of pulses in the call.

Myrberg's (1972) study of bicolor damselfish concentrated on the behavior and responses of males. Myrberg et al. (1986) focused on responses of females to playback of courtship chirps. This kind of field experiment relies heavily upon detailed observational data in order to optimize experimental design. For example, earlier work had shown that females tend to spawn at sunrise in the third quarter of the lunar cycle (Schmale 1981). In order to guarantee a reasonable number of ovulating females, the playbacks were limited to sunrise during the three peak days of spawning. The first playback experiment contrasted chirps of two males, one of which was 4 mm larger than the other. This larger male appeared to be more attractive to females, as he had more egg batches in his nest. The peak frequency of chirps from the larger male was 710 Hz, that from the smaller male 780 Hz, a frequency difference likely to be discriminable by fish. Of 21 females tested, 15 swam to the playback of the larger male and only two females swam towards the smaller male. A second playback series compared responses of females to chirps of conspecifics with either congeneric chirps or synthetic chirp-like sounds. All 14 of the females in this second playback series chose the chirps of conspecifics. These results show that females not only can use chirps to find a mating partner of the appropriate species, but also to select a mate from among several conspecifics.

Charles Darwin (1871) coined the term "sexual selection" to describe the evolutionary selection pressures for traits that are concerned with increasing mating success. There are two ways sexual selection can work. It can increase the ability of an animal to compete with a conspecific of the same sex for fertilization of a member of the opposite sex (intrasexual selection) or it can increase the likelihood that an animal will be chosen by a potential mate (intersexual selection). The evidence just discussed for bicolor damselfish would emphasize intersexual selection for the evolution of courtship chirps. In the bicolor damselfish, as in most animal species, it is the male sex that produces courtship displays and the female sex that selects a male for mating. The basic question for the evolution of displays by intersexual selection is why are females choosing particular features of the male display? Three factors that may influence female choice are: (1) correlations between a feature of the display and the quality of the male; (2) sensory bias of the female; and (3) positive feedback between a female preference and features of a male display. The first factor appears to be most important for the evolution of the chirps of bicolor damselfish, and discussion of the other two factors will be postponed for more appropriate species.

One important feature of quality in male fish appears to be body size, as females of many species tend to select larger males as mates (e.g., Schmale 1981, for bicolor damselfish). Myrberg et al. (1993) demonstrated a strong correlation between the body size of bicolor damselfish and the fundamental frequency of the chirp sound. Myrberg et al. (1993) suggested that differences in the peak frequency of chirps result from differences in the volume of the swim bladder. If females are selecting larger males that have chirps with the lowest peak frequency, then this will create a selection pressure for males to make the lowest chirps they can. However, if the minimum peak frequency a male can produce is constrained by the volume of his swim bladder and if the swim bladder volume correlates with body size, then males may be constrained to produce an honest advertisement of their body size. Similar cases of selection for low frequency indicators of body size are also known for a diverse array of vertebrates, and may involve fighting assessment between males as well as mate choice (e.g., Davies and Halliday 1978, for anurans; Clutton-Brock and Albon 1979, for a mammal). The reliability of the association between peak frequency and body size may have selected for females to attend to this frequency cue, even though fish are thought in general to be more sensitive to temporal features of their sounds.

Studies of communication in the oyster toadfish:, *Opsanus tau*, also include responses of females to the courtship sounds of males. Winn (1972) describes two primary acoustic signals from the oyster toadfish: the grunt and the boatwhistle. The grunt is produced during aggressive interactions, with more grunts produced in more intense interactions. Male toadfish produce the boatwhistle sound when they establish nests at the beginning of the breeding season. While grunts are usually produced in the context of aggressive interactions, boatwhistle sounds are produced at regular intervals by a male even when he is alone. Winn conducted a series of playback experiments in order to test whether males produce this call to attract females to lay eggs in their nests. Initial tests showed that toadfish seldom respond before evening, so Winn restricted the playbacks to this evening response time. He made pens and put an open can similar to a toadfish nest in each of the four corners of the pen. Speakers were attached to each of these cans. He then introduced a spawning male and female into the center of the pen and played boatwhistle sounds from the two of the four speakers. The other two silent cans were considered control cans. None of the females that had no eggs because they had already spawned entered any of the cans. Of the 66 males tested, none entered a control can and only four entered a can with boatwhistle sounds. However, of the 44 females with eggs, 15 entered boatwhistle cans, one entered a control can, and 28 did not enter a can. This suggests that females that are ready to spawn are attracted to the boatwhistle sound.

Features associated with effort or cost appear to be important in the evolution of the boatwhistle display. Boatwhistles are very loud and prolonged compared with most fish vocalizations. Fine (1978) suggests that the fundamental frequency of the harmonic part of the boatwhistle call is affected more by the rate of contraction of sound-producing muscles rather than by the volume of the swim bladder. If so, then the frequency of boatwhistles may not be a reliable indicator of the size of the displaying fish, as was suggested for the bicolor damselfish. If the fundamental frequency of the boatwhistle is driven by the rate of muscle contraction, then the rate of contraction may correlate with effort of the displaying fish. The fundamental frequency of boatwhistles varies from < 150 Hz early in the breeding season to > 250 Hz nearer the peak of the season, and duration shows a similar increase (Fine 1978). When a male toadfish sees a female or hears an indication that a female may be nearby, he increases his own display effort in order to attract the female. While most males produced boatwhistles at rates of about 7/min, they increase their calling rate when they hear playbacks of boatwhistles at rates greater than 12 boatwhistles/min (Winn 1972). Males also increase their rate of boatwhistle calls when a female comes within a meter or so (Gray and Winn 1961; Fish 1972). All of these responses of males indicate that females may select a mate based upon his display effort.

Ryan et al. (1990) argue that preexisting biases in the sensory systems of females may be an important factor in the evolution of displays by sexual selection. Most communication systems show a match between the frequency range of the signals and the receptors, but there appears to be a mismatch for boatwhistles which may have influenced some of the boatwhistles, acoustic characteristics. While male toadfish respond to playbacks of tone bursts in the range 180–400 Hz as if they were boatwhistles, they do not respond to sounds that contain the lower frequencies of 90–100 Hz (Winn 1972). This suggests that there is a lower limit in frequency below which boatwhistles become less stimulatory to males. This is particularly interesting because Fish and Offut (1972) and Fine (1981) present evidence that toadfish are most sensitive to sounds in the frequency range from 40 to 90 Hz, with sensitivity decreasing rapidly at higher frequencies (see Figure 5). If females have the same kind of bias for preferring boatwhistle sounds in a frequency range to which they are less sensitive, as was demonstrated by Winn (1972) for males, then this may force males to put more effort into the display, influencing some features of the boatwhistle, such as its extraordinary loudness.

Communication is often analyzed as an exchange of one signal and an immediate response. For example, Myrberg (1972) analyzed the functions of a damselfish display by comparing which displays immediately preceded or followed a particular display. This may be appropriate for many signals, but it does not seem appropriate for analyzing the boatwhistle which is produced spontaneously at regular intervals by males during the breeding season whether another animal is present or not. The boatwhistle of a toadfish is more likely to be followed by another boatwhistle from the same fish than by a response from a different fish. These kinds of signals have been called advertisement displays, and Winn (1972) suggests that the boatwhistle display is analogous to the songs of birds which are a well known reproductive advertisement display. The structure of advertisement displays cannot be understood simply in terms of an immediate response to one signal. One must consider more broadly how the signal is designed to manipulate choices of animals that hear the display (Krebs and Dawkins 1984). Advertisements are often produced in great quantity in order to modify the outcome of a choice that an animal may make over a long period of time. Advertisements are in general designed to be attention getting. For example, the boatwhistles of toadfish are loud, sustained, and repeated.

Two kinds of features that females might use in choosing a male have just been discussed. One is a direct correlation between the display and some physical attribute of male quality that will be important to the female. Female bicolor damselfish may choose a lower frequency as an indicator of large size of a male. Another kind of feature is related to the cost of the display. If producing the display is taxing or dangerous enough, the male who achieves the most intense display may be in the best condition. Female toadfish may use a variety of features to assess the level of effort a male is putting into his display.

5.3
Marine Mammals

5.3.1
Reproductive Advertisement Displays

Marine mammals also produce reproductive advertisement displays. The best known acoustic advertisement displays are called songs, which are usually defined as a sequence of notes that are repeated in a predictable pattern. Male seals of some species repeat acoustically complex songs during the breeding season. Songs are particularly common among seals that inhabit polar waters and that haul out on ice. The vocalizations of ringed seals, *Phoca hispida*, become more common along with agonistic behavior as the breeding season gets underway (Stirling 1973). The songs of bearded seals, *Erignathus barbatus*, are produced by sexually mature adult males (Ray et al. 1969). Bearded seal songs are heard frequently during the peak of the breeding season in May; but by July song is seldom heard. Male walruses, *Odobenus rosmarus*, also perform ritualized visual and acoustic displays near herds of females during their breeding season (Fay et al. 1981; Sjare and Stirling 1993). Males can inflate internal pouches that can produce a metallic bell-like sound (Schevill et al. 1966). Walruses make loud knock, whistle, and breathing sounds in air when they surface during these displays. When they dive, these males produce distinctive sounds underwater, usually a series of pulses followed by the gong- or bell-like sounds. Usually several males attend each female herd, so the relative roles of this display in intersexual vs. intrasexual behavior are not known (Sjare and Stirling 1993). Antarctic Weddell (*Leptonychotes wedelli*), seals also have extensive vocal repertoires and males repeat underwater trills (rapid alternations of notes) during the breeding season. Males defend territories on traditional breeding colonies. These trills have been interpreted as territorial advertisement and defense calls (Thomas et al. 1983). Whether females may also use them in selecting a mate is unknown.

The songs of humpback whales are the best understood advertisement display in the cetaceans. These songs sound so beautifully musical to our human ears that they have been commercial bestsellers. Figure 8 shows a spectrogram of a humpback song made in Hawaiian waters during a period when songs contained up to nine themes. The third theme was not included in this song, but was in the next song of this whale, as is indicated by the star in the spectrogram. Each theme is made up of repeated phrases or series of sounds lasting on the order of 15 s. Phrases of one theme repeat a variable number of times before a new theme is heard. In Figure 8, the phrase boundaries are marked by vertical lines. Humpbacks tend to sing themes in a particular order, and it often takes about 10 min before a singer comes back to the initial theme.

The observational key to our understanding of humpback song lies in the ability of biologists to find and follow a singing humpback whale. The first step is to use a small boat to monitor underwater sound in different areas within a breeding ground until the song of one whale is much louder than any other. Singers often surface to breathe once per song during a particular theme, and they *blow* or breathe during the silent intervals between notes (Tyack 1981). When the singer reaches this part of the song, it can be

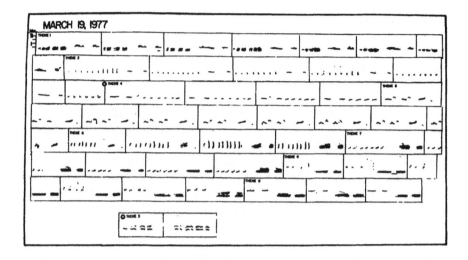

Fig. 8. Spectrogram of the song of a humpback whale, *Megaptera novaeangliae*. This song was recorded from a lone humpback on 19 March 1977 in the Hawaiian Islands. Each *line* represents 120 s, and this song took seven lines or 14 minutes before repeating the initial theme. The song at this time had up to nine themes in it. The start of each theme is marked on the spectrogram. Each theme is made up of repeated phrases, and the boundary between phrases is marked by vertical lines. This particular song did not include theme 3, but the next song in this sequence did, and is indicated at the bottom of the spectrogram. (reprinted from Payne et al. 1983, with permission)

located by careful visual scanning from the boat that is monitoring the song. The song gets fainter seconds before the singer breaks the surface to breathe, and this often allows one to identify one surfacing whale as the singer. When one approaches the singer, the sound of the song is often loud enough to be heard through the hull of the boat. Once a singer has been identified in this way, its sex can be determined by underwater photography of the genital slit (Glockner 1983) or by taking a small tissue sample for genetic determination of sex (Baker et al. 1991; Palsbøll et al. 1992). The continuous vocalizations and predictable surfacing behavior of singers allows them to be followed for periods of hours. These observations have shown that almost all singing humpbacks are lone adult males, and they repeat songs in bouts that can last for hours.

Interactions between whales can take so long and can occur over such great distances that special techniques are required to document them. Observation from elevated sites on land has been important for studying species, such as humpbacks, that concentrate near coastal waters. Observers on land can use a surveyor's theodolite to pinpoint the location of each whale surfacing and can communicate by radio with vessels that can obtain more detailed behavioral and acoustic data while they follow whale groups (Tyack 1981). These visual locations can also be linked to acoustic locations from arrays of hydrophones (Frankel et al. 1989). These techniques have documented responses of singing whales with groups nearly 10 km away and are well suited to observing whales for periods up to 10 h. The interactions of whales are so much slower than our own pace, that it is often necessary to make a plot of movement patterns and behavioral displays in order to make sense out of an interaction that took many hours to unfold.

These kinds of behavioral observations have shown that singing humpback whales appear motivated to join with nearby whales, and singers often swim towards other whales while still singing (Tyack 1981). Singing humpbacks usually stop singing when they join with other whales. Females have occasionally been observed to approach and join with a singing humpback; behavior associated with sexual activity has been observed after the female joins the singer (Tyack 1981). Aggressive behavior has often been seen after a singer joins with another male. Males may approach and join with a singer to initiate these competitive interactions. After a male has joined with a singer, the singer may then stop singing and leave, while the whale that joined may start singing himself. However, most groups of whales seem to avoid the potentially aggressive singers. While breeding males are not territorial on the breeding ground, song mediates both competitive interactions between males and some associations between males and females. Humpback song appears to play a role both in male—male competition for mates (intrasexual selection) and perhaps in female choice of a mate (intersexual selection).

There appears to be a strong force for vocal convergence in the songs of humpback whales on a breeding area at any one time, coupled with progressive change in all aspects of the song over time (Payne et al. 1983; Payne and Payne 1985). Recordings of humpback song made from different whales in the same breeding area at the same time are quite similar. At any one time, all of the singers within a breeding area sing the same themes in the same order, and the individual sounds that make up the song are quite similar. However, the song changes dramatically from month to month and from year to year. For example, in hundreds of recordings made later in the spring of 1977 and for years afterwards, theme three from Figure 8 was never heard after the end of March 1977. There is no indication that these changes in the song reflect changes in the message; the whales appear to be engaged in similar interactions even as the song changes.

The combination of complexity and progressive changes in humpback song suggest that sexual selection may select for complexity of the acoustic display per se. As Darwin (1871) pointed out for other species, we clearly may have to thank the aesthetic sensibilities of generations of female humpbacks for the complexity and musical features of the males' songs. Fisher (1958) explained the evolution of this kind of extraordinarily complex and beautiful secondary sexual character in terms of a *runaway process* of sexual selection. The runaway process emphasizes positive feedback between the female preference and the elaboration of a male display. Let us start by assuming that females have developed a preference for an acoustic display with a particular feature. This preference could arise because the display was correlated with some valuable male trait, the display could make the male more easy to find, or females could simply have some bias to respond preferentially to a particular stimulus. Whatever the origin of the preference, the tendency for females with stronger preferences to mate with males with the feature means that genes for the preference will covary with genes for the feature. If females on average select males with extreme development of the feature, then the next generation will have more sons with the exaggerated feature and more females with the preference. This creates a positive feedback loop, potentially producing a runaway process leading to extreme and exaggerated development of secondary sexual characters in males and preferences in females.

Male humpback whales appear to have two alternate strategies for gaining access to a female. They may sing in order to attract a female or they may join a group in which a female is already with one or more males. Males in these larger groups may fight for the position closest to the female in the group. During these fights, a set of sounds is heard that differ from song. These sounds are called social sounds. As with the grunts of toadfish, the more intense the aggressive interaction, the greater the number of social sounds heard (Silber 1986). Singing humpbacks may stop singing and make a beeline to join these groups from distances as far as 10 km (Tyack and Whitehead 1983). Groups of one to two adults without calves (and presumed to be males) also have been observed to join these competitive groups. It is unlikely that any cues other than acoustic ones such as the social sounds are detectable from such ranges.

Tyack (1983) conducted playback experiments to test whether the differential responses of different kinds of humpback social group to song or social sound were in fact mediated by these sounds. Techniques had already been established to follow each group and to monitor approach and avoidance as described above. The underwater sound projector and amplifier selected for this experiment were powerful enough to approach the source level of singing humpbacks, yet were portable enough to be deployed from 4–5 m boats. The two stimuli used were song and social sounds, and these were played back to lone singers, groups of one to two adults without a calf, groups of one to two adults with a calf, and competitive groups of three or more adults. While it was difficult to build large sample sizes for these playbacks, the results were consistent. During 9 out of the 16 social sound playbacks, singing whales or groups of one to two adults moved rapidly and directly to the playback boat, passing within meters of the underwater sound source. No whales rapidly approached the boat during the playbacks of song; in most of these, the target whale groups moved away. The results thus mirrored the responses observed when these sounds were produced by whales. Since two natural stimuli were used with no other control stimulus, one can only argue for differential response, not a highly specific response to each stimulus. Some of the critical features of this experiment were: (1) previous observation allowed prediction of responses; (2) predicted responses were strong and easy to score; and (3) limiting to two playback stimuli was crucial, given the difficulty of boosting sample size.

Other baleen whales also have been reported to produce reproductive advertisement displays. Bowhead whales, *Balaena mysticetus*, have been recorded producing songs as they migrate from their breeding grounds (Ljungblad et al. 1982). These whales spend their winter breeding season in Arctic waters such as the Bering Sea, where it is difficult to observe them, so we know little about behavior during the peak of the breeding season. While the individual sounds that make up bowhead songs are somewhat similar to those of humpbacks, the songs are simpler, lasting a minute or so. Like the songs of humpback whales, bowhead songs also change from year to year.

The long series of loud 20 Hz calls of finback whales also may be a reproductive advertisement display. They have a seasonal occurrence that closely matches the winter breeding season of these whales (Watkins et al. 1987). These 20 Hz series of finback whales are much simpler than the songs of bowheads and humpbacks. As has already been described, this appears to be a design feature for long-distance communication.

This matches what we know of the breeding pattern of finback whales. Humpback whales are thought to congregate in protected waters in the lee of tropical islands or banks for calving and breeding. A female in a breeding area can often hear many singers within a range of 10 km or so, in which most of the detail of humpback song is audible. Finback whales, on the other hand, are thought to disperse into tropical and temperate waters during their winter breeding season. While we actually know little of the spacing of finback whales during the breeding season, it appears likely that a female might have to listen for males over much longer ranges. This need for communication over long ranges may have created selection pressures for a simpler, more repetitive, loud and low frequency signal. If female finback whales had reasons for selecting large males and if low frequency vocalizations were a reliable indicator of size in finback males, then the low frequencies of these calls might also result from the kind of sexual selection described for bicolor damselfish and many other vertebrates. On the other hand, finback males are smaller on average than females, so selection for large size appears to be greater for females than males (Ralls 1976). In order to clarify these issues, we need better observations of vocalizing finbacks, responses of females to these 20 Hz series, and playback experiments varying frequency, source level, interval between pulses, etc.

Better evidence is needed on the role of reproductive advertisement vocalizations in mediating male—male competition and in mediating female choice for marine mammals. Real-time computerized acoustic source location allows direct measurement of the spacing of vocalizing males (Frankel et al. 1991). For species where females can also be tracked from their own vocalizations, this same technique could be used to follow females as they move through the field of advertising males. Data of this sort may help clarify the process of female choice and the acoustic and behavioral features which females may use to choose a mate. It will also be necessary to develop methods to identify which females are receptive. Female humpbacks may only be receptive for a short period of time around ovulation. Whaling data from the southern hemisphere indicates that nearly 50 % of adult females may be ovulating during one 10–day interval, with much lower rates of ovulation during most other parts of the breeding season (Chittleborough 1954). As with the toadfish, nonreceptive females are unlikely to approach a singing male. Playbacks to females should focus on receptive females, either during periods of peak ovulation or using a technique, such as a biopsy, to determine each subject's reproductive status.

While it is often difficult to apply sampling methods developed for visual observation of terrestrial animals (e.g. Altmann 1994) to marine bioacoustics, new instrumentation may also resolve some of these problems. Reliable and systematic sampling of vocal behavior will be improved by passive acoustic tracking of vocalizing animals, and sophisticated tags for animals that can sense vocalizations, the sound field of the animal, and other response measures. Animals that do not vocalize frequently and distinctively enough to allow acoustic tracking from their own vocalizations may be followed visually, or tracked after attachment of a tag. These techniques become much more powerful when used together to address specific scientific issues. For example, imagine tagging a female humpback as she swims into an area where all singers can be located acoustically. Suppose the tag attachment allows one to recover a biopsy that can later assess whether the female is receptive. Even a simple tag would allow one to track the female's movements through the field of singers, and more sophisticated tags might be able to record what she heard and her responses to different males.

5.3.2
Individual and Group Recognition with Special Emphasis on Sounds of Toothed Whales

Myrberg (1981) developed a general framework for analyzing communication in fish. He emphasized two kinds of intraspecific social interactions: finding and selecting mates and competition for mates or other resources. These are the kinds of interactions that have dominated the preceding section (Sect. 5.3.1). However, Myrberg also mentioned that a general framework must include the potential importance of social interactions directed toward the survival and reproduction of one's own offspring or more distantly related kin. If interactions with kin were important, they would create a selection pressure for some system for kin recognition. Blaustein and Waldman (1992) discuss evidence for kin recognition using chemical cues among amphibians, and similar chemical recognition of kin also occurs in some fish (Winberg and Olsén 1992). However, there is much less evidence for use of acoustic cues in kin recognition among fish, and Myrberg (1981) suggested that kin relationships among adults may be less important for fish than for some other animal groups, such as social insects, and some birds and mammals, including marine mammals. Many species with extensive parental care have a system for parent—offspring recognition. Parental care is limited in fish and rarely lasts for more than several weeks. Most models of the behavior of animals that aid one another through reciprocation also require some form of individual recognition. While the remarkable coordination of fish in a school clearly demonstrates impressive communicative mechanisms, there is little evidence of individual-specific relationships within such a school.

Myrberg and Riggio (1985) do provide evidence of individual recognition of chirp sounds in competitive interactions between male bicolored damselfish. As mentioned in Section 5.2, adult males in this species hold territories that are stable for months or years. This means that each territorial male will persistently be exposed to the displays of his neighbors. Myrberg and Riggio (1985) conducted playback experiments with a colony of bicolor damselfish on a small reef that contained five adult males. Chirp sounds were recorded from each of the males, and all five sets of chirps were played back to all five males. Each male showed a greatly reduced rate of competitive courtship displays to chirps from the nearest neighbor compared with all of the other males. However, this result might simply stem from increased habituation to the most familiar sound. In order to control for this, the sounds of the two nearest neighbors were played back, once from the neighbor's own territory and once from the territory of the other neighbor. Males responded much more strongly when the sound of a neighbor was played back from the wrong territory. This kind of playback design has yielded similar results in many other vertebrates, an effect known as the "dear enemy" phenomenon.

Many marine mammals do live in kin groups, and social interactions within these groups may have a powerful effect on fitness. The different structures of these cetacean societies create different kinds of problems of social living, and there appears to be a close connection between the structure of a cetacean society and the kinds of social communication that predominate in it. For example, stable groups are found in fish-eating killer whales, *Orcinus orca*, in the coastal waters of the Pacific Northwest, and these whales also have stable group-specific vocal repertoires. The only way a killer whale group, called a pod, changes composition is by birth, death, or rare fissions of very large groups (Bigg et al. 1987). Many of the calls of killer whales are stereotyped and stable over decades. These are called discrete calls. Each pod of killer whales has a group-specific repertoire of discrete calls that is stable for many years (Ford 1991). Each individual whale within a pod is thought to produce the entire call repertoire typical of that pod. Analysis of variation in call use within a pod suggests that some calls may be more common in resting groups, others more common in more active groups (Ford 1989). However, each discrete call in the pod's repertoire can be heard regardless of what the pod is doing. Different pods may share some discrete calls, but none share the entire call repertoire. The entire repertoire of a pod's discrete calls can thus be thought of as a group-specific vocal repertoire. Different pods may have ranges that overlap and may even associate together for hours or days before diverging. These group-specific call repertoires in killer whales are thought to indicate pod affiliation, maintain pod cohesion, and to coordinate activities of pod members.

Bottlenose dolphins (*Tursiops truncatus*) do not have stable groups as in resident killer whales, but rather live in a fission – fusion society in which group composition changes from hour to hour or even minute by minute. While dolphin groups are remarkably fluid, there may be strong and stable bonds between particular individuals. Some wild individual bottlenose dolphins show stable patterns of association for many years (Wells et al. 1987). This combination of highly structured patterns of association between individuals, coupled with fluid patterns of social grouping, argues that individual specific social relationships are an important element of bottlenose dolphin societies (Tyack 1986a). It is difficult to imagine how dolphins that share a bond could remain together without an individually distinctive acoustic signal. Caldwell and Caldwell (1965) demonstrated that each dolphin within a captive group produced an individually distinctive whistle. The Caldwells called these *signature whistles*, and they postulated that signature whistles function to broadcast individual identity. Initial studies of signature whistles in adult dolphins, primarily of isolated animals, suggested that well over 90 % of an individual's whistle repertoire was made up of its signature whistle (reviewed in Caldwell et al. 1990). Signature whistles were initially discovered in captive dolphins, but similar signature whistles have been documented in wild dolphins (Sayigh et al. 1990). In the wild, it is seldom possible to determine which dolphin makes a whistle when a group is swimming freely, but this can be accomplished by recording dolphins when they are restrained in a net corral. When a dolphin is held in the corral, one can attach a hydrophone to its head with a suction cup. This is not a very natural context for these animals, but it allows one to sample the results of normal whistle development among wild dolphins. Moreover, when dolphins are swimming freely immediately after being corralled, or when they are recorded immediately after release, they produce whistles very similar to those produced while they were restrained (Sayigh et al. 1990). Recordings of whistles from wild dolphins demonstrate that dolphins develop signature whistles by 1 – 2 years of age and these are stable for over a decade.

5.3.2.1
Mother—Infant Recognition

All mammalian young are born dependent upon the mother. Most need to suckle frequently, and many species depend upon the mother for thermoregulation and protection from parasites and predators. Most mammals have a vocal system for regaining contact when mother and offspring are separated. Colonially breeding seals often face difficult mother–young location and recognition problems. In many otariid seals, a mother leaves her young pup on land in a colony of hundreds to thousands of animals, feeds at sea for a day or more, and then must return to find and feed her pup. Among Galapagos fur seals, *Arctocephalus galapagoensis*, pups spend more time calling during their first day of life than later, and mothers learn to recognize the calls of their young within the first day of life (Trillmich 1981). Mothers give pup contact calls as early as during birth. Later, mothers can signal with a pup-attraction call to a pup that is moving away. When a mother returns from feeding at sea, she comes up on the beach giving pup-attraction calls. Her own pup usually seems to recognize her call and approaches. If a pup approaches to suckle, the mother sniffs the pup for a final olfactory check. If it is not her offspring, she almost always rejects the pup, a rejection which can cause injury or occasionally death to the pup (Trillmich 1981). There is thus a strong incentive for both mother and pup to recognize each other correctly. Playback experiments of pup attraction calls indicate that 10-12-day-old pups prefer their mother's call, and this recognition persists until they become independent at more than 2 years of age (Trillmich 1981).

The young of many dolphin and other odontocete species are born into groups comprised of many adult females with their young, and they rely upon a mother–young bond that is even more prolonged than that of otariids. Many of these species have unusually extended parental care. For example, both sperm whales (*Physeter macrocephalus*) and pilot whales (*Globicephala macrorhynchus*) suckle their young for up to 13–15 years (Best 1979; Kasuya and Marsh 1984). Bottlenose dolphin calves typically remain with their mothers for 3–6 years (Wells et al. 1987). These dolphin calves are precocious in locomotory skills, and swim out of sight of the mother within the first few weeks of life (Smolker et al. 1993). Calves this young often associate with animals other than the mother during these separations. This combination of early calf mobility with prolonged dependence would appear to select for a mother–offspring recognition system in bottlenose dolphins. In the following paragraphs and Section 5.3.3, the terms "dolphin", "mother", and "calf" will be used to refer to the bottlenose dolphin, *Tursiops truncatus*.

Dolphin mothers and young use signature whistles as signals for individual recognition. Observations of captive dolphins suggest that whistles function to maintain contact between mothers and young (McBride and Kritzler 1951). When a dolphin mother and her young calf are forcibly separated in the wild, they whistle at high rates (Sayigh et al. 1990); during voluntary separations in the wild, the calf often whistles as it is returning to the mother (Smolker et al. 1993). Experimental playbacks have demonstrated that mothers and offspring respond preferentially to each others' signature whistles even after calves become independent from their mothers (Sayigh 1992).

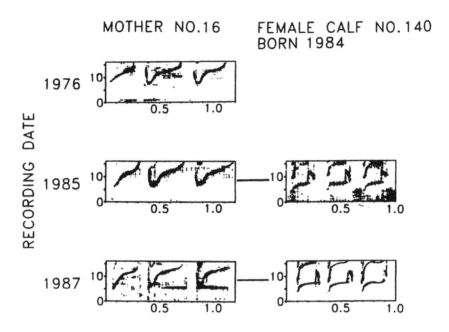

Fig. 9. Spectrograms of signature whistles from one wild adult female bottlenose dolphin recorded over a period of 11 years and of her daughter at 1 and 3 years of age (Reprinted with permission from Fig. 2 in Sayigh et al. 1990). Note the stability of both signature whistles. The *x-axis* indicates time in seconds and the *y-axis* indicates frequency in kilohertz on the spectrogram

Dolphins do not just use whistles for mother–infant recognition. Calves show no reduction in whistling as they wean and separate from their mothers. While adult males are not thought to provide any parental care, they whistle just as much as adult females. Bottlenose dolphins may take up to 2 years to develop an individually distinctive signature whistle, but once a signature whistle is developed, it appears to remain stable for the rest of the animal's lifetime (Figure 9; Caldwell et al. 1990; Sayigh et al. 1990; Sayigh et al. 1995). These results suggest that signature whistles continue to function for individual recognition in older animals.

5.3.3
Vocal Learning Among Marine Mammals

Few animals outside of several families of birds have been shown to modify the acoustic structure of their vocal repertoire based upon what they hear. One source of evidence for vocal learning is imitation of novel sounds. If an animal can imitate a sound that is not normally part of its repertoire, then it learned to modify its normal vocalizations to match the model. Given the lack of evidence for vocal learning among terrestrial nonhuman mammals, there are a surprising number of reports of marine mammals imitating manmade sounds: Hoover, a harbor seal (*Phoca vitulina*) at the New England

Aquarium, imitated human speech well enough to have a recognizable New England accent (Ralls et al. 1985). Captive bottlenose dolphins from many aquaria have been reported to learn to imitate computer-generated tones and pulses. Logosi, a beluga whale at the Vancouver Aquarium, was able to imitate his own name (Eaton 1979). Sperm whales (*Physeter macrocephalus*) have been heard to respond so precisely to a depth sounder that their pulses produce false targets on the depth recorder (Backus and Schevill 1962). These reports include both seals and whales, which have different terrestrial ancestors. Presumably, this means that both groups independently evolved this skill.

Vocal learning may play a particularly important role in the production of individually distinctive calls in marine animals. Many birds and mammals rely upon individually distinctive social relationships within their societies, and many of these species recognize individual differences in the vocalizations of different animals. Slight variations in the vocal tracts of terrestrial animals lead to predictable differences in the voices of individuals. These involuntary characteristics of voice are not likely to be as reliable for diving animals, however. The vocal tract is a gas-filled cavity, and gases halve in volume for every doubling of pressure as an animal dives. Since different parts of the vocal tract are more or less elastic, changes in volume will lead to changes in shape. These depth-induced changes in the vocal tract are likely to outweigh the subtle developmental differences that lead to voice differences. If diving animals rely upon individually distinctive calls, they may need to create them by learning to modify acoustic features under voluntary control, such as the frequency modulation of whistles. This analysis suggests one reason why vocal learning may be more common in marine than terrestrial mammals.

5.3.3.1
Imitation of Signature Whistles

The bottlenose dolphin, *Tursiops truncatus*, is the marine mammal species whose imitative skills are best known. Captive bottlenose dolphins of both sexes are highly skilled at imitating synthetic pulsed sounds and whistles (Caldwell and Caldwell 1972; Herman 1980). Bottlenose dolphins may imitate sounds spontaneously within a few seconds after the first exposure (Herman 1980), or after only a few exposures (Reiss and McCowan 1993). Dolphins can also be trained using food and social reinforcement to imitate manmade whistle-like sounds (Evans 1967; Richards et al. 1984; Sigurdson 1993). The frequency modulated whistle-like sounds that are most commonly used as stimuli for imitation research are similar to signature whistles. My own interest in dolphin whistles was stimulated by the apparent contrast between the evidence that each dolphin tended to produce one individually distinctive whistle and evidence that dolphins were so skilled at imitating manmade whistle-like sounds. Might dolphins be using imitation in their own natural system of communicating by whistles? In order to address this question, it was necessary to develop a method to identify which dolphin makes a sound when animals are interacting together. I have developed several techniques to solve this problem. The first device, called a *vocalight*, is a small hydrophone connected to a display of light-emitting diodes (LEDs) (Tyack 1985). For every doubling of signal intensity (increase of 3 dB), another diode lights up. There are ten LEDs, so the device responds over a range of 30 dB. The vocalight is small and can be attached to a dolphin's head with a suction cup. In order to identify which dolphin produces which whistle, one puts a vocalight with a different color of LEDs on each dolphin. Then one broadcasts underwater sound in air. When one hears a whistle, one looks to see which color of vocalight lit up the most.

A dolphin tends to produce its own signature whistle when it is isolated. However, by using the vocalight, I found that when a captive dolphin is interacting with members of its group, it can also produce precise imitations of the signature whistles of the other dolphins. Imitation of signature whistles is also found in wild dolphins that share strong social bonds (Sayigh et al. 1990), such as mothers and calves or pairs of adult males who may remain together in a coalition for years (Connor et al. 1992).

Studying the function of cognitive skills such as whistle imitation is difficult and often requires both controlled experiments in the laboratory and careful observation in the natural setting in which these skills evolved (Tyack 1993). Experiments on vocal labeling in dolphins demonstrate that dolphins have the cognitive abilities required to use imitated whistles as names. Captive dolphins can be taught to imitate manmade sounds upon command (Richards et al. 1984). In the vocal labeling paradigm, when the trainers gave a dolphin the command to imitate a sound, they would simultaneously play the model sound and hold up an object associated with that sound. For example, for the tune "MARY HAD A LITTLE LAMB", they might show a frisbee, and for "ROW, ROW, ROW YOUR BOAT", they might show a pipe. After the dolphin got used to this, the trainers started occasionally to show the object but not to play the model sound. In order to respond with the right sound, the dolphin had to remember which sound was associated with which object. After sufficient training, the dolphin succeeded in learning how to label each manmade object with an arbitrary manmade tune.

The Richards et al. (1984) vocal labeling experiments suggest that imitated signature whistles might function as similar vocal labels, and that a dolphin may imitate the whistles of particular individuals in order to name them. In order to test whether imitated signature whistles function as names, one must work with a group of three or more dolphins. It is difficult to use the vocalight with groups of more than two animals. When one hears a whistle, one must be able to observe vocalights on all of the animals in the group in order to determine which one lit up the most and therefore which dolphin made the whistle. I therefore developed a small datalogging computer that is also small enough to attach to a dolphin with suction cups (Tyack and Recchia 1991). The computer is mounted in an underwater housing with a hydrophone, preamplifier, and filter. It logs the level and frequency of sound 20 times a second for up to 45 min. As many data loggers as dolphins in the tank are synchronized at the start of a session using a signal from a video time generator. This same generator time-stamps a video recording of the session. A video camera mounted over the dolphin pool provides a video record of behavior, and underwater audio signals from hydrophones in the pool are recorded on hi-fi audio channels of the VCR. Once the session is over, one downloads data from each logger into a personal computer. These data are analyzed after the session using a program that reads data from each data logger. When one finds a time of interest on the videotape, one indicates this time to the program which plots the levels of each data logger and the frequency at the loudest logger. The data logger with the loudest signal is presumed to be on the whistling dolphin.

Neither the vocalight nor the data logger techniques are particularly well suited for use with free-ranging dolphins. Several techniques are currently under development. One scheme is to use a more sophisticated data logger, which can actually record the entire signal of interest. This can either be recovered later or linked to a digital radio telemetry transmitter. Other promising techniques include localizing where sounds are originating using a two-dimensional beamformer or acoustic location system (e.g. Freitag and Tyack 1993) and overlaying the sound source location on an underwater or overhead video image of animals in view.

5.4
Sound Classification

Section 4 on hearing suggests that different species of animal are likely to perceive sounds in different ways. This is due both to differences in receptors and in central auditory processing. For example, timing cues seem more important than frequency in fish, while frequency coding is an important element of mammalian hearing, and mammals are better than fish at frequency discrimination. All good ethologists understand this idea that each animal has its own sensory world. However, when we study animal signals, we initially must rely upon our own senses and instruments to detect and classify them. The development of instruments is particularly important for marine bioacoustics, since we must rely upon electronic detectors and since so many of the signals are outside our own hearing range. The classical approach for analyzing sounds in bioacoustics is to listen to recordings and to perform spectrographic analysis on detected sounds. A sound spectrogram plots the energy of sounds as a function of frequency and time. There is a time window associated with the frequency analysis, and the choice of time and frequency settings has a great impact on what features are apparent and how they appear (Watkins 1967). The way in which spectrographic analysis encodes the energy in different frequencies has some rough analogies to the way in which the cochlea detects energy at frequencies high enough that the auditory system cannot follow the waveform. Spectrograms thus have the advantage that they can present acoustic information in visual figures that may roughly match the way sounds are processed by the peripheral auditory system in mammals (also see discussions by Gerhardt, Clements, and Beeman, this Volume). The humble waveform plot of actual pressure values as a function of time can also be extremely useful, especially for viewing and analyzing temporal features of a signal. This is particularly relevant for those animals which appear to rely primarily on timing features for recognizing conspecific signals. The basic point is that vocalizations should be presented using displays that parallel how the sounds are processed by the auditory system of the species that receives the signal.

When biologists set out to characterize the vocal repertoire of a species or an individual, they typically make lots of recordings, listen to them, and begin to recognize and name classes of signals. Spectrographic analysis can help aid the ear in such recognition, but often the biologist only makes spectrograms of a few examples of each sound type. More detailed data on the repertoire are usually presented as counts of the different signal types. There are several problems and pitfalls with this approach. These will be discussed in the next few paragraphs.

Humans cannot directly sense many features that are easily detected by other animals. We must be careful to use the proper instruments to detect and faithfully record, for example, the high frequency clicks of a porpoise or the low frequency calls of a blue whale (see Pye and Langbauer, this Volume). Studies of visual observations of animal behavior put strong emphasis on avoiding attentional biases and inter-observer reliability (Altmann 1974). Classical bioacoustic analyses are vulnerable to both problems. Humans are excellent at creating categories out of continuous variables and our categories may not match those of the species under study. If we set out to make categories, we may miss graded or continuous elements of a communication system. There are several alternatives to the subjective categorization of sounds. One approach is to measure well-defined acoustic features from a set of sounds and then to analyze them using multivariate statistics. For example, Clark (1982) performed this kind of analysis on the calls of southern right whales, and found that the calls did not fall into the discrete categories suggested by earlier analyses, but formed a structured continuum. Many studies have extracted a few acoustic features such as duration, and minimum and maximum frequency for statistical analysis. These features have been selected for ease of measurement, not necessarily because of their salience to the animals. Measurements of extreme minima and maxima are not very reliable, being modified by signal to noise and spectrographic settings. Modern digital signal processing facilitates making more robust measurements, measuring more features, and selecting features for their relevance to the animals.

A variety of promising signal processing techniques have been developed for the detection and classification of transient signals. Mellinger and Clark (1993) developed a spectrogram correlator filter which compares the spectrogram of a transient signal with a stored spectrographic model, and this had an excellent error rate of only 3.6 % when classifying the endnote from songs of bowhead whales (Potter et al. 1994). Potter et al. (1994) also used *neural net* processing to classify calls of bowhead whales, which decreased the error rate to only 22 out of 1475 calls for a rate of 1.5 %. The neural net has two advantages over the linear detector. The model for the spectrogram filter was synthesized by a human, while the neural net is able to calculate a better spectrogram model using a subsample of the data set for training. The neural net is also better able to reject interfering calls which share features with the desired call.

Another approach for classifying sounds is to make an explicit model of which acoustic features are most salient to an animal. For example, most analyses of signature whistles focus upon specific features such as frequency changes in the fundamental frequency of the whistle rather than upon the entire spectrogram. While the frequencies of spontaneous whistles are relatively stable, the timing of whistles changes by up to 30 % or more (Tyack 1986b). Buck and Tyack (1993) developed a method to compare similarity in frequency contour of whistles while allowing timing to vary by a factor of up to 2. This kind of similarity index makes explicit assumptions that the pattern of frequency change of the fundamental is a critical feature for whistles. These assumptions can be tested by comparing the goodness of fit of the index with responses to naturalistic playbacks or psychophysical tests of how dolphins perceive the similarity of different signals.

All of the methods described above still rely upon the human listener to define what constitutes the unit of analysis for a signal. These decisions can be arbitrary, yet are very important for later conclusions. For example, when people compare whale song with bird song, they are often amazed that whale song can last for over ten minutes while most bird songs last 1 s or so. At first glance, this would seem to indicate that whale song is more complex than bird song. However, the definition of song may differ between whales and birds. When the wren sings AAABBB, we say it is repeating the same song as it moves through its song repertoire. When the whale sings AAABBB, we say it is repeating phrases from theme to theme as it completes its song. I would argue that the appropriate comparison is either between the bird song and the humpback phrase or between the bird's song repertoire and the humpback song. Many birds sing a structured series of songs that may last many minutes. Song bouts in whales and birds can last up to hours, so there may not be a great difference in the durations of the two displays.

The comparison of whale phrases with bird song raises questions about how an outside observer can parse a sequence of animal vocalizations into the appropriate units. An example from human communication will illustrate the point that one cannot analyze sequences of vocalizations without making decisions about the proper signal unit. A critical feature of human speech is the way it is chunked into different levels of organization, from phonemes to words and sentences. While we humans effortlessly parse phonemes in speech, the acoustic boundaries between phonemes are not obvious. In spoken speech, the features of one phoneme may depend upon and leak into the next phoneme, even if these phonemes are from two different words (e.g.,Öhman 1966; see also discussion by Rubin and Vatikiotis-Bateson, this Volume). It has proven very difficult to develop schemes to detect phonemes and words from spectrograms of the speech signal. A major focus of linguistics has been to study the rules by which phonemes and words may be combined to make meaningful utterances. There have been attempts to define this kind of syntactic organization in animal communication, but these usually assume that whole utterances form the proper parallel with the phoneme. We do not really know whether or how animals "chunk" their vocal output.

There is also an important statistical issue concerning the unit of analysis of animal signals. Many statistical analyses of communication signals treat each signal as an independent event. When signals occur in a series, they may be serially dependent. This means that one signal may follow another more often than expected by chance. While this kind of serial correlation can inform us about the structure of signals and interactions, the serial dependence of observations can inflate the sample size for some statistical analyses (Kramer and Schmidhammer 1992).

Both humans and many animals detect graded signals categorically (Harnad 1987), but there is no guarantee that we will draw the perceptual boundaries in the same place as our study animals. Even close relatives of Japanese macaques find it difficult to categorize Japanese macaque calls which are easily discriminated by the species which makes the call (Zoloth et al. 1979). If such close relatives differ in how they process species-specific vocalizations, how much less likely are humans to match the appropriate patterns of distantly related animals living in completely different habitats, such as under the sea? Clearly any human- or computer-generated categorization of vocalizations will need to be validated by testing with the species producing the calls.

In addition, variation in the developmental plasticity of vocalizations will also affect how we interpret the vocal repertoire of a species or an individual. For example, when ethologists start to study a species, they traditionally construct a catalog of species-specific displays. Most research in marine invertebrates or fish suggests that each individual does inherit species-specific display patterns with a structure that is not modifiable through experience to any significant degree. However, we have just seen that marine mammals may learn vocalizations used for individually distinctive or group-specific vocal repertoires. They may even add new vocalizations to their repertoire throughout their lifetime, as in the imitated signature whistles, which raises questions about the applicability of the model of a fixed vocal repertoire.

5.5
Acoustic Localization and Telemetry Open New Windows on Communication Among Marine Animals

A primary obstacle to progress in the study of social communication in marine animals has been the difficulty of identifying which animal within an interacting group produces a sound underwater. Biologists who study terrestrial animals take it for granted that they can identify which animal is vocalizing. They can use their own ears to locate the source of a sound and to direct their gaze. Most terrestrial animals produce a visible motion associated with the coupling of sound energy to the air medium. Mammals and birds open their mouths when vocalizing, many insects produce visible stridulation motions, and frogs inflate their throat sacs. Once their gaze has been directed to the sound source, terrestrial biologists can correlate movements associated with sound production with the sound they hear to confirm which animal is vocalizing. The simplicity of this process should not obscure how important it is for research on communication. Without this ability, researchers can scarcely begin to tease apart the pattern of signal and response that inform us about a system of animal communication.

Humans are not able to locate sounds underwater in the same way they locate airborne sounds. Furthermore, while many fish and invertebrates produce visible motions coordinated with sound production, whales and dolphins seldom do. Some dolphin sounds are coordinated with a visible display, like the so, called "jaw clap" which occurs at the same time as an open mouth display. Dolphins also occasionally emit a stream of bubbles while vocalizing. But these special cases may bias the sample and are not common enough to allow systematic analysis. The need for some technique to identify which cetacean produces which sound during normal social interaction has been discussed for over three decades. The following three different approaches have emerged:

1. An electro-acoustic link between animals isolated in two tanks
2. Passive acoustic location of sound sources using an array of hydrophones
3. Telemetry of information about sound production from each animal in a group

Several investigators have attempted to study communication between isolated captive dolphins using an electronic acoustic link between two pools (Lang and Smith 1965; Burdin et al. 1975; Gish 1979). However, this approach has several serious drawbacks.

The electronic reproduction of dolphin sounds may be discriminably different from natural sound, and it is next to impossible to control for this problem in electronic acoustic link experiments. Even if dolphins accept the acoustic quality of the link, the sounds emanate from an underwater sound projector rather than from another dolphin. As soon as an animal approaches the source and can inspect the projector, it is likely to respond differently than if the source was another dolphin. Furthermore, in order to study the social functions of vocalizations, one would hope to study what roles they play in social interactions. However, the isolated dolphins are able to interact only acoustically.

The second technique, acoustic location of vocalizing animals, is a promising method for identifying which animal is producing a sound. It involves no manipulation of the animals, just placement of hydrophones near them. In some applications, animals may vocalize frequently enough and be sufficiently separated that source location data may suffice to indicate which animal produces a sound. Tracks of continuously vocalizing finback whales, *Balaenoptera physalus*, were obtained in the early 1960s using bottom mounted hydrophones made available to geologists by the US Navy (Patterson and Hamilton 1964; an example of data from these hydrophones is shown in Figure 2). Hydrophones placed on the sea floor to record seismic activity have also been used more recently to track blue and finback whales (McDonald et al. 1995).

The US Navy has devoted considerable resources to using bottom mounted hydrophones in order to locate ships and to track them. These sophisticated systems have recently been used to locate and track whales over long ranges, including one whale tracked for >1700 km over 43 days (Figure 10; Costa 1993). These arrays have proven capable of detecting whales at ranges of hundreds of kilometers, as was predicted by the acoustic models described in Payne and Webb (1971) and Spiesberger and Fristrup (1990). For biologists used to digesting the trickle of information available on the vocalizations of baleen whales, data from these arrays gushes forth as if from a fire hydrant. Thousands of vocalizations may be detected per day, some of them only vaguely similar to sounds identified to species in the earlier literature. One set of problems arises from dealing with this volume of data, but problems just as serious remain with keeping the analysis rooted in basic biological questions. Early studies of whale vocalizations were plagued by the potential confusion that a sound recorded in the presence of one species might be produced by an animal of another species just over the horizon or under the water. Ground truthing of species identification is critical for much of these data, and questions will remain about new sounds not compared with a visual record, for one must compare the source location with a visual record of the animal's locations in order to identify which animal produces a sound.

Several different approaches have been used to locate the sounds of animals over shorter ranges from locations where one could look for the vocalizing animal(s). Watkins and Schevill (1972) devised a four hydrophone array that could rapidly be deployed from a ship. This array has been used to locate vocalizing finback whales, right whales (*Eubalaena glacialis*), sperm whales (*Physeter macrocephalus*), and several species of dolphins. This array is not rigid, and the hydrophone locations must be calibrated using underwater noise sources called pingers. When the array was first used near sperm whales, the pingers were low enough in frequency to be detected by the whales, which immediately stopped vocalizing, reducing the value of the array (Watkins and Schevill

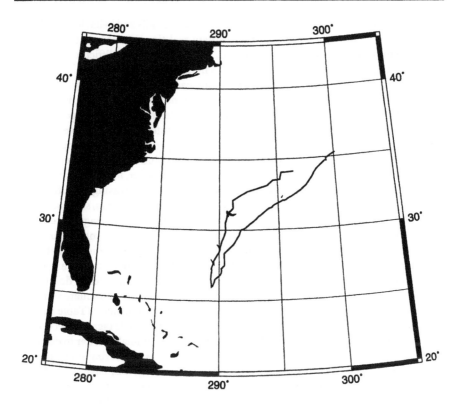

Fig. 10. Track of vocalizing blue whale, *Balaenoptera musculus*, made using the US Navy's Sound Surveillance System (SOSUS) arrays. (Courtesy of Christopher Clark, Laboratory of Ornithology, Cornell University)

1975). Since that time, the array has been deployed with higher frequency pingers and has provided important data on the coda repertoires of individual sperm whales and coda exchanges between different sperm whales (Watkins and Schevill 1977). The array has not been so useful for identifying which dolphin produces a sound (Watkins and Schevill 1974). These animals move rapidly and often swim so close to one another that source locations are not enough by themselves to identify which animal produces a sound. Locations are only calculated back in the laboratory, and it has proved difficult to record the location of different individuals in sufficient detail to allow correlation with the acoustic source locations. Freitag and Tyack (1993) present a different approach to locating dolphin sounds, but in this application it was also difficult to relate acoustic locations to visual records of animal locations. Clark (1980) developed a device that measures the phase difference for a vocalization of a southern right whale (*Eubalaena australis*) arriving at different hydrophones in a relatively small rigid array. This device gives source bearings in real-time and was deployed just off a coastal observation site where it was used successfully to correlate right whale sounds and behavior. Clark et al. (1986) have used a computer in the field to obtain real-time source locations of vocalizing bowhead whales (*Balaena mysticetus*), but it also proved difficult to associate visual sightings with acoustic source locations in this study.

The third technique does not require locating each animal within a group. If each animal carries a telemetry device which records or transmits acoustic data while on the animal, then the repertoires of each individual can be determined. The biggest problem for telemetry has been deciding how to transmit the information effectively without disturbing the animals. Evans and Sutherland (1963) proposed the development of a radio telemetry device to broadcast sounds from a dolphin's head. This technique is limited to applications where the antenna remains above water, for the conductivity of seawater limits electromagnetic transmission, with losses increasing with frequency (Mackay 1968). The use of sound for underwater telemetry is common, since sound has favorable propagation characteristics underwater. Sonic telemetry tags have proved extremely valuable for tracking fish, seals, and some whales (e.g., Amlaner and MacDonald 1980). However, there are serious limitations to the use of sonic telemetry with dolphins and possibly other cetaceans. If the sounds of the telemetry device are audible to the animal, they may interfere with normal behavior. Many species of dolphin can hear frequencies as high as 150 kHz. Frequencies higher than this attenuate so rapidly in sea-water that they are not particularly effective for sonic telemetry (Table 1). These problems with telemetry have led biologists to develop recoverable tags that record data while on an animal, but that need to be recovered from the animal in order to download the data. Recently, biologists have had successful programs recovering such tags from seals, porpoises, and baleen whales. Recoverable acoustic tags may have scientific uses well beyond identifying vocalizations. Figure 11 shows acoustic and dive data sampled from an elephant seal as it swam in Monterey Bay. The tag was able to monitor both the acoustic stimuli heard by the seal and the acoustic signatures of the seal's breathing and heart beat.

6
Acoustic Impact of Human Activities on Marine Animals

Since marine mammals and humans both produce and use sound underwater, there is potential competition for channel space (Green et al. 1994). Transient sounds of biological origin may dominate the noise for manmade sonars in some regions (Urick 1983). Taking the animals' point of view, Payne and Webb (1971) have raised concerns about the potential masking effects of shipping noise on whale communication. Sound generated by commercial shipping is the dominant source of underwater noise over the entire globe in the frequency range from about 20 to 500 Hz, elevating the average ambient by tens of dB (Figure 12; Urick 1983). Judging by the similar frequency range of vocalizations of baleen whales, baleen whales are likely to be sensitive at these frequencies. This is also the primary range of hearing in most marine fish (Figure 5). Myrberg (1980) points out that noise from rough seas or shipping traffic may reduce the effective range of the courtship chirps of bicolor damselfish from 9 m at sea state 1 with light shipping traffic to only 1 m when shipping traffic is heavy. This carries the potential

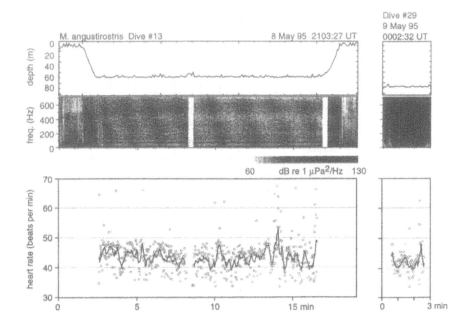

Fig. 11. Data recorded from tag deployed on an elephant seal. *Mirounga angustirostris*, as it swam across Monterey Bay, California. *Top* Depth of seals dive measured with a pressure sensor. *Middle* Spectrogram of noise sensed from a hydrophone on tag during this dive. Darker gray represents higher noise levels as indicated on the horizontal gray scale bar; breathing noises made by seal at surface (*dark areas below 200 Hz at start of record*); flow noise from rapid swimming as seal starts its descent (*broader frequency band of dark gray*). As soon as seal reaches bottom of its dive at nearly 60 m, flow noise is greatly reduced; noise from ship passing overhead of seal (*middle curved bands of light and dark gray*); data gaps as tag was writing data from memory to a disk drive (two *narrow white vertical lines*). Acoustic record from seal was also able to detect low frequency sounds of its heart beat. *Bottom* Estimates of heart rate of diving seal made from detecting this acoustic signature of heart beat. (Courtesy of William Burgess, Monterey Bay Aquarium Research Institute)

for disrupting reproductive activity. Environmental noise has been shown not only to mask communication signals, but also to cause physiological stress responses, hearing loss, and interference with normal activities in humans and other terrestrial animals.

Little attention has focussed upon the impact of noise on marine invertebrates or fish, but some research has focused upon endangered species and marine mammals because of the greater legal protection they enjoy under United States law. For example, since lease of offshore areas for oil exploration and production requires environmental impact statements in the USA, the federal Minerals Management Service has funded a number of studies attempting to quantify the effects of noise that results from oil industry activities. More is known about whale responses to these stimuli than to any other human noise source. For example, one sound playback study involved observation of > 3500 migrating gray whales, *Eschrichtius robustus*, and showed statistically significant responses to playback of industrial noise (Malme et al. 1984). These whales were tracked from elevated shore observation sites using the same theodolite method described for singing humpbacks. This is one of few ways to observe whales without

AVERAGE DEEP-SEA NOISE

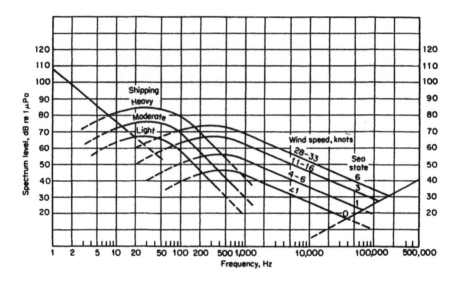

Fig.12. Average deep-sea noise levels. Noise below about 20 Hz primarily stems from geological activity. Shipping noise dominates the ambient noise from 20 to 200 Hz. Above this frequency, noise stems primarily from wind and waves. (from Urick 1983, used with permission)

having to worry about whether the observation platform might be altering the animals' behavior. Whales slowed down and started altering their course away from the sound source at ranges of 1–3 km, where the estimated level of the playback source was near the local ambient noise level. This resulted in an avoidance reaction — an increase in the distance between the whales and the source at their closest point of approach (Figure 13). Half of the migrating gray whales avoided exposure to the continuous playback stimuli at received levels $\geq$ 117–123 dB re 1 µPa. The 50 % avoidance level for airgun pulses used in seismic surveys was 170 dB. The ranges predicted for a 50 % probability of avoidance were 1.1 km for one of the continuous playback stimuli, a drillship, and 2.5 km for the seismic array.

There is little reason to expect that oil industry activities are more likely to harm whales than other activities. The acoustic impacts of the oil industry upon whales have been highlighted primarily for political reasons, such as federal government involvement in leasing offshore tracts to the oil industry. Other human activities, such as ships, underwater explosions, military sonar, and acoustic tomography may generate enough noise to have similar potential impacts. These impacts may be particularly important if they inhibit the recovery of marine mammal populations which were depleted in the past by human hunting or whaling. Humans may have to be more careful now than in the past to avoid polluting the habitat of wild animals by noise as well as chemical contaminants. Unfortunately, we know very little about what effect noise has on the hearing and behavior of most marine species. Even for well-studied species, it is unwise to extrapolate to contexts beyond those studied. For example, the migrating gray whales were watched as they were exposed to a novel and relatively faint source. They might not show the same avoidance response to a more distant and louder source after they habituated to repeated exposure.

In order to protect humans against exposure to noise in the workplace, the US government pays little attention to annoyance or behavioral disturbance, but instead focuses on preventing hearing loss (EPA 1974). Safe levels of noise are determined in experiments where the hearing of a subject is tested before and after exposure to a well-defined noise source. When the level and duration of the noise exposure is just enough to yield a detectable decrement in hearing ability, this is called a *temporary threshold shift* (TTS) for hearing. While this TTS does no harm in itself, it is assumed that chronic repeated exposure to levels that cause TTS are likely to lead to permanent hearing loss. Therefore, regulations limit exposure to sound beyond the levels and durations that cause TTS. So many marine animals depend so heavily on sound that there is reason to be concerned that permanent hearing loss might affect biologically significant activities. Ideally one might want to define standards for exposure to underwater noise in order to prevent a negative impact, particularly on the recovery of endangered marine animal populations. Unfortunately there is not one marine animal for which we can specify that a specific noise exposure leads to hearing loss. Furthermore, noise might disrupt biologically significant activities of animals at levels lower than those required to cause hearing loss. Until both hearing loss and behavioral disruption are better defined, it will not be possible to regulate noise exposure in a way that protects marine animals while minimizing the burden on human seagoing activities. Marine bioacoustics thus clearly is not just a fascinating area of basic scientific research, but also is central to protecting our marine environment from habitat degradation by noise pollution.

Playback Control

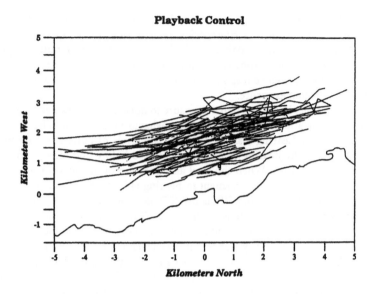

Production Platform Playback

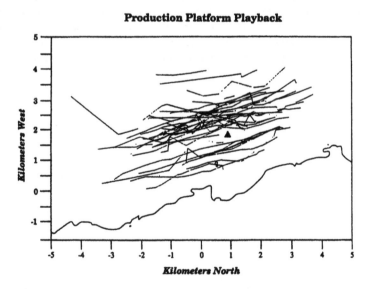

Fig. 13. Plot of tracks of migrating gray whales during control observations (*top*) and during playback of the noise of an offshore oil industry production platform (*bottom*). The *black triangle* at approximately 1 km north and 2 km west marks the location of the playback vessel. (Adapted from Malme et al. 1984, with permission)

References

Altmann J (1974) Observational study of behavior: sampling methods. Behaviour 49: 227–265

Amlaner CJ, MacDonald DW (1980) A handbook on biotelemetry and radio tracking: proceedings of an international conference on telemetry and radio tracking in biology and medicine. Pergamon, Oxford

Andersen S (1970) Auditory sensitivity of the harbour porpoise, *Phocoena phocoena*. In: Pilleri G (ed) Investigations on Cetacea, vol 3. Benteli, Berne, pp 255–259

Aroyan JL (1996) Three dimensional numerical simulation of biosonar signal emission and reception in the common dolphin. PhD Thesis, University of California at Santa Cruz, Santa Cruz

Astrup J, Møhl B (1993) Detection of intense ultrasound by the cod *Gadus morhua*. J exp Biol 182: 71–80

Au WWL (1980) Echolocation signals of the Atlantic bottlenose dolphin (*Tursiops truncatus*) in open waters. In: Busnel R-G, Fish JF (eds) Animal sonar systems. Plenum, New York, p 251

Au WWL (1993) The sonar of dolphins. Springer Verlag, Berlin Heidelberg New York

Au WWL, Moore PWB (1984) Receiving beam patterns and directivity indices of the Atlantic bottlenose dolphin *Tursiops truncatus*. J Acoust Soc Am 75: 255–262

Au WWL, Moore PWB (1990) Critical ratio and critical bandwidth for the Atlantic bottlenose dolphin. J Acoust Soc Am 88: 1635–1638

Au WWL, Floyd RW, Haun JE (1978) Propagation of Atlantic bottlenose dolphin echolocation signals. J Acoust Soc Am 64: 411–422

Au WWL, Carder DA, Penner RH, Scronce BL (1985) Demonstration of adaptation in beluga whale echolocation signals. J Acoust Soc Am 77: 726–730

Au WWL, Pawloski JL, Nachtigall PE, Blonz M, Gisiner RC (1995) Echolocation signals and transmission beam pattern of a false killer whale (*Pseudorca crassidens*). J Acoust Soc Am 98: 51–59

Backus R, Schevill WE (1962) Physeter clicks. In: Norris KS (ed) Whales, dolphins, and porpoises. University of California Press, Berkeley, p 510

Baerends GP, Baerends-van Roon JM (1950) An introduction to the study of the ethology of cichlid fishes. Behaviour, (Suppl) 1: 1–243

Baker CS, Lambertsen RH, Weinrich MT, Calambokidis J, Early G, O'Brien SJ (1991) Molecular genetic identification of the sex of humpback whales (*Megaptera novaeangliae*). Rep Int Whaling Comm, Spec Issue 13, IWC, Cambridge, p 105

Batzler WE, Pickwell GV (1970) Resonant acoustic scattering from gas-bladder fish. In: Farquhar GB (ed) Proc Int Symp on Biological sound scattering in the ocean. Govt Printing Office, Washington, DC

Best PB (1979) Social organization in sperm whales, *Physeter macrocephalus*. In: Winn HE, Olla BL (eds) Behavior of marine animals, vol. 3. Cetaceans. Plenum, New York

Bigg MA, Ellis GM, Ford JKB, Balcomb KC (1987) Killer whales – a study of their identification, genealogy and natural history in British Columbia and Washington State., Phantom Press, Nanaimo

Blaustein AR, Waldman B (1992) Kin recognition in anuran amphibians. Anim Behav 44: 207–221

Buck J, Tyack PL (1993) A quantitative measure of similarity for *Tursiops truncatus* signature whistles. J Acoust Soc Am 94: 2497–2506

Bullock TH, Ridgway SH (1972) Evoked potentials in the central auditory systems of alert porpoises to their own and artificial sounds. J Neurobiol 3: 79–99

Burdin VI, Reznik AM, Skornyakov VM, Chupakov AG (1975) Communication signals of the Black Sea bottlenose dolphin. Sov Phys Acoust 20: 314–318

Caldwell MC, Caldwell DK (1965) Individualized whistle contours in bottlenosed dolphins (*Tursiops truncatus*). Science 207: 434–435

Caldwell MC, Caldwell DK (1972) Vocal mimicry in the whistle mode by an Atlantic bottlenosed dolphin. Cetology 9: 1–8

Caldwell MC, Caldwell DK, Tyack PL (1990) A review of the signature whistle hypothesis for the Atlantic bottlenose dolphin, *Tursiops truncatus*. In: Leatherwood S, Reeves R (eds) The bottlenose dolphin: recent progress in research. Academic Press, San Diego, p 199

Chittleborough RG (1954) Studies on the ovaries of the humpback whale, *Megaptera nodosa* (Bonnaterre) on the western Australian coast. Aust J Mar Freshwater Res 5: 35–63

Clark CW (1980) A real-time direction finding device for determining the bearing to the underwater sounds of southern right whales, (*Eubalaena australis*). J Acoust Soc Am 68: 508–511

Clark CW (1982) The acoustic repertoire of the southern right whale, a quantitative analysis. Anim Behav 30: 1060–1071

Clark CW, Ellison WT, Beeman K (1986) Acoustic tracking of migrating bowhead whales. Proc Oceans '86, 23–25 Sept 1986, Washington DC, p 341

Clarke GL, Denton EJ (1962) Light and animal life. In: Hill MN (ed) The sea vol. 1, Interscience, New York, p 456

Clay CS, Medwin H (1977) Acoustical oceanography. Wiley, New York

Clutton-Brock TH, Albon SD (1979) The roaring of red deer and the evolution of honest advertisement. Behaviour 69: 145–169

Connor RC, Smolker RA, Richards AF (1992) Aggressive herding of females by coalitions of male bottlenose dolphins (*Tursiops* sp.). In: Harcourt AH, de Waal FBM (eds) Coalitions and alliances in humans and other animals. Oxford University Press, Oxford, p 415

Coombs S, Popper AN (1979) Hearing differences among Hawaiian squirrelfish (family Holocentridae) related to differences in the peripheral auditory system. J Comp Physiol A 132: 203–207

Costa DP (1993) The secret life of marine mammals. Oceanography 6: 120–128

Cranford TW (1992) Functional morphology of the odontocete forehead: implications for sound generation. PhD Thesis, University of California at Santa Cruz, Santa Cruz

Darwin C (1871) The descent of man and selection in relation to sex. J Murray, London

Davies NB, Halliday TR (1978) Deep croaks and fighting assessment in toads, *Bufo bufo*. Nature (Lond) 391: 56–58

Dunning DJ, Ross QE, Geoghegan P, Reichle JJ, Menezes JK, Watson JK (1992) Alewives avoid high-frequency sound. North Am J Fish Manage 12: 407–416

Eaton RL (1979) A beluga whale imitates human speech. Carnivore 2: 22–23

EPA (1974) Information on levels of environmental noise requisite to protect public health and welfare with an adequate margin of safety. National Technical Information Service, PB–239 429

Evans WE (1967) Vocalization among marine mammals. In: Tavolga WN (ed) Marine bioacoustics, vol. 2. Pergamon, Oxford, p 159

Evans WE, Sutherland WW (1963) Potential for telemetry in studies of aquatic animal communication. In: Slater LE (ed) Bio-telemetry. Pergamon, Oxford, p 217

Everest FA, Young RW, Johnson MW (1948) Acoustical characteristics of noise produced by snapping shrimp. J Acoust Soc Am 20: 137–142

Ewing M, Worzel JL (1948) Long-range sound transmission. Geol Soc Am Mem 27

Fay FH, Ray GC, Kibal'chich AA (1981) Time and location of mating and associated behavior of the Pacific walrus, *Odobenus rosmarus divergens* Illiger. In: Fay FH, Fedoseev GA (eds) Soviet-American cooperative research on marine mammals. NOAA Tech Rep NMFS Circ 12, vol 1. Pinnipeds, Washington, DC, p 89

Fay RR, Feng AS (1987) Mechanisms for directional hearing among nonmammalian vertebrates. In: Yost WA, Gourevitch G (eds) Directional hearing. Springer, Berlin Heidelberg New York p 179

Fay RR (1988) Hearing in vertebrates: a psychophysics databook. Hill-Fay, Winnetka, Illinois

Fine ML (1978) Seasonal and geographical variation of the mating call of the oyster toadfish *Opsanus tau*. Oecologia (Berl) 36: 45–47

Fine ML (1981) Mismatch between sound production and hearing in the oyster toadfish. In: Tavolga WN, Popper AN, Fay RR (eds) Hearing and sound communication in fishes. Springer, Berlin, Heidelberg, New York, p 257

Fish JF (1972) The effect of sound playback on the toadfish. In: Winn HE, Olla BL (eds) Behavior of marine animals, vol. 2. Vertebrates. Plenum, New York, p 386

Fish JF, Offut GC (1972) Hearing thresholds from toadfish, *Opsanus tau*, measured in the laboratory and field. J Acoust Soc Am 51: 1318–1321

Fisher RA (1958) The genetical theory of natural selection. Dover, New York

Ford JKB (1989) Acoustic behavior of resident killer whales (*Orcinus orca*) off Vancouver Island, British Columbia. Can J Zool 67: 727–745

Ford JKB (1991) Vocal traditions among resident killer whales (*Orcinus orca*) in coastal waters of British Columbia. Can J Zool 69: 1454–1483

Frankel AS, Clark CW, Herman LM, Gabriele CM, Hoffhines MA, Freeman TR, Patterson BK (1989) Acoustic location and tracking of wintering humpback whales (*Megaptera novaeangliae*) off South Kohala, Hawaii. Proc 8th Biennial conf on the Biology of marine mammals, Pacific Grove, California

Frankel AS, Clark CW, Herman LM, Gabriele CM, Hoffhines MA, Freeman TR (1991) Spacing function of humpback whale song. Proc 9nth Biennial conf on the Biology of marine mammals Chicago, Illinois

Freitag L, Tyack PL (1993) Passive acoustic localization of the Atlantic bottlenose dolphin using whistles and clicks. J Acoust Soc Am 93: 2197–2205

Frisch K von (1938) The sense of hearing in fish. Nature 141: 8–11

Gish SL (1979) A quantitative description of two–way acoustic communication between captive Atlantic bottlenosed dolphins (*Tursiops truncatus* Montagu). PhD Thesis, University of California at Santa Cruz, Santa Cruz

Glockner DA (1983) Determining the sex of humpback whales (*Megaptera novaeangliae*) in their natural environment. In: Payne RS (ed) Communication and behavior of whales. AAAS Sel Symp. Westview Press, Boulder, p 447

Goodson AD, Kastelein RA, Sturtivant CR (1995) Source levels and echolocation signal characteristics of juvenile harbour porpoises (*Phocoena phocoena*) in a pool. In: Nachtigall PE, Lien J, Au WWL, Read AJ (eds) Harbour porpoises – laboratory studies to reduce bycatch. De Spil, Woerden, p 41

Goold JC, Jones SE (1995) Time and frequency domain characteristics of sperm whale clicks. J Acoust Soc Am 98: 1279–1291

Gray GA, Winn HE (1961) Reproductive ecology and sound production of the toadfish, *Opsanus tau*. Ecology 42: 274–282

Green DM, DeFerrari HA, McFadden D, Pearse JS, Popper AN, Richardson WJ, Ridgway SH, Tyack PL (1994) Low-frequency sound and marine mammals: current knowledge and research needs. National Academy Press, Washington DC

Harnad S (1987) Categorical perception. Cambridge University Press, Cambridge

Hawkins AD (1981) The hearing abilities of fish. In: Tavolga WN, Popper AN, Fay RR (eds) Hearing and sound communication in fishes. Springer, Berlin Heidelberg New York, p 109

Hawkins AD (1993) Underwater sound and fish behavior. In: Pitcher TJ (ed) Behavior of teleost fishes. Fish and Fisheries Series 7. Chapman and Hall, London, p 129

Hawkins AD, Myrberg AA Jr (1983) Hearing and sound communication under water. In: Lewis B (ed) Bioacoustics: a comparative approach. Academic Press, New York, p 347

Hazlett BA, Winn HE (1962) Sound production and associated behavior of Bermuda crustaceans (*Panulirus, Gonodactylus, Alpheus*, and *Synalpheus*). Crustaceana (Leiden) 4: 25–38

Helweg DA, Roitblat HL, Nachtigall PE, Hautus MJ (1996) Recognition of aspect-dependent three-dimensional objects by an echolocating Atlantic bottlenose dolphin. J Exp Psychol Anim Behav Processes 22(1): 19–31

Herman LM (1980) Cognitive characteristics of dolphins. In: Herman LM (ed) Cetacean behavior: mechanisms and functions. Wiley-Interscience, New York, p 363

Jacobs DW, Tavolga WN (1967) Acoustic intensity limens in the goldfish. Anim Behav 15: 324–335

Johnson CS (1966) Auditory thresholds of the bottlenosed porpoise (*Tursiops truncatus* Montagu). US Naval Ordnance Test Station, Tech Publ 4178: 1–28

Kamminga C (1988) Echolocation signal types of odontocetes. In: Nachtigall PE, Moore PWB (eds) Animal sonar: processes and performance. Plenum, New York, p 9

Kamminga C, van der Ree AF (1976) Discrimination of solid and hollow spheres by *Tursiops truncatus* (Montagu). Aquat Mammals 4: 1–9

Kamminga C, Wiersma H (1981) Investigations on cetacean sonar II. Acoustical similarities and differences in odontocete sonar signals. Aquat Mammals 8: 41–62

Kasuya T, Marsh H (1984) Life history and reproductive biology of the short-finned pilot whale, *Globicephala macrorhynchus*, off the Pacific coast of Japan. Rep int Whal Comm Spec Issue 6: 259–310

Kenyon TN (1994) The significance of sound interception to males of the bicolor damselfish, *Pomacentrus partitus*, during courtship. Environ Biol Fishes 40: 391–405

Ketten DR (1994) Functional analyses of whale ears: adaptations for underwater hearing. IEEE Proc on Underwater acoustics 1. Brest, France, pp 264–270

Kramer MW, Schmidhammer J (1992) The chi-squared statistic in ethology: use and misuse. Anim Behav 44: 833–841

Krebs JR, Dawkins R (1984) Animal signals: mind reading and manipulation. In: Krebs JR, Davies NB (eds) Behavioural ecology: an evolutionary approach, 2nd edn. Blackwell, Oxford, p 380

Lang TG, Smith HAP (1965) Communication between dolphins in separate tanks by way of an electronic acoustic link. Science 150: 1839–1844

Ljungblad DK, Thompson PO, Moore SE (1982) Underwater sounds recorded from migrating bowhead whales, *Balaena mysticetus*, in 1979. J Acoust Soc Am 71: 477–482

MacGinitie GE, MacGinitie N (1949) Natural history of marine animals. McGraw-Hill, New York

Mackay RS (1968) Bio-medical telemetry. Wiley, New York, p 359

Malme CI, Miles PR, Clark CW, Tyack P, Bird JE (1984) Investigations of the potential effects of underwater noise from petroleum industry activities on migrating gray whale behavior. Phase II: January 1984 migration. Bolt Beranek and Newman Report No 5586 submitted to Minerals Management Service, US Dept of the Interior, Washington

McBride AF, Kritzler H (1951) Observations on pregnancy, parturition, and postnatal behavior in the bottlenose dolphin. J Mammal 32: 251–266

McDonald MA, Hildebrand JA, Webb SC (1995) Blue and fin whales observed on a seafloor array in the northeast Pacific. J Acoust Soc Am 98: 712–721

Mellinger DK, Clark CW (1993) Bioacoustic transient detection by image convolution. J Acoust Soc Am 93: 2358

Møhl B, Andersen S (1973) Echolocation: high frequency component in the click of the harbour porpoise (Phocoena phocoena L.). J Acoust Soc Am 54: 1368–1372

Moulton JM (1956) Influencing the calling of sea robins (Prionotus spp.) with sound. Biol Bull 111: 393–398

Moulton JM (1957) Sound production in the spiny lobster Panulirus argus (Latreille). Biol Bull 113: 286–295

Murchison AE (1980) Detection range and range resolution of echolocating bottlenose porpoise (Tursiops truncatus). In: Busnel R-G, Fish JF (eds) Animal sonar systems. Plenum, New York, p 43

Myrberg AA Jr (1972) Ethology of the bicolor damselfish, Eupomacentrus partitus (Pisces: Pomacentridae): a comparative analysis of laboratory and field behavior. Anim Behav Monogr 5: 197–283

Myrberg AA Jr (1980) Fish bio-acoustics: its relevance to the 'not so silent world.' Environ Biol Fishes 5: 297–304

Myrberg AA Jr (1981) Sound communication and interception in fishes. In: Tavolga WN, Popper AN, Fay RR (eds) Hearing and sound communication in fishes. Springer, Berlin Heidelberg New York, p 395

Myrberg AA Jr, Riggio RJ (1985) Acoustically mediated individual recognition by a coral reef fish (Pomacentrus partitus). Anim Behav 33: 411–416

Myrberg AA Jr, Spires JY (1980) Hearing in damselfishes: an analysis of signal detection among closely related species. J Comp Physiol 140: 135–144

Myrberg AA Jr, Mohler M, Catala JD (1986) Sound production by males of a coral reef fish (Pomacentrus partitus): its significance to females. Anim Behav 34: 913–923

Myrberg AA Jr, Ha SJ, Shamblott MJ (1993) The sounds of bicolor damselfish (Pomacentrus partitus): predictors of body size and a spectral basis for individual recognition and assessment. J Acoust Soc Am 94: 3067–3070

Nelson DA, Marler P (1990) The perception of birdsong and an ecological concept of signal space. In: Stebbins WC, Berkley MA (eds) Comparative perception. vol II. Complex signals. Wiley, New York, p 443

Nestler JM, Ploskey GR, Pickens J, Menezes J, Schilt C (1992) Responses of blueback herring to high-frequency sound and implications for reducing entrainment at hydropower dams. North Am J Fish Manage 12: 667–683

Nolan BA, Salmon M (1970) The behavior and ecology of snapping shrimp (Crustacea: Alpheus heterochelis and Alpheus normanni). Forma Funct 2: 289–335

Norris KS (1968) The evolution of acoustic mechanisms in odontocete cetaceans. In: Drake ET (ed) Evolution and environment. Yale University Press, New Haven, p 297

Offut CG (1968) Auditory response in the goldfish. J Aud Res 8: 391–400

Offut CG (1970) Acoustic stimulus perception by the American lobster, Homarus. Experientia Basel 26: 1276–1278

Öhman S (1966) Coarticulation in VCV utterances: spectrographic measurements. J Acoust Soc Am 39: 151–168

Palsbøll, PJ, Vader A, Bakke I, El-Gewely MR (1992) Determination of gender in cetaceans by the polymerase chain reaction. Can J Zool 70: 2166–2170

Patterson B, Hamilton GR (1964) Repetitive 20 cycle per second biological hydroacoustic signals at Bermuda. In: Tavolga WN (ed) Marine bioacoustics. Pergamon, Oxford, p 125

Payne KB, Payne RS (1985) Large scale changes over 19 years in songs of humpback whales in Bermuda. Z Tierpsychol 68: 89–114

Payne KB, Tyack P, Payne RS (1983) Progressive changes in the songs of humpback whales. In: Payne RS (ed) Communication and behavior of whales. AAAS Sel Symp. Westview Press, Boulder, p 9

Payne RS, Webb D (1971) Orientation by means of long range acoustic signaling in baleen whales. Ann NY Acad Sci 188: 110–141

Popper AN, Platt C (1993) Inner ear and lateral line. In: Evans DH (ed) The physiology of fishes. CRC Press, Boca Raton, p 99

Potter JR, Mellinger DK, Clark CW (1994) Marine mammal call discrimination using artificial neural networks. J Acoust Soc Am 96: 1255–1262

Ralls K (1976) Mammals in which females are larger than males. Q Rev Biol 51: 245–276

Ralls K, Fiorelli P, Gish S (1985) Vocalizations and vocal mimicry in captive harbor seals, *Phoca vitulina*. Can J Zool 63: 1050–1056

Ray C, Watkins WA, Burns JJ (1969) The underwater song of *Erignathus* (bearded seal). Zoologica 54: 79–83+3 plates

Rayleigh Lord (1945) The theory of sound. Dover, New York

Reiss D, McCowan B (1993) Spontaneous vocal mimicry and production by bottlenose dolphins (*Tursiops truncatus*): evidence for vocal learning. J Comp Psychol 107: 301–312

Richards DG, Wolz JP, Herman LM (1984) Vocal mimicry of computer-generated sounds and vocal labeling of objects by a bottlenosed dolphin, *Tursiops truncatus*. J Comp Psychol 98: 10–28

Richardson WJ, Greene CR Jr, Malme CI, Thomson DH (1996) Marine mammals and noise. Academic Press, New York

Ryan MJ, Fox JH, Wilczynski W, Rand AS (1990) Sexual selection for sensory exploitation in the frog *Physalaemus pustulosus*. Nature 343: 66–67

Sayigh LS (1992) Development and functions of signature whistles of free-ranging bottlenose dolphins, *Tursiops truncatus*. PhD Thesis, MIT/WHOI Joint Program, WHOI 92-37, Woods Hole, Massachusetts

Sayigh LS, Tyack PL, Wells RS, Scott MD (1990) Signature whistles of free-ranging bottlenose dolphins, *Tursiops truncatus*: stability and mother-offspring comparisons. Behav Ecol Sociobiol 26: 247–260

Sayigh LS, Tyack PL, Wells RS (1993) Recording underwater sounds of free-ranging bottlenose dolphins while underway in a small boat. Mar Mammal Sci 9: 209-213

Sayigh LS, Tyack PL, Wells RS, Scott MD, Irvine AB (1995) Sex difference in whistle production in free-ranging bottlenose dolphins, *Tursiops truncatus*. Behav Ecol Sociobiol 36: 171–177

Schein H (1977) The role of snapping in *Alpheus heterochaelis* Say, 1918, the big-clawed snapping shrimp. Crustaceana 33: 182–188

Schevill WE, Lawrence B (1949) Underwater listening to the white porpoise (*Delphinapterus leucas*), Science 109: 143–144

Schevill WE, Lawrence B (1950) A phonograph record of the underwater calls of *Delphinapterus leucas*. Woods Hole Oceanogr Inst Reference No 50–1

Schevill WE, Backus RH, Hersey JB (1962) Sound production by marine animals. In: Hill MN (ed) The sea, vol. 1. Interscience, New York, p 540

Schevill WE, Watkins WA, Ray C (1966) Analysis of underwater *Odobenus* calls with remarks on the development and function of the pharyngeal pouches. Zoologica 51: 103–106+5 plates and phonograph disk.

Schmale MC (1981) Sexual selection and reproductive success in males of the bicolor damselfish *Eupomacentrus partitus* (Pisces: Pomacentridae). Anim Behav 29: 1172–1184

Schusterman RJ (1980) Behavioral methodology in echolocation by marine mammals. In: Busnel R-G, Fish JF (eds) Animal sonar systems. Plenum. New York, p 11

Shockley RC, Northrop J, Hansen PG, Hartdegen C (1982) SOFAR propagation paths from Australia to Bermuda: comparisons of signal speed algorithms and experiments. J Acoust Soc Am 71: 51–60

Sigurdson J (1993) Whistles as a communication medium. In: Roitblat HL, Herman LM, Nachtigall P (eds) Language and communication: comparative perspectives. Erlbaum, Hillsdale, New Jersey, p 153

Silber GK (1986) The relationship of social vocalizations to surface behavior and aggression in the Hawaiian humpback whale (*Megaptera novaeangliae*). Can J Zool 64: 2075–2080

Sjare B, Stirling I (1993) The breeding behavior and mating system of walruses. In: Proc 10 Biennial Conf on the Biology of marine mammals, Galveston, Texas, p 10

Smolker RA, Mann J, Smuts BB (1993) Use of signature whistles during separation and reunions by wild bottlenose dolphin mothers and infants. Behav Ecol Sociobiol 33: 393–402

Spanier E (1979) Aspects of species recognition by sound in four species of damselfishes, Genus *Eupomacentrus* (Pisces: Pomacentridae). Tierpsychol 51:301–316

Spiesberger JL, Fristrup KM (1990) Passive localization of calling animals and sensing of their acoustic environment using acoustic tomography. Am Nat 135: 107–153

Stabentheimer A (1988) Correlations between hearing and sound production in piranhas. J Comp Physiol A 162: 67–76

Stanton TK, Chu D, Wiebe PH (1995) Acoustic scattering characteristics of several zooplankton groups. ICES J Mar Sci 53:289– 295

Stebbins WC (1983) The acoustic sense of animals. Harvard University Press, Cambridge

Stirling I (1973) Vocalization in the ringed seal (*Phoca hispida*). J Fish Res Board Can 30: 1592–1594

Surlykke AM (1988) Interactions between echolocating bats and their prey. In: Nachtigall PE, Moore PWB (eds) Animal sonar: processes and performance. Plenum, New York, p 635

Tappert FD (1977) The parabolic approximation method. In: Keller JB, Papadakis JS (eds) Wave propagation and underwater acoustics. Springer, Berlin Heidelberg New York

Tavolga WN (1962) Mechanisms of sound production in the ariid catfishes *Galeichthys* and *Bagre*. Bull Am Mus Nat Hist 124: 1–30

Tavolga WN (1964) Marine bioacoustics, 2 vols. Pergamon, Oxford

Tavolga WN, Popper AN, Fay RR (1981) Hearing and sound communication in fishes. Springer, Berlin Heidelberg New York

Thomas JA, Zinnel KC, Ferm LM (1983) Analysis of Weddell seal (*Leptonychotes weddelli*) vocalizations using underwater playbacks. Can J Zool 61: 1448–1456

Thompson PO, Friedl WA (1982) A long term study of low frequency sounds from several species of whales off Oahu, Hawaii. Cetology 45: 1–19

Thompson RKR, Herman LM (1975) Underwater frequency discrimination in the bottlenose dolphin (1–140 kHz) and the human (1–8 kHz). J Acoust Soc Am 57: 943—948

Tinbergen N (1951) The study of instinct. Oxford University Press, Oxford

Trillmich F (1981) Mutual mother–pup recognition in Galpagos fur seals and sea lions: cues used and functional significance. Behaviour 78: 21–42

Tyack P (1981) Interactions between singing Hawaiian humpback whales and conspecifics nearby. Behav Ecol Sociobiol 8: 105–116

Tyack P (1983) Differential response of humpback whales to playbacks of song or social sounds. Behav Ecol Sociobiol 13: 49–55

Tyack P (1985) An optical telemetry device to identify which dolphin produces a sound. J Acoust Soc Am 78: 1892–1895

Tyack P (1986a) Population biology, social behavior, and communication in whales and dolphins. Trends Ecol Evol 1: 144–150

Tyack P (1986b) Whistle repertoires of two bottlenosed dolphins, *Tursiops truncatus*: mimicry of signature whistles? Behav Ecol Sociobiol 18: 251–257

Tyack P (1993) Why ethology is necessary for the comparative study of language and communication. In: Roitblat H, Herman LM, Nachtigall PE (eds) Language and communication: comparative perspectives. Erlbaum, Hillsdale, New Jersey, p 115

Tyack P, Whitehead H (1983) Male competition in large groups of wintering humpback whales. Behaviour 83: 132–154

Tyack PL, Recchia CA (1991) A datalogger to identify vocalizing dolphins. J Acoust Soc Am 90: 1668–1671

Urick RJ (1983) Principles of underwater sound. McGraw-Hill, New York

USRD (1982) Underwater electroacoustic standard transducers. Standards Section, Transducer Branch, Underwater Sound Reference Department, Naval Res Lab, Orlando

Watkins WA (1967) The harmonic interval: fact or artifact in spectral analysis of pulse trains. In: Tavolga WN (ed) Marine bioacoustics, vol. 2. Pergamon, Oxford, p 15

Watkins WA, Schevill WE (1972) Sound source location by arrival-times on a non-rigid three-dimensional hydrophone array. Deep-Sea Res 19: 691–706

Watkins WA, Schevill WE (1974) Listening to Hawaiian spinner porpoises, *Stenella* cf. *longirostris*, with a three-dimensional hydrophone array. J Mammal 55: 319–328

Watkins WA, Schevill WE (1975) Sperm whales (*Physeter catodon*) react to pingers. Deep-Sea Res 22: 123–129

Watkins WA, Schevill WE (1977) Sperm whale codas. J Acoust Soc Am 62: 1485–1490

Watkins WA, Tyack P, Moore KE, Bird JE (1987) The 20–Hz signals of finback whales (*Balaenoptera physalus*). J Acoust Soc Am 82: 1901–1912

Wells RS, Scott MD, Irvine AB (1987) The social structure of free-ranging bottlenose dolphins. Curr Mammal 1: 247–305

Wenz GM (1964) Curious noises and the sonic environment in the ocean. In: Tavolga WN (ed) Marine bioacoustics. Pergamon, Oxford, p 101

Winberg S, Olsén KH (1992) The influence of rearing conditions on the sibling odour preference of juvenile Arctic charr, *Salvelinus alpinus* L. Anim Behav 44: 157–164

Winn HE (1972) Acoustic discrimination by the toadfish with comments on signal systems. In: Winn HE, Olla BL (eds) Behavior of marine animals, vol. 2: Vertebrates. Plenum, New York, p 361

Winn HE, Marshall JA (1963) Sound producing organ of the squirrelfish, *Holocentrus rufus*. Physiol Zool 36: 34–44

Zoloth SR, Petersen MR, Beecher MD, Green S, Marler P, Moody DB, Stebbins W (1979) Species–specific perceptual processing of vocal sounds by monkeys. Science 204: 870–872

CHAPTER 7

Ultrasound and Infrasound

J. D. PYE AND W. R. LANGBAUER, JR.

1
Introduction

Ultrasound and infrasound differ from "ordinary" sounds in three distinct ways that influence all the considerations of this chapter. The first and most obvious characteristic of these sound types is that, by definition, they are "extreme" frequencies that fall outside the normal response curve for the human ear (see Figure 1) and are therefore inaudible. Ultrasound, which includes biologically significant sounds ranging from 15 kHz or so up to 200 kHz, is too high in frequency. Infrasound, effectively extending downwards from about 20 to 0.1 Hz or less, is too low in frequency. In both cases, therefore, it is necessary to use special instruments merely to detect the signals, which only increases the fascination of studying them. When the appropriate technology is applied, it becomes possible to observe phenomena that may be quite common among nonhuman species, but have previously been unknown.

The second characteristic that sets ultrasound and infrasound apart from other acoustic signals is wavelength. Wavelength is inversely proportional to frequency, and ultrasound therefore has short wavelengths while infrasound has long ones (see Table 1). This particular characteristic is significant because of the way sound waves interact with solid objects, including vocal and auditory organs. Animals can only emit or receive sound directionally if they are larger than, or at least comparable in size with, the wavelengths involved. They must, therefore, exploit these two bands in very different ways. While ultrasound reflects strongly from even small objects and can be highly directional for both large and small animals, infrasound has neither of these characteristics.

The third critical aspect of these signals is that high-frequency sound is strongly absorbed by air, whereas very low frequencies propagate with little loss. Figure 2 shows that attenuation, expressed in dB/m, approximates a function of frequency raised to the power of 1.3. This effect occurs independently of the purely geometric "spreading loss" expressed in the *inverse-square law* of sound attenuation, which applies equally at all frequencies (at least in open, homogeneous media). Thus, ultrasound is effective only over short distances while infrasound can be used for communication over very long ranges. A similar relationship occurs underwater, but there the losses are much less. For a given frequency, attenuation over 200 m of seawater or 2 km of freshwater is comparable with that occurring in 1 m of air (see also Tyack, this Volume).

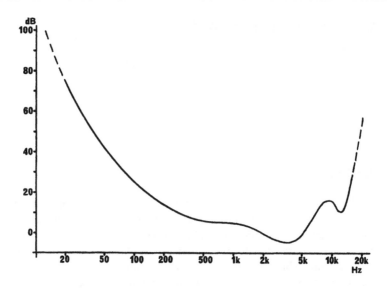

Fig. 1. Sensitivity of a typical human ear, showing the lowest detectable sound level in decibels re 20 µPa as a function of sound frequency. (Compiled from various data sources)

Table 1. Illustrative wavelengths of ultrasound and infrasound signals. Values of 340 m/s and 1500 m/s were used to calculate sound velocity in air and water, respectively

Ultrasound			Infrasound		
Frequency (kHz)	Wavelength (mm)		Frequency (Hz)	Wavelength (m)	
	Air	Water		Air	Water
200	1.7	7.5	100	3.4	15
100	3.4	15	0	17	75
20	17	75	1	340	1500
10	34	150	0.1	3400	15000

2
A Brief Historical Overview

The first suggestion that animals might not share the range of human sensitivity to sounds appears to have been made by William Wollaston in 1820. Wollaston noted that some insect songs are inaudible to certain people and speculated that there might be others that were inaudible to all humans. The matter was raised again by Sir John Lubbock in 1879. He showed experimentally that ants can see some light in the ultraviolet

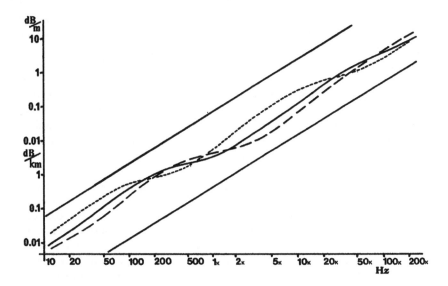

Fig. 2. Attenuation of sound in air in decibels per meter travelled as a function of sound frequency at 20°C and relative humidities of 20% (*dotted line*), 50% (*solid line*), and 100% (*dashed line*). (Drawn from data in Evans and Bass 1972)

range while being blind to red light. He also suspected, but was unable to prove, that ants produce ultrasonic communication signals. Prophetically, Lubbock wrote, "Our range is, however, after all, very limited, and the universe is probably full of music which we cannot perceive...If any apparatus could be devised by which the number of vibrations produced by any given cause could be lowered so as to be brought within the range of our ears, it is probable that the result would be most interesting."

Unfortunately, the technology of Lubbock's day was inadequate for this task. A later, well-founded attempt by Lutz (1924) to build a high-frequency sound detector failed despite help from the Westinghouse Company laboratories, and nearly 60 years more went by before Lubbock's prophesy was proved true. In contrast, Sir Francis Galton was readily able to demonstrate ultrasonic sensitivity in a range of mammals (especially in the cat family) before the turn of the century (Galton 1883). He mounted a tunable, high-pitched "Galton whistle" on the end of a walking stick, using a rubber tube to blow it. Then he went "through the whole of the Zoological Gardens" holding his whistle "as near as is safe to the ears of animals" and observed their reactions to its sound.

The occurrence of biological infrasound was perhaps first, but incorrectly, suggested by Sir Hiram Maxim. He proposed that bats navigate by echolocation, using the 10 to 14 Hz waves emanating from their beating wings (Maxim 1912). It is now clear that these low frequencies do not radiate from such small sources and would not produce clear echoes from significant objects. However, Hartridge suggested in 1920 that bats might instead use echoes of vocal ultrasound, and this ultrasound was first detected by Pierce and Griffin in 1938. Pierce, a physicist, produced the first effective detectors for ultrasound in air and

and also discovered a variety of ultrasonic communication signals in insects, which he later studied for their own sake (Pierce 1948). The pioneering work that finally established echolocation by bats is described in a classic account by Griffin (1958).

The first accurate observations of biological infrasound occurred in the world's oceans in the early 1950s. At that time, pure 20 Hz signals approximately 1 s in duration were detected in the North Atlantic and were soon found in other oceans as well. Initial speculation as to the origin of these sounds included artificial sources (e.g., Russian submarines or other boats) and whale heartbeats. Schevill et al. (1964) eventually linked these sounds to fin whales (*Balaenoptera physalus*) and convincingly argued that they were not due to heartbeats but were produced by some other mechanism, which is still not well understood.

The discovery of these signals was made possible by the rapid advances in electronics that occurred during the Second World War. The ability of these new devices to expand human perception of the world was a marvel to researchers. Describing their recordings of fin whale sounds, Patterson and Hamilton (1964) wrote with a touch of wonder that "It is quite incredible to see these high level signals reproduced on a direct writing oscillograph and yet be unable to hear them on an 18-inch...loudspeaker". Even though the requisite technology has been available for over 30 years, only recently have terrestrial animals been found to produce calls with infrasonic components (e.g., elephants studied by Payne et al. 1986). This slow pace of discovery may be more reflective of our anthropocentric expectations than any paucity of other "infrasound-capable" animals.

3
Sources of Ultrasound and Infrasound

Natural sources of ultrasound in air are few, but include the rustling of leaves, crunching of gravel underfoot, snapping of twigs, and splashing of water. Artificial sources include vehicle brakes and some electronic sensors. Furthermore, because of scattering and atmospheric attenuation, these sounds do not propagate far and their ambient levels are generally low. Overall, the world is rather quiet at ultrasonic frequencies. There are also various accidental ultrasounds of biological origin, such as the whistling of birds' feathers in flight, the fricatives and sibilants of human speech, and nasal "fluting". Functional signals, however, are emitted by a variety of small mammals, by many odontocete cetacea, and by several groups of insects. Ultrasonic sensitivity seems to be of even wider incidence, especially among mammals, but, significantly, not in the great apes or most other primates. A broad review of ultrasonic phenomena in biology is given by Sales and Pye (1974). Comprehensive treatment of echolocation is provided by two symposia, edited by Busnel and Fish (1980), and by Nachtigall and Moore (1988), respectively. These publications contain full bibliographies and in general only more recent references will be listed here.

While there is typically little ambient ultrasound, the opposite is true for infrasound. The world is full of low-frequency sound sources and, since these sounds propagate so well, this richness results in a high background level of infrasound. Some infrasonic phenomena, for example the rush of wind past the ears, are for all intents and purposes just noise. But other sources of infrasound can theoretically be used by animals for orientation (Kreithen 1980, 1983; Sand and Karlsen 1986, Arabadzhi 1990). These include weather phenomena (e.g., thunder), earthquakes, the interaction of wind and topography, highway noise, and ocean waves (see Kreithen 1980, 1983 for excellent reviews). Atlantic cod (*Gadus morrhua*, Sand and Karlsen 1986), cuttlefish (*Sepia officinalis*), octopus (*Octopus vulgaris*), and squid (*Loligo vulgaris*, Packard et al. 1990), pigeons (*Columba livia*, Kreithen and Quine 1979; Schermuly and Klinke 1990), guinea fowl (*Numida meleagris*, Theurich et al. 1984), and elephants (*Elephas maximus*, Heffner and Heffner 1982) can all perceive infrasound, some down to 0.05 Hz! However, how these species can localize sound of such low pitch remains problematical. It is interesting to note that many of the animals either home or migrate over long distances. In addition, a number of species are now known to produce infrasonic signals, presumably for communication. These include fin whales (Schevill et al. 1964), Asian (*Elephas maximus*) and African (*Loxodonta africana*) elephants (Payne et al. 1986; Poole et al. 1988; Langbauer et al. 1991), and the capercaillie, an Eurasian grouse (*Tetrao urogallus*, Moss and Lockie 1979).

4
Echolocation

As stated in Section 2, the echolocation calls of insectivorous bats were among the first ultrasonic signals to be predicted and detected. From the early 1950s on, the history of discovery unfolded approximately as follows (see, for instance, Sales and Pye 1974; Nachtigall and Moore 1988 for reviews). The earliest discoveries were made by Griffin and his collaborators, who were studying species of Vespertilionidae, the only family of bats found in New England. In Germany, Möhres then established that bats in the Old World Rhinolophidae family use a strikingly different form of echolocation. A third mode of echolocation was soon found in the *Rousettus* genus of the Old World Pteropidae family of fruit bats (suborder Megachiroptera). It is not yet clear whether any other pteropids can use similar acoustic guidance, although some, such as *Eonycteris*, roost deep in dark caves. However, it does seem virtually certain that all 18 or so currently recognized families of the other suborder of bats, the Microchiroptera, rely on ultrasonic, vocal echolocation for flight guidance, although probably to varying extents (see Nachtigall and Moore 1988). The only families for which such signals have not been documented through recording and analysis are the Furipteridae and the Thyropteridae. Ultrasonic pulses have, however, been observed in members of these groups as well, using oscilloscopes, detectors, or both.

It is now clear that although there is considerable variety in the structure of bat calls, they fall into the three functional types that were described in the early studies. It is also becoming increasingly evident that many bats can flexibly emit different kinds of signals under various conditions, presumably in order to obtain different information about their surroundings, to optimize signal operation, or to counter the countermeasures employed by their prey. It is therefore no longer possible to assume that the echolocation behavior of any one species is stereotyped, or to suppose that recordings made under only one set of conditions will necessarily typify that species.

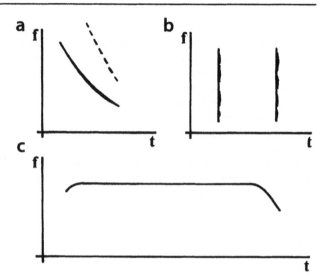

Fig. 3. Three different
frequency patterns used by
different bats for ultrasonic
echolocation. a; *Upper left*
the frequency sweep of a
vespertilionid (the weak
second harmonic is dotted).
b, *Upper right* the double
click of a *Rousettus* fruit bat
c, *Bottom* the long, CF pulse
of a horseshoe bat

The first, and possibly most common type of echolocation pulse in bats is a short, loud, deep-frequency sweep, as produced by most Vespertilionidae species (shown in Figure 3a). These pulses are typically emitted during cruising flight, at a rate of 10 to 15/s that is correlated with the respiratory and wingbeat cycles. During interception of prey, landing, or obstacle-avoidance maneuvers, the pulse rate rises to 100 to 200/s, the pulses shorten and are greatly reduced in intensity, and the frequency of each pulse becomes lower with reduced sweep extent. This vocalizing pattern is often called the "buzz", because of the sound produced by a bat detector in response to the high calling rates involved.

Radar theory suggests that deep frequency sweeps, as used in *chirp* radar and sonar, are ideal for measuring the propagation time of echoes and thus the distance to a target (its range). Bats using similar frequency-modulated calls (*FM bats*) therefore also seem to be operating in the polar coordinates of direction and range. Behavioral discrimination tests, notably by Simmons and his collaborators, have generally confirmed this proposal (see Simmons 1993). However, other studies have produced situations in which bats have either performed badly or apparently achieved results far better than expected. These outcomes present puzzles that remain to be resolved and the associated controversies are summarized by Pollak (1993) and Simmons (1993).

Pulse timing, and thus range estimation, depend on the bandwidth of the emitted signals and a very wide bandwidth must necessarily include some ultrasonic frequencies. Some FM bats produce calls that sweep over a range of 50 kHz or more. Bandwidth can also be increased by the inclusion of harmonics, and other species emit a short sweep that covers a smaller frequency range but includes several strong harmonics. Indeed, it seems probable that pulses of this kind (and perhaps of longer duration) represent the "ancestral and primitive" form of bat calls. This way of increasing bandwidth may have a disadvantage, in that it produces small uncertainties in timing and range that can be avoided by using "advanced", pure, single-component sweeps. It may be significant that most bats that use pulses with multiple harmonics suppress the fundamental frequency of these signals, leaving for instance, only harmonics 2 to 4.

Another way of producing a wide-bandwidth sound is to emit a very brief impulse, or click, which may otherwise be almost structureless. A drawback in this case is that, due to its short duration, the pulse's energy is limited by its peak amplitude. This is the kind of signal emitted by *Rousettus* bats. Individuals in these species do it nonvocally by clicking their tongues to produce a pair of very brief (e.g., 100 μs) impulses with a well-defined interval of 20 to 30 ms between pulses (see Figure 3b). Sound frequencies within each click range from below 10 kHz to over 60 kHz, but, strangely, most measures of hearing in *R. aegyptiacus* suggest that auditory sensitivity in this species peaks rather sharply at around 12 kHz (reviewed by Pye and Pye 1988). In members of *R. amplexicaudatus*, mid-brain evoked potentials can be recorded in response to frequencies that extend to over 50 kHz. Analogous potentials have not been recorded in *R. aegyptiacus*.

The significance of the "double-click" is unknown, although this signal resembles the audible double-clicks used by most species of *Aerodramus* (also known as *Collocalia*), which comprise the majority of echolocating birds. The human tongue can produce a similar "cluck" by a sudden downward movement from the palate and this sound also consists of a click pair with a highly stereotyped interval of similar magnitude. Odontocete cetacea, such as dolphins, also use clicks, which can be of such brief duration that their spectra extend to over 150 kHz. The prey of dolphins are much larger than those of bats, which should obviate the need to produce ultra-high-frequency signals, but wavelengths are over four times greater underwater than in air. The clicks are commonly emitted in pairs and repetition rates may reach several hundred per second in brief bursts. The reason why clicks may be better-suited to the underwater environment of cetaceans, while in air most bats use structured vocal pulses, has been discussed by Zbinden (1985-1986).

Doppler-mode operation is a third method of echolocation. This form is most commonly found in the Rhinolophidae and Hipposideridae bats, but also occurs in some Mormoopidae and perhaps occasionally in other families. It is characterized by pulses of long duration and constant frequency (*CF*), with consequent narrow bandwidths — although brief, terminal FM sweeps are also commonly present (see Figure 3c). Such pulses appear to be poorly suited for measuring the time delay between the original call and its echo, but provide an excellent means of detecting *Doppler shifts* — frequency changes in echoes that occur due to differences in the relative flight velocities and directions of a bat and its target. These changes, of course, represent the first derivative of range and can also be obtained from successive range measurements made at known time intervals. Similarly, integration of velocity yields the change in range over a given time, and the remaining distance can be deduced from other data (such as changes in intensity levels).

That CF bats do indeed exploit the Doppler shift has been most clearly demonstrated by Schnitzler and colleagues, who discovered the phenomenon called *Doppler compensation* (discussed by Nachtigall and Moore 1988). A CF bat demonstrates Doppler compensation when, in response to an approaching target, it lowers the pitch of its pulses so that the echoes return (after an upward Doppler shift) at a fixed frequency. Neuweiler and others have shown that a species using this strategy is likely to exhibit an auditory system that is very sharply tuned to this frequency, with major parts of the neural elements of its cochlea and higher-level auditory system being devoted to detecting and processing energy in this narrow spectral band. Theoretically, the accuracy of velocity measurement in such a system is necessarily dependent on the product of pulse duration and frequency (and hence of the number of waves in each pulse). As the use of higher-frequency sounds allows velocity to be established requisitely more quickly than if lower-frequencies are used, there is once again good reason for ultrasonic signals to be used.

One special case of echolocation should be mentioned. While the short wavelength of an ultrasonic signal allows very small objects to be distinguished, there are some cases, such as underwater navigation in large basins, where only objects of relatively large size need to be located. In these instances, lower-pitched sound can be effective and, indeed, some man-made sonar navigation systems use acoustic signals as low as a few hundred Hertz in such applications. There has been speculation that some whale calls might serve the same function (George et al. 1989).

4.1
Countermeasures Shown by Prey Species

Remarkable though echolocation may be, it is not an ideal way to find prey as they can be alerted by the searching signals and take evasive action. Work in this area was pioneered by Roeder and his colleagues, who found ultrasound-sensitive tympanic organs on the thorax of moths in the Arctiidae — Noctuidae group, on the abdomen of Geometridae moths, and in modified labial-labral mouthpart structures in certain Sphingidae moths (see Nachtigall and Moore 1988). Ultrasound sensors have also been found on the abdomen of moths of the family Pyralidae, on the forewings of some lacewings (order Neuroptera) and, most recently, in an unpaired, mid-ventral organ of *Mantis* (Yager and Hoy 1989). Other insects that use sound for intraspecific communication, such as crickets (Gryllidae) and bush crickets (Tettigoniidae), may also rely on their hearing to evade bats.

Because of these acoustic countermeasures, insectivorous bats may not always employ the most sensitive and accurate echolocation behavior available to them. The optimum system in this case is not one that detects the most insects, but rather is one that allows the bat to catch the most prey. One "counter-countermeasure" may therefore be to reduce the intensity of pulse emissions, since passive detection of the echolocation sounds by the prey then suffers more than does echo-detection by the bat. Another method is to use extremely high frequencies that the prey seem to find difficult to detect. Conversely, Rydell and Arlettaz (1994) have shown that some bats may employ exceptionally low frequencies (10 to 12 kHz) that fall below the hearing-range in moths. Such counter-countermeasures must surely account for much of the diversity to be found among the echolocation systems of bats.

Many tiger moths (Arctiidae) and their relatives (the Ctenuchidae) are themselves able to emit a barrage of ultrasonic clicks by stressing a pair of corrugated, air-backed sclerites on the metathorax. This measure is used as a "last-ditch" strategy when a bat is closing in using its buzzing "attack sequence". It has been suggested that the moth sound may merely startle the bat. However, this click may also mask the echo produced by the moth's body, thereby upsetting the bat's ability to locate the prey, or it may act as an *aposematic* (warning) signal, indicating that the moth is chemically distasteful. The data on this question are still equivocal, although Surlikke and Miller (1985) produced good evidence for the warning function in one case. Of course, none of the three possible functions outlined here are mutually exclusive. Other moths is also known to make noises, but these sounds have not yet been linked to predation by bats.

5
Ultrasonic Communication

Echolocation has been described here in some detail because it involves the animal in some highly technical problems whose precise solutions demand the use of ultrasound. But ultrasound is also used widely for intraspecific, social communication. Strangely, rather little is known about vocal communication in bats, although such signaling is certainly prevalent and important. A review of this topic has been provided by Fenton (1985).

When Pierce produced the first ultrasound detector in the 1930s, he soon experienced hitherto unknown signals from insects (see Pierce 1948) — as predicted more than 110 years before by Wollaston. Other signals, previously thought to be weak, or audible only to certain people under quiet conditions, were found to have significantly more energy at higher frequencies. Bush crickets, which are perhaps the "most ultrasonic" of the singing insects, are the prime examples of species producing such signals, but Pierce also studied the sounds of crickets (gryllidae), grasshoppers (Acrididae), and cicadas (Cicadidae). All these groups have subsequently attracted attention from researchers, showing that the occurrence of ultrasonic components in communication signals, coupled with ultrasonic hearing capabilities, is far from unusual. This general outcome should not be surprising for two reasons. First, insects make their sounds mechanically using skeletal elements of small size and requisitely high resonant frequencies. Second, these animals are very unlikely to be able to effectively radiate energy at longer wavelengths. It is significant that two insects that do radiate low-frequency energy both use accessory structures. Mole crickets (*Gryllotalpa vinae*), dig exponential horn-shaped burrows, while the tree cricket (*Oecanthus burmeisteri*) cuts a hole in a large leaf and uses it as an extended baffle (see Pye 1979).

Study of ultrasonic communication by mammals other than bats started slowly, but was greatly accelerated by work on rodents — first by Noirot and then by Sales and others beginning in about 1966 (reviewed by Sales and Pye 1974). Ultrasonic distress calls of mouse pups left outside the nest elicit maternal retrieval and prevent the mother from exhibiting the aggression normally shown to organisms close to the nest. This behavior occurs widely and is perhaps universal in the altricial young of the Muridae and Cricetidae families. In the adult life of rodents, ultrasonic signals are found in a variety of social situations, where they apparently play important roles in courtship and mating, aggression and (in rats) submission, and even in exploration. The spectra of these signals are nicely matched by a (second) peak in auditory sensitivity in the appropriate frequency range. The ultrasounds are not produced by vocal vibration, but appear to be a form of whistling, produced in the vocal tract that can jump instantaneously to structured vocalizations (squeaks) with much lower fundamental frequencies.

The use of high frequencies in these sounds may have much to do with the need for sound localization, but absorption by air and scattering by vegetation and other objects may also make such signals difficult for a predator to detect and home in on from a longer distance. Underwater sound is considered in detail by Tyack (this Volume), but it may be pointed out here that, despite their large size, odontocete cetacea both emit and hear ultrasound. In addition to using echolocation clicks that extend to very high frequencies, these animals also produce a wide variety of other signals. Many of these sounds are clearly used for communication, although it is possible that some may be used for echolocation in the open sea.

6
Infrasonic Communication

As mentioned in the introduction, infrasound is little degraded or attenuated by the environment, and is therefore well-suited for long-distance communication. Underwater, these distances are potentially immense. Using a number of reasonable assumptions, Payne and Webb (1971) calculated that in the days before low-frequency shipping noise, a fin whale situated at the depth of minimum sound velocity (the "SOFAR" layer) could broadcast calls that were audible to conspecifics up to 5000 km away! This outcome is possible not only because of the longer wavelength and lesser attenuation of sound in water, but also because water that is different from surrounding layers in temperature and pressure (and thus density) can act as a sound channel, resulting in cylindrical rather than spherical spreading of sound. Although *audible distances* for infrasound in air are orders of magnitude less than in water, they are still considerably better than can be achieved with sounds that humans can hear. Thus, Langbauer et al. (1991) have shown that elephants can perceive the infrasonic calls of conspecifics over distances of at least 4 km.

It is important to note that all animals that are known to produce infrasonic calls have natural histories where the ability to communicate over long distances appears to confer a distinct advantage, often involving reproduction. Fin whales, for instance, migrate over vast distances every year to reach favored breeding grounds (Brown 1954). Mature male and female elephants live apart and yet must find one another to mate during the very brief and infrequent times when an individual female is in estrus — on average, such periods last four days and only occur once every four years (Moss and Poole 1983). Female elephants make a distinctive, low-frequency estrus call, and male elephants have been shown to walk silently more than 1.5 km toward a loudspeaker that is playing recordings of this sound (Langbauer et al. 1991). The male capercaillie, the grouse mentioned in Section 3, produces its low-frequency call at a display ground in spring, presumably to attract hens that fly in to mate (Moss and Lockie 1979). Like the case of the calling female elephant, it is to the grouse's advantage to be able to attract potential mates from a large area.

In addition to long-distance communication, infrasound also appears to be useful for broadcasting calls in thick forest or shrub lands. At 1 kHz, sounds propagating through forest or over grass can be subject to attenuation that is as much as 23 dB per 100 m greater than that predicted by the inverse-square law (Eyring 1946, Dneprovskaya et al. 1963) — although this *excess attenuation* is not necessarily linear (see Michelsen 1978; Wiley and Richards 1978, 1982; Hopp and Morton, this Volume). Even greater attenuation is predicted for higher frequency sound. Low-frequency sound (i.e., less than 100 Hz) is predicted to show little, if any, excess attenuation in these environments (Eyring 1946; Aylor 1971; Marten and Marler 1977; Marten et al. 1977; Waser and Waser 1977; Wiley and Richards 1978; Langbauer et al. in press). In grassy savannahs or woodlands, for example, Asian elephants should be able to perceive infrasonic calls better than their more "normally pitched" *trumpeting* calls when the distance between animals is as little as 100 to 300 m. A researcher interested in finding new biological sources of infrasound might thus be wise to specifically investigate animals that either live in dense habitats, or have life histories where long-distance communication would be an advantage.

7
Physical Properties of Ultrasound and Infrasound

The physical properties of ultrasound and infrasound, mentioned briefly in the introduction, have important consequences for both research techniques and the nature of the phenomena being studied (see, for instance, Pye 1993). These implications will be reviewed in turn.

7.1
Atmospheric Attenuation

The absorption of sound by air rises rapidly with increasing frequency. Although negligible at infrasonic frequencies, absorption in the ultrasonic spectrum can reach as much as 10 dB/m at 200 kHz in humid air (see Figure 2). This effect can therefore be much more important than the inverse-square law in governing signal levels occurring at different distances from the source. The bat *Cloeotis*, for example, might use its narrowband echolocation signals of just over 200 kHz to observe a small target that moves from 1m to a distance 2 m away. It will experience the usual reduction in echo intensity of 12 dB per doubling of distance due to the inverse fourth-power law for echoes, but an additional reduction of 20 dB or more may occur due to energy absorption by air. The total loss of 32 dB may then render a given target undetectable — as the echoes fade it will "disappear" into the "murky" air.

Because energy absorption effects are so strongly frequency-dependent, any propagation pathway acts as a high-cut filter whose slope is related to the distance traveled by the sound (see Stoddard, this Volume, for an explanation of filtering). One implication is that broadband signals are "distorted" by distance and a bat using such sounds might therefore have problems in correlating particular echoes with the original calls that produced them. Unfortunately, there are two problems in applying precise corrections to allow for this effect. First, the degree of absorption is clearly dependent on the amount of water vapor in the air, and hence on both temperature and relative humidity (which are easier to measure). The relationship is not a simple one, however, for damp air is known to absorb the most energy at frequencies above 60 kHz, whereas drier air absorbs more than damp air at frequencies below 30 kHz (as shown in Figure 2).

Second, a degree of confusion has entered the literature, apparently due to the inaccuracy of estimates of absorption proposed by Pohlman (see Lawrence and Simmons 1982). Pöhlman's estimates are apparently too high by a factor of 2 (in dB/m). Unfortunately, Griffin (1971) based a seminal paper on these figures — although he expressed some reservations about them. Pye (1980) quoted the same figures, while also noting suspicions about their accuracy. Lawrence and Simmons (1982) checked the evidence very carefully and made some new measurements of their own, concluding that Pöhlman's values were too high. These original estimates should therefore be interpreted with caution, although they are not far wrong if regarded as the round-trip attenuation values for echoes.

The safest available approach to calculating energy absorption in air is to consult the detailed tables drawn up by two authoritative groups who not only agree both on theory and empirical results, but have also worked and published together. The tables and graphs of Evans and Bass (1972) for air at 20 ° C provide estimates based on 22 values of relative humidity that range from 0 to 100%, and frequencies covering a span of 12 kHz to 1 MHz. Bazley (1976) gives values at 18 relative humidities ranging from 10 to 95%, at every 2 ° C between 0 to 30 °C, at frequencies of 500 Hz to 100 kHz (as well as some data concerning the effects of reduced atmospheric pressure). Pye (1971), noted that fine water droplets suspended in air appear to absorb ultrasound very strongly, perhaps due to deformation resonance of the droplets. This effect may be a reason why bats appear to avoid patches of dense mist.

7.2
Wavelength-Related Effects

The very long wavelengths of infrasound mean that it is effectively reflected only by very large objects. This property renders infrasound unsuitable for use in echo-location. However, requisitely little attenuation occurs due to scattering by objects in the environment, which further increases the effectiveness of infrasound for long-distance communication. In contrast, the very short, millimeter-long wavelengths of airborne ultrasound mean that interactions such as reflection, diffraction, and interference affect these signals even on a small scale. Both larger and smaller bats may therefore experience and exploit phenomena that are unfamiliar to us at the much longer, "audio" wavelengths.

7.2.1
Directionality of Ultrasound

Although the details differ, the directionality of acoustic energy radiated from a vibrating plate, a parabolic reflector, and a long, tapered horn all increase as the diameters of these sources become greater than the equivalent of one wavelength of the radiated sound. *Beamwidth*, a measure of the directionality of radiated energy, is usually characterized as the observed angle between the *half-power* directions — points at which signal strength is 3 dB below the amplitude occurring along the beam axis (see Figure 4). As a rule of thumb, this value is about 1 radian (e.g., 60°) at a diameter of one wavelength, and it decreases by a factor equivalent to the ratio of wavelength to diameter with increasing frequency. Due to the reciprocity relationship, the same directionality will occur in frequency-related sensitivity when these energy-radiating elements are used as receivers (e.g., if the vibrating plate is a microphone diaphragm).

Wavelength-dependent directionality in reflected ultrasound is clearly of special interest in cases, like those of echolocating animals, in which echoes are reflected back towards the source and the emitting and receiving mechanisms are physically close together. Both the applicable theory and actual empirical results are complicated, and depend on the size, shape, and orientation angle (*aspect*, or angle of "view") of the target. A number of simple target shapes have been analyzed by Bowman et al. (1987) and by Neubauer (1986). The simplest case is that of a spherical target, because orientation is then invariant. *Rayleigh scattering*, in which echo intensity is proportional to the inverse-fourth-power of wavelength, occurs when the circumference of the sphere is less than the wavelength of incident energy. When the circumference is equivalent to more than ten wavelengths, echo intensity is almost independent of wavelength. Figure 5 shows a third effect that occurs between the areas illustrating the first two relationships. In this intermediate location, the *Mie region*, echo intensity fluctuates rapidly as wavelength changes.

Pye (1980) used a much-published function for the reflection of electromagnetic waves from an energy-conducting sphere as a model for understanding how sound is reflected from a hard sphere. This analogy is not entirely accurate, as electromagnetic echoes show a strong peak when the circumference of the sphere is equal to the wavelength of incident energy, whereas sound echoes approach a "plateau" value more gradually. Of course, this simplified case is highly artificial in that no natural target is entirely spherical, but it does point to some useful and important principles. Specifically, if a target is small in comparison with the wavelength of an incident sound, its shape is rather unimportant and is well-approximated by a sphere. The amplitude of the echoes that result is greatly enhanced if the energy includes higher frequencies and, thus, shorter wavelengths. The need for insectivorous bats to use ultrasound is clear from this relationship alone. When the target is comparable in size with, or a little larger than the wavelength, echo intensity varies greatly with both wavelength and orientation for aspherical targets. The complexities that therefore arise for FM bats that use wideband calls with wavelengths that vary rapidly within each pulse may be imagined, but are difficult to assess formally.

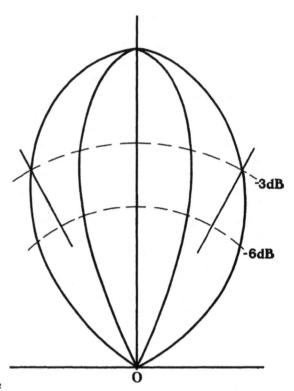

Fig. 4. Two examples of beam width (equivalent to directional sensitivity) shown as polar diagrams. Signal amplitude in each direction is shown by the distance of the *line* from the origin, O. The simplest single measure is the angle between the half-power (-3 decibels) directions, which is here twice as large for the wider beam as for the narrower beam

7.2.2
Sound-Collecting Effects

The next topic to be considered in this context is the sound-collecting property of a horn, which can serve as a model of a *pinna* (the external, fleshy portion of the typical mammalian ear). Again, the effects are wavelength-dependent, with significant outcomes occurring only if the horn (or pinna) in question is equal to or larger than the wavelength of incident sound. Based on this consideration alone, then, bats clearly have more "technical options" for achieving significant sound-collecting effects than do larger species. A 113-kHz signal, for instance, has a wavelength of only 3 mm and can be readily accommodated by a correspondingly small pinna. Even the highest-frequency signal that a human can produce, for instance a soprano singer's 1045-Hz, high-C, has a significant wavelength of 325 mm. Applying theoretical characterizations of sound collectors developed by Fletcher and Thwaites (1979) and Fletcher (1992), Coles et al.

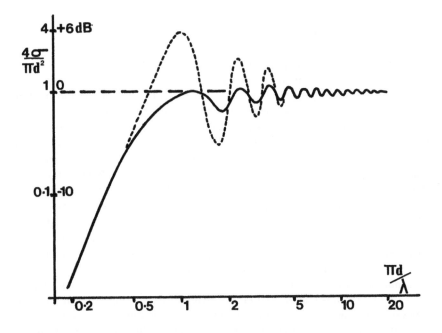

Fig. 5. Backscatter from a sphere. The strength of the echo returning to the source is shown on both a relative decibel scale and also as the area (σ) of an equivalent flat, circular target in relation to the actual cross-section of the sphere. Wavelength (λ) is shown in terms of the circumference of the sphere. The *solid line* represents sound; the *dotted line* shows the somewhat different curve for electromagnetic waves

(1989) applied these models to the long-eared bat (*Plecotus auritus*). Their measurements were in good agreement with theoretical expectations, showing an amplitude gain of up to 23 dB at 9 - 20 kHz decreasing to 0 dB at about 4kHz and at 70 kHz .

7.2.3
Ground Attenuation Effects on Infrasound

There is one additional wavelength-dependent source of attenuation that may, in some cases, affect infrasonic signals. Based on theoretical considerations, Ingard (1953) predicted that a peak of increased attenuation due to boundary interference with the ground (*ground attenuation*) occurs at heights corresponding to 1.5 to 2 wavelengths. For a 30-Hz signal, this value corresponds to a height of 6 to 8 m. On this basis, one would predict that any birds using infrasound for communication should call from close to the ground rather than from an aerial perch. In the case of the capercaillie, this is indeed what occurs.

8
Sound Localization

The pinna of a single ear can provide cues to the location of a sound source if it acts as a "large horn". However, binaural hearing provides a range of reliable cues to the direction of incident energy, including the relative time of arrival, phase, and amplitude of the sound at each ear. The effectiveness of these methods depends largely on the relationship between a particular animal's interaural distance and the wavelength of the sound in question. Overall, then, each works better for high-frequency than for low-frequency signals. For small animals, differences in the time of arrival may be too small to be helpful, but transient characteristics of a wide-bandwidth signal, for instance related to the sound's onset and offset, can be quite sharp. Binaural localization is also aided by *head-shadowing* effects, in which the head (or body, in the case of insects) has a stronger attenuating effect on shorter wavelengths than on longer wavelengths as the sound travels to the ear further from the sound source. As the ears of mammals act independently, these effects may explain why many of these animals, especially small ones, need to hear ultrasound. The same may also apply to most insects.

In contrast to mammals, the ears of some insects (most notably crickets) and many vertebrates (for instance, anurans, lizards, and birds) act as a single system, coupled by an internal, interaural, acoustic pathway. This *pressure gradient detector* arrangement (see Lewis 1983) can produce sharp nulls in well-defined directions at certain very low frequencies and thus provide a basis for directionality at relatively longer wavelengths. The need for ultrasound sensitivity is thus reduced and is indeed absent in vertebrates other than mammals.

At infrasonic frequencies, all the localization methods discussed above are ineffective because wavelengths are so long that, for most animals, the sound signal at each ear is nearly identical. It may be marginally possible for an elephant to produce a shadowing effect on a conspecific call by spreading its ears, but the problem facing infrasound-sensitive birds is much more acute. The wavelength of a 0.05 Hz sound in air is 6.8 km, while the distance between a bird's ears is in the order of a few centimeters. Kreithen and Quine (1979) put forth the intriguing possibility that birds might use the Doppler shift to determine the direction of an infrasonic sound source. They found that in the laboratory, pigeons can detect a 5% change in sound frequency. A pigeon first flying towards and then away from a sound source can generate a total frequency shift of up to 12% if it flies at the reasonable rate of 20 m/s, making such a detection strategy at least theoretically feasible.

9
Directional Sound Emission

Just as with localization, sound emission can more easily be made directional at higher frequencies. In echolocation, it is inefficient to radiate energy in directions other than towards the target(s) of interest, since echo intensities will not then be maximized. (It is sometimes advantageous to restrict the radiation of sound to certain preferred directions in intraspecific communication as well.) Many bats form their mouths into horn-like shapes that can produce half-power (i.e., -3 dB) beamwidths of about 1 radian. There is a clear analogy between this strategy and the passive megaphones used by humans before the development of electronics technology.

An alternative method is used by many bats that emit echolocation signals through their nostrils rather than their mouths. In this case, interference effects between the two sound sources can impart inherent directionality to the signal produced, at least in any plane passing through both sources (there should be no directionality in the median, or vertical, plane). In *nasal-emitting* bats, the nostrils are surrounded by a structure of folded skin, called the noseleaf. This varied and sometimes elaborate organ appears to influence sound directionality in further ways, probably by diffraction, and possibly also by reflection and horn action.

Thus, *Rhinolophus ferrumequinum*, like many noseleaf CF bats, has nostrils spaced just one-half-wavelength apart. However, measurements by Schnitzler and Grinnell (1977) showed that resulting beamwidth is only half the value expected based on two-source interference effects, and that there are strong constraints in the vertical, median plane. Hartley and Suthers (1988) showed that in *Carollia perspicillata*, whose calls show frequency sweeps and harmonics, there is again vertical beam-shaping and that pulse distortion away from the principal axis does not degrade the inherent timing accuracy for range measurement. Many bats show rapid changes in noseleaf shape that may allow them to voluntarily alter radiation patterns. Pye (1988) suggested that some noseleaf bats may be able to steer their sound beams by introducing phase differences between the nostrils.

Finally, other parts of an animal may influence vocal directionality. For example, the enormous ears of *Plecotus* are held forward in flight, over upward-facing and sound-emitting nostrils. Even the wings of flying bats may have some effect, but no data have been published.

10
Instrumentation and Techniques

The inaudibility of ultrasound and infrasound means that special attention must be given to the instrumentation, which is essential even for the detection of signals. The fidelity of recordings or of synthesized signals cannot be judged by ear, as it can for most sounds, and some technical understanding is essential if pitfalls are to be avoided. The techniques involved are also rather unconventional compared with most methods in bioacoustics (see, for instance, Pye 1992).

10.1
Microphones

Almost all microphones used for ultrasound are of the *solid-dielectric capacitance* type first developed in 1954. In this design, a thin insulating film, metalized on one side, is stretched over a metal backplate and charged by a steady "polarizing" voltage. Nowadays, the film is usually Mylar, 3 to 10 µm thick with aluminium on its outer, grounded surface. The backplate is usually a sintered metal disc (to give compliance) on an insulating mount, adjustable to vary membrane tension. Polarizing voltages of 100 V or more are used, depending on film thickness. Higher voltages provide greater sensitivity, but eventually the film "saturates", and may even break down and spark over. Unfortunately, Mylar cannot retain a permanent charge as an *electret* foil.

Small, commercial electret microphones are, however, used in cheaper ultrasound detectors and for infrasonic radio-telemetry techniques. These require no polarizing voltage and have an internal field-effect transistor that can be connected so as to reduce the output *impedance* (resistance in AC circuits) or, alternatively, to amplify the signals (see Pye 1983). These "lapel" microphones are really intended only for use at audio frequencies and their response declines rapidly below about 5 Hz and above 10 to 15 kHz. However, the noise level of these instruments also falls off steeply at higher frequencies, so that an amplifier with a compensatory "high-boost" response can provide a reasonably flat characteristic that is adequate for many ultrasound purposes.

For a variety of reasons, the sensitivity of a solid-dielectric microphone slowly fluctuates. For precise work, especially if intensity measurement is required, one of the miniature (for ultrasound) or 0.5-inch (for infrasound) *air-gap-dielectric capacitance* microphones must be used. These instruments are expensive, demand high-quality power supplies, and are mechanically vulnerable since the thin metal diaphragm is supported only at its edge. But they are accurately and reliably calibrated, showing frequency responses that may be flat up to 150 kHz or more.

All capacitance microphones have a high output impedance that must be matched by the input of the first amplifier used with them. If the microphone is to be used with a cable, this amplifier must be placed at the microphone end as a *headstage*. Modern field-effect transistors are cheap and have very high input impedances that are ideal for this application. One alternative that has been used as the microphone for some simple ultrasound detectors is the narrowband *piezoelectric* or *Gulton* transducer (in which pressure fluctuations induce electrical variation directly). This instrument is cheap, rugged, sensitive, and undemanding, but its response is centred on 45 kHz and extends no more than 10 kHz above and below this point. More sophisticated piezoelectric hydrophones are commonly used for underwater applications (see Tyack, this Volume).

For ultrasound, microphone sensitivity can be increased by the use of horns, but doing so results in increased directionality. Most solid-dielectric microphones are inherently quite directional because of their size, but smaller microphones can usefully take a horn with an aperture of 3 cm or so. Used for ultrasound, larger horns and parabolic reflectors are not subject to the frequency-response problems that plague their use in the audio band, but have a directionality that is generally too great to handle. Even if the sound source is fixed and in a known direction, atmospheric attenuation effects mean that operating distances cannot be greatly extended in this way.

For infrasound, with its low excess-attenuation, operating range is hardly ever a problem. This apparent advantage is actually unfortunate since, due to the long wavelength of low-frequency sound, it is practically impossible to design aids to increase microphone sensitivity and directionality by "focusing" the signal. A parabolic reflector designed to be effective at 20 Hz, for instance, would need to be at least 15 m (50 ft) in diameter.

Another special consideration for infrasound is eliminating background noise, especially that related to wind. Wind noise is sometimes called "pseudosound", because it is due to localized turbulence that causes nonpropagating changes in pressure. If a microphone element is located at some distance from these localized pressure changes, the noise is much reduced. This is the purpose of windscreens, which are constructed in such a way as to keep turbulence as far as possible away from the microphone element while allowing coherent sound waves to propagate through the screen. In general, the larger the windscreen the better, and the cheap "puff-ball" windscreens that are provided with many microphones are usually inadequate for recording in the field. A very inexpensive and effective alternative is to stretch nylon pantihose over a wire mesh cylinder (first suggested to W.R. Langbauer by D.R. Griffin). Rubber bands are then used to suspend the microphone in the middle of the cylinder. Care must be taken to stretch the nylon tightly and to tie off the loose ends tightly in order to avoid wind-generated, "luffing" noises.

10.2
Recorders

Specialized *instrumentation recorders* have traditionally been used to record ultrasound or infrasound for later playback or analysis. For analyzing ultrasound, amplified but otherwise unaltered signals are recorded at high tape speeds and subsequently replayed at much lower speed, thereby reducing the pitch of the sounds to an audible level. The procedure is reversed for infrasound, with signals being recorded at low speeds and subsequently speeded up in order to raise their pitch. The *dynamic range* of such recorders is considerably lower than for most audio-frequency tape-recorders and at best reaches about 40 dB (the maximum undistorted signal level attainable with respect to background noise). This limit reflects the need to record and replay at different speeds, which makes the use of frequency compensation impracticable, as a separate equalizing filter would be needed for every combination of record and playback speeds. As a result, a flat recording characteristic is used and dynamic range cannot be optimized.

It should also be noted that tape-recorder specifications should not be interpreted like those of a linear amplifier. For a tape-recorder, signals above the specified upper frequency limit are not simply represented at lower-than-normal amplitudes, but rather generate distortion artifacts at "new" frequencies. Silver and Halls (1980) demonstrated this effect when resolving a controversy involving recordings made with two different sets of equipment, one of which was operated beyond its bandwidth specification.

In addition, when recording infrasound, care must be taken that the instrument does not include a high-pass filter that will eliminate the very frequencies one is concerned with. Such a filter is almost universal on audio recorders, where it is used to exclude wind and mechanical noise that would degrade recordings of music or other signals that are audible to humans. Most instrumentation-recorders have a means of selecting any one of several different industry-standard high-pass filters (known, for instance, as *A*- or *C-weighting scales*), or of eliminating high-pass filtering altogether (the *flat*, or *linear* response setting). Always ensure that a flat response is selected.

It has to be admitted that even instrumentation recorders have severe drawbacks, although these are mainly logistic in nature. The instruments are large, heavy, and expensive. They consume a lot of power and, as the Ni-Cd batteries typically used for fieldwork operate for only a few hours before needing a recharge, it is not feasible to stray very far from mains electricity. The recorders operate at transport speeds of 30 ips (76 cm/s) and 60 ips (152 cm/s) when recording frequencies of up 150 kHz and 300 kHz, respectively, so that a reel of tape lasts only a few minutes. Finally, recorded results are not available until the tape has been rewound and played at a different speed. The use of instrumentation recorders is therefore limited to recording and analysis, and is in no sense a detection technique. It is often difficult to relate the sounds that are recorded to the events they represent, since any commentary recorded by the observer becomes unintelligible at the altered playing speeds.

There is a further, more serious drawback to using instrumentation recorders in *direct mode* to record infrasound. While these instruments can typically record sounds down to 2 to 3 Hz, they can only play back sounds above about 25 Hz. This discrepancy occurs because the voltage induced in the playback head of a tape-recorder is proportional to the rate of change of magnetic flux (corresponding to the frequency of the recorded signal) present in the playback head-gap. At low frequencies and normal playback speed, the rate of change of the signal across the playback head-gap is very small and the resulting output is so low in amplitude that accurate reproduction is impossible.

While this limitation does not seriously affect signal analysis, since the sounds can be speeded up to raise their pitch to appropriate levels, it is a major problem if one wishes to play infrasonic signals back to animals to observe their responses. This problem can be resolved in two ways. First, a recorder with *FM-mode* capabilities can be used. Here, a high-pitched (e.g., 10 kHz) carrier tone is modulated by the input signal arriving from the microphone. This modulation causes the frequency of the carrier to fluctuate above and below its nominal value, with the amount of modulation (in Hz) corresponding to the amplitude of that signal. It is this modulated signal that is actually recorded. Thus, it is a high-frequency signal that is read from the tape on playback, at an amplitude that is sufficient for accurate reproduction. The modulated signal is then passed through demodulating circuits to recover the original input, which can, for instance, be analyzed or amplified and sent to a loudspeaker. A further advantage of FM-mode recording is that it can extend down to *DC level* (corresponding to 0 Hz), while direct-mode recording, as stated earlier, has a lower limit of 2 to 3 Hz. However, one pays for this advantage in decreased high-frequency performance — FM-recorders often have an upper frequency limit of less than 1 kHz.

In the last several years, digital tape-recorders have also become available. These instruments seem to offer numerous advantages for infrasonic work — they are lightweight, relatively inexpensive, use small, convenient cassette tapes, and are technically capable of recording and playing back sounds down to DC levels. However, there are at least two things to be aware of when buying a digital recorder for infrasound applications. First, there is often high-pass circuitry in the input line, and second, early models exhibited unexplained distortion at frequencies below about 15 Hz. While this technology will definitely set the standard in future work in bioacoustics, it is critical to carefully check both the frequency response and distortion of any particular machine being considered for purchase.

10.3
Loudspeakers

Ultrasonic loudspeakers designed for use in air are currently almost invariably scaled-up versions of solid-dielectric capacitance microphones (with a diameter of 5 cm or more). They need a high voltage-swing to operate and a correspondingly high polarizing voltage if frequency-doubling and the appearance of spurious even-order harmonics in the output are to be avoided.

The design of infrasonic loudspeakers is more problematical, due simply to the fact that the volume of air that must be moved to produce a sound at a given level varies as the third power of the wavelength (see Figure 6). As a result, a million times as much air is moved in generating a 20 Hz sound as in producing a 2 kHz signal. It is therefore difficult to achieve high sound-pressure levels (e.g., above about 100 dB at 1 m) with commercially available equipment, unless a great many transducers and very large cabinets are used. However, at least for frequencies as low as 14 to 16 Hz, one effective servo-driven loudspeaker is available. In this design, a low-inertia, AC motor is driven by the input signal voltage, reversing direction when the voltage reverses polarity. The motor is connected to the loudspeaker's diaphragm membranes through driverods, and in following the AC input signal, is able to move the membranes through very large excursions. These speakers can produce sound pressure levels up to 112 dB at 1 m for 16 Hz tones, with minimal distortion.

Other options that exist for producing signals in this frequency range at moderate sound pressure levels (for instance, 90 to 100 dB sound pressure level, SPL, at 1 m) include *transmission-line* loudspeakers and custom-designed, *bass-reflex subwoofers* (following the design rules of Theile and Small, as summarized by Rossing 1980). For infrasound at very low frequencies (i.e., from about 0.1 to 10 Hz), the volume of air that must be moved is extremely high. However, Kreithen and Keaton (1974) got around this problem by using a sealed pressure chamber and generating infrasound using loudspeaker cones modified to act as bellows-pumps that alternately raised and lowered the overall air pressure of the chamber. Of course, this solution is most convenient when small animals that fit into requisitely small chambers are being studied — Kreithen and Keaton were working with pigeons. However, given an appropriate outlay of funds, the same principle could be applied to larger chambers and larger animals.

10.4
Ultrasound Detectors

The first successful ultrasound detectors were of the *superheterodyne* type (see Pierce 1948). In this design, a narrow band of frequencies (e.g., 10 kHz wide) from the microphone signal is amplified, converted to a higher frequency for additional amplification, converted again to an audible frequency, and finally used to drive a loudspeaker or headphones. This seemingly elaborate arrangement has a number of technical advantages, most of which are shared with radio receivers and are therefore discussed in textbooks on radio. J.D. Pye made his first detector from a portable radio in 1959, only later discovering Pierce's designs.

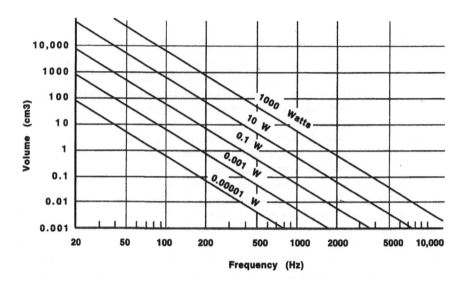

Fig. 6. The volume of air moved by a loudspeaker as a function of sound power and frequency .
(After Rossing, 1983)

The majority of commercial detectors now use superheterodyne circuits or incorporate this technology as an available mode of operation. Superheterodyne design offers
a quick, sensitive, and reliable way to measure frequencies, including detection and
analysis of harmonics. To do so with maximum accuracy, one needs to know whether
the tuning scale is calibrated to read correctly based on a given output tone or for *zero-
beat* (in which one seeks a silent spot occurring between two output tones as the instrument is tuned). In practice, however, calibration-related measurement errors that might
occur, for instance in measuring the frequency of pulses emitted by a CF bat, are likely
to be equalled or exceeded in magnitude by Doppler-shift effects occurring as the animal
flies to and fro.

One drawback to superheterodyne detectors is that they miss signals occurring outside the frequency band to which they happen to be tuned. As a result, several ways have
been devised to produce a broadband response, thereby allowing the user to listen "at
all frequencies at once". An early *envelope-detector* design produced a signal that outlined the shape of each ultrasound pulse by following the peaks of their individual cycles.
A drawback in this case is that while the intensity and repetition rate of the input are
captured, frequency information is lost. Longer sounds, like those produced by CF bats,
do not produce clear output, and a continuous input signal is not registered at all. This
type of detector is no longer widely used, and because it currently offers little economic
advantage over superheterodyne circuitry, is found only in the very cheapest instru-

ments. A subsequent, *panoramic heterodyne* arrangement allowed simultaneous tuning across the whole spectrum, but the frequency information provided by these detectors was slight or even confusing.

These designs gave way to the *frequency divider*, which was first developed by Andersen and Miller (1977). In this approach, incoming waveform cycles are counted as the signal voltage swings across the zero, or baseline level. A new waveform is then generated at some fixed (e.g., 1/10) or selectable ratio (e.g., 1/4, 1/8, 1/16, or 1/32) of the original frequency, and is given the amplitude characteristics of the input signal. The new waveform emerges as a lower-frequency, audible version of the original sound that retains both its frequency and amplitude patterns and can be recorded on audio-level cassette-tape.

Although this scheme may seem ideal, it has two significant drawbacks. First, the new waveform tends to track the frequency of the most intense energy component of the input signal and the instrument's performance may therefore be disturbed by the presence of multiple harmonics. Second, some bat pulses are of such short duration that that they cannot be effectively transformed to a lower frequency. In response to a 1 ms, 40 kHz signal, for instance, a frequency divider at a setting of one-tenth produces only four waveform cycles. Nevertheless, the frequency-divider design is deservedly popular, especially when complemented by superheterodyne-based detection, with both outputs being available for "stereo" input to a cassette tape-recorder. When the output of a frequency divider is recorded, the ratio value should be set so as to produce the highest possible frequency that does not exceed the upper limit of the tape-recorder. This approach provides the best representation of the frequency pattern of the input signal, even though a different setting may provide a better-sounding signal when it is heard through a loudspeaker.

In a different approach, called *automatic frequency scanning*, the detector "searches" for signals through continuous frequency-response changes. This automatic tuning process sweeps repeatedly across the entire frequency-band of interest, stopping only when sounds are detected. This approach saves the investigator the tedium of persistently turning the detector's tuning knob by hand, but does not overcome the inherent problem of detecting signals that might be occurring simultaneously at different frequencies.

Finally, a few points of qualification for assessing detector characteristics should be considered. One commercial superheterodyne detector uses a narrowband piezoelectric microphone that works well at 35 to 55 kHz. However, the specifications for this instrument claim a tuning band of 20 to 120 kHz, with corresponding markings appearing on its tuning dial! Despite this deception, this detector provides a very useful response and is sensibly designed. Most FM bats use calls with high-energy levels or second harmonics within the actual sensitivity range of the detector and at short range even the suppressed fundamental of the high-frequency calls of some CF bats can be detected. The output signal of this instrument is quite loud, which some users mistakenly take to indicate high inherent sensitivity. However, true sensitivity is indicated by the level of the weakest signal that the instrument can detect. The user can easily test the relative sensitivity of a detector by walking away from an ultrasound source and judging the maximum distance at which the instrument still produces an output signal. In this particular case, the instrument in question is comparable with other superheterodyne detectors at 40 to 50 kHz, but is hopelessly inferior outside these limits. In general, increases in output amplitude are "paid for" in faster consumption of battery power, and loudspeaker volume should therefore not be higher than necessary.

Two points should be considered in using detectors based on the frequency-divider approach. First, it is possible when shaping the amplitude of the divided signal to suppress its smallest intensity fluctuations. However, while this technique seems to increase the signal-to-noise ratio of the output, the apparent improvement is spurious. Weak signals, for instance, will be lost altogether. Although the frequency content of the output waveform will also be changed by such amplitude shaping, this effect is inconsequential due to the second point — that the only data of value in this waveform are its fundamental frequency and peak amplitude values. While circuitry capable of producing a pure, sinusoidal output waveform has been developed, using it does not appear to justify the extra cost and power consumption involved. It is, however, important to note that in analyzing recordings made using a frequency-divider detector, no attention whatsoever should be paid to energy that it may produce at harmonics of the output signal's fundamental frequency.

10.4.1
Time-expansion methods

In using any of the detectors described in Section 10.4, some information about the original signal is lost. As discussed by Beeman and others (this Volume), simultaneous description of the spectral and temporal properties of a signal is inherently limited by the product of the time and frequency ranges involved (the *time-bandwidth product*). Since ultrasound detectors work in real-time with a narrow spectral bandwidth, this product is reduced by a factor of 10 or more compared with the original signal, so information that is potentially available is thereby "rejected". In choosing a particular detection method, therefore, the observer should take into account what information is required and what can be dispensed with. Due to the inherent limitations of any given technique, some detectors are designed to offer two complementary methods that might even be used simultaneously.

A radically different approach to this general problem, however, is to extend the time-scale of the recorded signal in order to maintain the time-bandwidth product and thereby represent the entire signal in full detail. Since about 1960, instrumentation recorders have been used in this manner, as explained in Section 10.2. This technique is well-established and can give excellent results if used by informed operators. However, because of the delay that necessarily occurs between recording and playback, it is only suited to signal analysis and not to signal detection.

Since 1985, a different method has been under development, using computer memory instead of tape for signal storage. In this approach, signals from the microphone are digitized and recorded in RAM chips. In this form, they can immediately be played through a digital-to-analog converter at any desired rate for listening or recording on audio cassette-tape. Because the instruments developed for this purpose give almost instantaneous results, they are often called *memory detectors* (e.g., Pye and Mutere 1985-1986). These detectors are small, much cheaper than high-speed tape-recorders and can operate for tens of hours on a few small alkaline power cells. Although ideal design specifications for these instruments have yet to emerge, and may well depend on the requirements of each application, the method is extremely flexible in principle and can easily be customized. This flexibility is perhaps most evident in the instrument developed by Maries et al. (1989), which incorporates a programmable microprocessor that controls all its functions. Memory sizes ranging from 64 kB to 1 mB have been used in these instruments, where the latter can record for up to 5 s using a 100 kHz bandwidth (and pro rata for other bandwidths).

As is the case for tape-recorders, signals whose frequencies exceed the specified upper limit for memory detectors cause problems for interpreting the signal that is recorded (see chapters by Beeman, Clements, and Stoddard, this Volume, for discussion of this *aliasing* problem). So far, the instruments have all used 8-bit resolution, which provides a dynamic range of 42 dB. This performance is slightly better than is characteristic of the best instrumentation-recorders, and the resulting signal quality is easily retained using a good analog cassette-recorder and type-III (metal) tape. This hybrid approach, involving both digital and analog equipment, is very convenient, although recording the signal directly in digital form on a compatible recorder or computer is also feasible.

An obvious drawback of using the time-expansion approach in the field, however, is that recording must cease during the extended playback periods. Playing a 5 sec signal at one-tenth the speed for listening or storage purposes takes 50 s, an interval during which all sorts of interesting things might happen. This problem can be minimized in at least two ways. First, the detector may be set to play back automatically at the end of each signal. In this way, for instance, it is occasionally possible to record every pulse produced by a single cruising bat in the silent intervals occurring between these signals (i.e., if the duty cycle of the bat's call is less than 10% and playback is at one-tenth the speed). Second, the observer may choose to set the detector to record and erase continuously, monitoring the signals using a separate instrument and playing back pulses of interest manually.

In each of these approaches, the output signal from a frequency divider can be recorded on one track of a tape, to give a continuous but simplified record of all signals and their time relations, while a memory detector provides high-fidelity records of selected samples on the other track. The high-speed recordings used historically were, in any case, highly wasteful in that only small samples from these tapes could ultimately be analyzed. It is much more effective to do such sampling at the time of recording, which has now become possible using RAM recording. A memory detector can also be made to play back repeatedly, allowing the observer to conduct on-the-spot assessment of recorded signals using a loudspeaker or an oscilloscope. If the memory sample is then restricted, this combination of instruments virtually becomes a "memory scope". This technique is especially effective if the oscilloscope is used with a period- or frequency-meter.

10.5
Infrasound Detectors

While the construction of infrasound detectors is possible in theory, such instruments have so far been little used in research. There are several interconnected reasons for this apparent underutilization of available technology. First, as mentioned above, the infrasonic spectrum tends to be rather noisy while the ultrasonic band contains little ambient noise. Because of this rather high level of ambient background energy, it is tricky to set the threshold sensitivity of a detector so that it responds to the sounds of interest while not being "swamped" by ongoing noise. Second, all biological infrasound discovered so far is used for communication and not for echolocation. This difference from ultrasonic signals makes the occurrence of infrasonic events relatively less predictable and further confounds the problem of setting the detector level appropriately. Finally, since infrasound usually occurs in the context of complex, on-going behavioral sequences, the most effective research strategy is often to make video- and audio-recordings of study animals rather than relying on detectors and on-the-spot data coding. These records can then be carefully analyzed in the laboratory in order to correlate the infrasonic signals with specific behaviors as accurately as possible. This process is not always straightforward — in the absence of overt visual signs of calling it is often impossible to determine which animal made a particular call. Since the identity of the caller influences the responses of the listeners in practically any social animal, this difficulty means that deciphering an infrasonic communication system is far from simple.

However, the preceding discussion is not meant to be discouraging, as none of these problems is insurmountable. The study of biological infrasound is a young field and there are probably many interesting phenomena waiting to be discovered. In fact, one application in which a memory detector might be even more suitable for detection and recording of infrasound than ultrasound is in using *time-compression* to speed up playback of a recorded signal and thereby greatly decrease the "down-time" involved in sampling of sounds in the field (although this approach does not yet seem to have been tried). A detector with a 1 MB memory, for example, would be sufficient to hold 1000 s (i.e., 16.5 min) of a 500 Hz bandwidth signal, while downloading it in 50 s using playback up to 10 kHz. Two such instruments used together could operate on a continuous basis, leaving no gaps in the resulting infrasonic record.

A second promising infrasound-detection method is to listen to the undemodulated output of an FM device. Wilcox (e.g., Jackson and Wilcox 1990) has used this technique to listen to the low-frequency vibrations caused by spiders moving on their webs by placing a transducer directly on the web itself. While this signal is not a form of sound, per se, such vibrations are also subject to interference by wind noise and other random events. Nonetheless, after a period of ear training, it was possible for the researchers to distinguish the distinctive warbles of the carrier-tone caused by the spider's movements from the less coherent and more random fluctuations caused by ambient noise. Although Wilcox built his own FM device, tapping the output of a commercial FM recorder just after the modulator stage would be a trivial task for a competent electronics technician. This output could then be amplified and monitored over headphones. One caveat is that it would be necessary to ensure that the carrier frequency was of sufficiently low pitch as to be readily audible. A carrier tone of less than 10 kHz, for instance, would be much easier to hear than one of higher frequency.

As animals that are capable of producing infrasound tend to be large, there is typically an opportunity to fit them with even rather sizeable radio collars, in order to both detect and record their signals. Langbauer et al. (in review) constructed such collars to simultaneously monitor the locations and vocalizations of free-ranging elephants. Each contained a walkie-talkie activated by the elephant's voice, thereby ensuring that transmission of vocal data occurred only when the animal made a powerful low frequency call, as well as a radio beacon that locate the animal when it was not calling. The calls broadcast from the elephants were automatically recorded by a computer located at a distant base-station. The system was easy to develop, requiring only simple modifications to commercially available equipment. A valuable advantage of using such radio collars is that, by having a microphone directly attached to each animal in a group, one can use the amplitude and time of arrival of a call at each collar (which a computer can measure with a resolution of 1 ms or less) to determine which animal made any given sound.

11
Summary

Ultrasound and infrasound are significantly different from acoustic energy that is audible to humans in the three fundamental aspects of frequency (they are higher and lower, respectively), wavelength (which is shorter and longer, respectively), and propagation loss (very high and very low, respectively). Although both these sound types are inherently difficult even to detect, let alone study, work on each has been on-going for some time and has been greatly aided by advances in electronics technology. While infrasound is broadcast widely, sources of ultrasound fade rapidly in a typical acoustic environment. The lack of resulting background noise in the very high-frequency ranges is an important aspect of setting the stage for effective use of such signals in echolocation, a major biological application of ultrasound. Among bats, three major forms of such echolocation signaling are found, including deep-frequency sweeps, broadband clicks, and Doppler-mode operation. Each presents a unique set of functional problems and many bat species are therefore likely to be able to flexibly employ more than one type. A problem that arises is that prey species hunted through echolocation come to exhibit countermeasures, which in turn gives rise to counter-countermeasures in bats.

Ultrasonic communication sounds are produced by a variety of animals, including insects, small mammals, and cetaceans, which are likely to be taking advantage of the potential localization cues inherent to high-frequency acoustic energy. Infrasonic communication signals are more strongly marked by relative immunity to propagation loss and are therefore typically found in species, such as whales and elephants, where long-distance communication appears to be an important aspect of typical life history, most notably in reproduction. The physical properties of ultrasound and infrasound that are most important to consider include atmospheric (and to a lesser extent, ground-related) attenuation, wavelength-related effects, and directionality in both reception and emission.

Specialized instrumentation and data-related techniques are a particularly prominent aspect of conducting scientific work involving ultrasound and infrasound. The need to understand both the problems that arise and the technological solutions that have been developed extends to every aspect of recording and storing these signals, including use of microphones, recorders, loudspeakers, and signal detectors. For each instrumentation and methodological approach, the researcher needs to explicitly consider a host of trade-offs that are inevitably involved. Recent developments in digital technology have significantly improved the effectiveness of detection and sampling of ultrasonic signals in the field, and hold great promise for infrasound applications as well. The authors will be happy to advise on commercial sources of the instruments described in this chapter. Please write to W.R. Langbauer about topics related to infrasound and to J.D. Pye concerning ultrasound.

References

Andersen BB, Miller LA (1977) A portable ultrasonic detection system for recording bat cries in the field. J Mammal 58: 226-229

Arabadzhi VL (1990) Migration of birds and infrasound. Biofizika 35: 361-362

Ayler D (1971) Noise reduction by vegetation and ground. J Acoust Soc Am 51: 197-205

Bazley EN (1976) Sound absorption in air at frequencies up to 100 kHz. Acoustics report Ac74. National Physical Laboratories, Teddington, UK

Bowman JJ, Senior TBA, Uslenghi PLE (eds) (1987) Electromagnetic and acoustic scattering by simple shapes. Radiation Laboratory, Univ Michigan, Ann Arbor

Brown SG (1954) Dispersal in blue and fin whales. Discovery Rep 26: 355-384

Busnel RG, Fish JF (eds) (1980) Animal sonar systems. Plenum Press, New York

Coles RB, Guppy A, Anderson ME, Schlegel P (1989) Frequency sensitivity and directional hearing in the gleaning bat, Plecotus auritus (Linnaeus 1758). J Comp Physiol A 165: 269-280

Dneprovskaya IA, Iofe VK, Levitas FI (1963) On the attenuation of sound as it propagates through the atmosphere. Sov Phys Acoust 8: 235-239

Evans JB, Bass HE (1972) Atmospheric absorption of sound as a function of frequency and relative humidity. Rep no WR 72-2, Wyle Laboratories, Huntsville, Alabama

Eyring CF (1946) Jungle acoustics. J Acoust Soc Amer 18: 257-270

Fenton MB (1985) Communication in the chiroptera. Indiana University Press, Bloomington

Fletcher NH (1992) Acoustic systems in biology. Oxford University Press New York

Fletcher NH, Thwaites S (1979) Physical models for the analysis of acoustical systems in biology. Q Rev Biophys 12: 25-65

Galton F (1883) Inquiries into human faculty and its development. Dent, London and Dutton, New York

George JC, Clark C, Carroll GM, Ellison WT (1989) Observations on the ice-breaking and ice-navigation behavior of migrating bowhead whales (Balaena mysticetus). Arctic 42: 24-30

Griffin DR (1958) Listening in the dark. Yale Univ Press, New Haven; republ (1974) Dover Publications, NY; (1986) Comstock/Cornell Univ Press, Ithaca New York

Griffin DR (1971) The importance of atmospheric attenuation for the echolocation of bats (Chiroptera). Anim Behav 19: 55-61

Hartley DJ, Suthers RA (1988) Directional emission and time precision as a function of target angle in the echolocating bat Carollia perspicillata. In: Nachtigall PE, Moore PWB (eds) Animal sonar: processes and performance. Plenum Press, New York

Hartridge H (1920) The avoidance of objects by bats in their flight. J Physiol 54:54-57

Heffner SR, Heffner EH (1982) Hearing in the elephant (Elephas maximus): absolute sensitivity, frequency discrimination, and sound localization. J Comp Physiol Psychol 96: 926-944

Ingard U (1953) A review of the influence of meteorological conditions on sound propagation. J Acoust Soc Am 25: 405-411

Jackson RR, Wilcox RS (1990) Aggressive mimicry, prey-specific predatory behaviour and predator — prey interactions of *Protia fimbriatta* and *Euryttus* sp. jumping spiders from Queensland, Australia. Behav Ecol Sociobiol 26: 111-120

Kreithen ML (1980) Detection of sound and vibration by birds. In: Proc 2nd Conference on Abnormal animal behavior. prior to earthquakes. Open file Rep 80 — 453, US Geol Surv, Menlo Park, California

Kreithen ML (1983) Strategies of bird orientation: a tribute to WT Keeton. In: Aspey WP, Lustick SI (eds) Behavioral energetics: vertebrate costs of survival. Ohio State University Press, Columbus, p 3

Kreithen ML, Keaton WT (1974) Detection of changes in atmospheric pressure by the homing pigeon, *Columba livia*. J Comp Physiol A 89: 73-82

Kreithen ML, Quine DB (1979) Infrasound detection by the homing pigeon: a behavioral audiogram. J Comp Physiol A 129: 1-4

Langbauer WR Jr, Charif R, Osborn F (in press) Transmission of low-frequency sound in three African environments and its relevance to long distance elephant communication. Behav Ecol Sociobiol (in press)

Langbauer WL Jr, Payne K, Charif R, Rapaport L, Osborn F (1991) African elephants respond to distant playback of low-frequency calls. J Exp Biol 157: 34-46

Langbauer WL Jr, Powell SP, Martin R (in review) A radiotelemetry system for monitoring location and vocalizations of elephants. J Wildl Manage

Lawrence BD, Simmons JA (1982) Measurements of atmospheric attenuation at ultrasonic frequencies and the significance for echolocation by bats. J Acoust Soc Am 71: 585-590

Lewis B (1983) Directional cues for auditory localization. In: Lewis B (ed) Bioacoustics: a comparative approach. Academic Press, London, NY, p 233-257

Lubbock J (1879) Observations on the habits of ants, bees and wasps: VI Ants. J Linn Soc Lond Zool 14: 607-626

Lutz FE (1924) Insect sounds. Bull Am Mus Nat Hist 50: 333-372

Maries K, Pye JD, Coppen D (1989) A microprocessor-controlled "memory" bat detector. In: Hanak V, Horacek I, Gaisler J (eds) European bat research 1987. Charles University Press, Prague, p 287-290

Marten K, Marler P (1977) Sound transmission and its significance for animal vocalization I: temperate habitats. Behav Ecol Sociobiol 2: 271-290

Marten K, Quine D, Marler P (1977) Sound transmission and its significance for animal vocalization II: tropical forest habitats. Behav Ecol Sociobiol 2: 291-302

Maxim HS (1912) A new system for preventing collisions at sea. Cassel, London, NY (also see Sci Amer, July 27, 1912)

Michelsen A (1978) Sound reception in different environments. In: Ali BA (ed) Perspectives in sensory ecology. Plenum, New York, p 345-373

Montgomery HC (1932) Do our ears grow old? Bell Lab Rec 10: 311-313

Moss C, Poole JH (1983) Relationships and social structure of African elephants. In: Hinde R (ed) Primate social relationships: an integrated approach. Blackwell Scientific Publ Oxford, p 315-325

Moss R, Lockie I (1979) Infrasonic components in the song of the capercaillie (*Tetrao urogallus*). Ibis 121: 94-97

Nachtigall PE, Moore PWB (eds)(1988) Animal sonar: processes and performance. Plenum Press, NY

Neubauer WG (1986) Acoustic reflection from surfaces and shapes. Nav Res Lab, Washington, DC

Packard A, Karlen HE, Sand O (1990) Low frequency hearing in cephalopods. J Comp Physiol A 166: 501-505

Patterson B, Hamilton GR (1964) Repetitive 20 cycles per second biological hydroacoustic signal at Bermuda. In: Tavolga WN (ed) Marine bioacoustics: symposium at Lerner Marine Lab, 1963. Macmillan, New York

Payne K, Langbauer WR Jr, Thomas E (1986) Infrasonic calls of the Asian elephant (*Elephas maximus*). Behav Ecol Sociobiol 18: 297-301

Payne RS, Webb D (1971) Orientation by means of long-range acoustic signalling in baleen whales. In: Adler HE (ed) Orientation: sensory basis. Ann NY Acad Sci 188: 110-141

Pierce GW (1948) The songs of insects. Harvard University Press, Cambridge

Pierce GW, Griffin DR (1938) Experimental determination of supersonic notes emitted by bats. J Mammal 19: 454-455

Pollak GD (1993) Some comments on the proposed perception of phase and nanosecond time disparities by echolocating bats. J Comp Physiol A 172: 523-531

Poole JH, Payne K, Langbauer WR Jr, Moss C (1988) The social context of some very low-frequency calls of African elephants. Behav Ecol Sociobiol 22: 385-392

Pye JD (1971) Bats and fog. Nature 229: 572-574

Pye JD (1979) Why ultrasound? Endeavour 3:57-62

Pye JD (1980) Echolocation signals and echoes in air. In: Busnel RG, Fish JF (eds) Animal sonar systems. Plenum Press, NY and London, p 309-353

Pye JD (1983) Techniques for studying ultrasound. In: Lewis B (ed) Bioacoustics: a comparative approach. Academic Press, London, NY, p 39-65

Pye JD (1988) Noseleaves and bat pulses. In: Nachtigall PE, Moore PWB (eds) Animal sonar: processes and performance. Plenum Press, NY and London, p 791-796

Pye JD (1992) Equipment and techniques for the study of ultrasound in air. Bioacoustics 4: 77-88

Pye JD (1993) Is fidelity futile? The "true" signal is illusory, especially with ultrasound. Bioacoustics 4: 271-286

Pye JD, Mutere FA (1985-6) Recording bat sounds by new techniques. Myotis 23/24: 245-248

Pye JD, Pye A (1988) Echolocation sounds and hearing in the fruit bat, *Rousettus*. In: Stephens SDG, Prasansuk S (eds) Measurement in hearing and balance, Karger, Basel Switzerland, p 1

Rossing TD (1980) Physics and psychophysics of high-fidelity sound, part III. Physics Teach 18: 426

Rossing TD (1983) The science of sound. Addison-Wesley, Reading, p 365

Rydell J, Arlettaz R (1994) Low frequency echolocation enables the bat *Tadarida teniotis* to feed on tympanate insects. Proc Roy Soc B 257: 175-178

Sales GD, Pye JD (1974) Ultrasonic communication by animals. Chapman and Hall, London

Sand O, Karlsen HE (1986) Detection of infrasound by the Atlantic cod. J Exp Biol 125: 197-204

Schermuly L, Klinke R (1990) Infrasound sensitive neurones in the pigeon cochlear ganglion. J Comp Physiol A 166: 355-363

Schevill WE, Watkins WA, Backus RH (1964) Underwater sounds of cetaceans. In: Tavolga WN (ed) Marine bioacoustics. Pergamon Press, New York, p 147

Schnitzler HU, Grinnell AD (1977) Directional sensitivity of echolocation in the horseshoe bat, *Rhinolophus ferrumequinum*, I. Directionality of sound emission. J Comp Physiol A 116: 51-61

Silver SC, Halls JAT (1980) Recording the sounds of hydropsychid larvae — a cautionary tale. J Comp Physiol A 140: 159-161

Simmons JA (1993) Evidence for perception of fine echo delay and phase by the FM bat, *Eptesicus fuscus*. J Comp Physiol A 172: 533-47

Surlikke A, Miller LA (1985) The influence of arctiid moth clicks on bat echolocation; jamming or warning? J Comp Physiol A 156: 831-843

Theurich M, Langner G, Scheich H (1984) Infrasound responses in the midbrain of the Guinea fowl. Neurosci Lett 49: 81-86

Waser PM, Waser MS (1977) Experimental studies of primate vocalization: specializations for long-distance propagation. Z Tierpsychol 43: 239-263

Wiley RH, Richards DG (1978) Physical constraints on acoustic communication in the atmosphere: implications for the evolution of animal vocalizations. Behav Ecol Sociobiol 3: 69-94

Wiley RH, Richards DG (1982) Adaptations for acoustic communication in birds: sound transmission and signal detection. In: Kroodsma DE, Miller EH, Ouellet H (eds) Acoustic communication in birds. Vol 1. Academic Press, New York, p 131

Wollaston WH (1820) On sounds inaudible by certain ears. Philos Trans R Soc Lond 110: 306-314

Yager DD, Hoy RR (1989) The cyclopean ear: a new sense for the praying mantis. Science 231: 649-772

Zbinden K (1985-6) Echolocation pulse design in bats and dolphins. Myotis 23-24: 195-200

CHAPTER 8

Measuring and Modeling Speech Production

P. RUBIN AND E. VATIKIOTIS-BATESON

1
Introduction

In human communication, the speech system is specialized for rapid transfer of information (Liberman et al. 1967; Mattingly and Liberman 1988). Significant events in the acoustic signal can occur in an overlapped or parallel fashion due to the coproduction of speech gestures. One result is that aspects of the signal corresponding to different linguistic units, such as consonants and vowels, often cannot be isolated in the acoustic stream. One way to help tease apart the components of the speech signal is to consider the physical system that gives rise to the acoustic information: The acoustic encoding of phonetic information is viewed in light of the flexibility inherent in the production apparatus, particularly the human supralaryngeal vocal tract, in which individual articulators or groups of articulators can function semi-independently. In this chapter we review this approach. First, we show how the analysis of speech acoustics has benefited by treating the sound production system as one in which the contributions of physical acoustic sources and physiologically determined filters are combined. We then discuss how acoustic diversity has resulted in a desire to find articulatory simplicity. In the process, we review some of the methods used to examine articulatory activity, and also describe in detail a particular attempt at modeling the coordination of the speech articulators. Finally, we consider some recent attempts to explore the links between production, perception, and acoustics in a dynamic-systems approach and in connectionist models. Where possible, recent trends in the field have been exemplified by projects involving ourselves and our colleagues. Although articulation in most animals is simpler than human speech production, the methods we describe are also applicable in this domain.

In order to consider the details of acoustics and production, it is necessary first to briefly describe the system being studied. Figure 1 shows a schematic view of the human sound production system. Note the extent of the anatomy involved in the production process (Ladefoged 1975; Borden and Harris 1984; Lieberman and Blumstein 1988). There are a variety of ways to produce sound. One method involves using the air pressure provided by the lungs to set the elastic vocal folds into vibratory motion. The larynx converts the steady flow of air produced by the subglottal system into a series of puffs, resulting in a quasi-periodic sound wave. Aperiodic sounds are produced by allowing air to pass through the open glottis into the upper airway (the supralaryngeal vocal tract) where localized turbulence can be produced at constrictions in the tract. A third method involves producing transient clicks and pops by rapid release of the articulatory

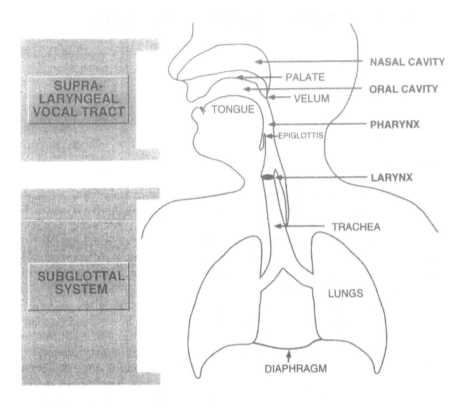

Fig. 1 The human speech production system (see text for details)

closure (Ladefoged 1975). Here the sound sources arise from the local changes in the vocal tract and do not require air pressure from the subglottal system.

2
The Acoustic Theory of Speech Production and Acoustic Analysis

2.1
The Source-Filter Model

Acoustic speech output in humans and many nonhuman species is commonly considered to be the combined outcome of a source of sound energy (e.g., the larynx) modulated by a transfer (filter) function determined by the shape of the supralaryngeal vocal tract. This combination results in a shaped spectrum with broadband energy peaks. This model is often referred to as the "source-filter theory of speech production" and stems from the experiments of Johannes Müller (1848) in which a functional theory of phonation was tested by blowing air through larynges excised from human cadavers.

"Müller...noticed that the sound that came directly from the larynx differed from the sounds of human speech. Speechlike quality could be achieved only when he placed over the vibrating cords a tube whose length was roughly equal to the length of the airways that normally intervene between the larynx and a person's lips. The sound then resembled the vowel [uh], the first vowel in the word *about...*" (Lieberman 1984, p. 131). In this model, the source of acoustic energy is at the larynx — the supralaryngeal vocal tract serves as a variable acoustic filter whose shape determines the phonetic quality of the sound (Fant 1960).

When the larynx serves as a source of sound energy, voiced sounds are produced by a repeating sequence of events. First, the vocal cords are brought together (adduction), which temporarily blocks the flow of air from the lungs and leads to increased subglottal pressure. When the subglottal pressure becomes greater than the resistance offered by the vocal folds, they open again. The folds then close rapidly due to a combination of factors, including their elasticity, laryngeal muscle tension, and the Bernoulli effect. If the process is maintained by a steady supply of pressurized air, the vocal cords continue to open and close in a quasiperiodic fashion. As they open and close, puffs of air flow through the glottal opening. The frequency of these pulses determines the fundamental frequency (F_0) of the laryngeal source and contributes to the perceived pitch of the produced sound. An example of the spectrum of the result of such glottal air flow is plotted at the top left of Figure 2. Note that there is energy at the fundamental frequency ($F_0 = 100$ Hz) and at the harmonics of the fundamental, and that the amplitude of the harmonics falls off gradually. The bottom of Figure 2 shows the comparable case for a fundamental frequency of 200 Hz. The rate at which the vocal folds open and close

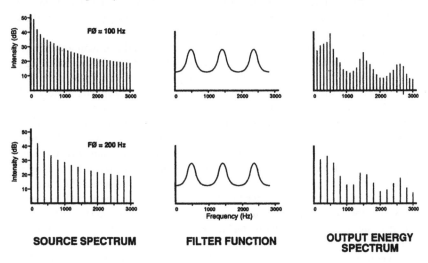

SOURCE SPECTRUM   FILTER FUNCTION   OUTPUT ENERGY SPECTRUM

Fig. 2. The source-filter model of speech production. The source spectrum represents the spectrum of typical glottal air flow with a fundamental frequency of 100 Hz. The filter, or transfer, function is for an idealized neutral vowel /ə/, with formant frequencies at approximately 500, 1500 and 2500 Hz. The output energy spectrum shows the spectrum that would result if the filter function shown here was excited by the source spectrum shown on the *left*

during phonation can be varied in a number of ways and is determined by the tension of the laryngeal muscles and the air pressure generated by the lungs. The shape of the spectrum is determined by details of the opening and closing movement, and is partly independent of fundamental frequency. In normal speech, fundamental frequency changes constantly, providing both linguistic information, as in the different intonation patterns associated with questions and statements, and information about emotional content, such as differences in speaker mood. In addition, the fundamental frequency pattern contributes to the naturalness of utterance production. This effect can be illustrated by creating a synthetic version of a natural utterance in which the spectral properties are left largely unchanged while the normally varying fundamental is replaced with one of constant frequency.

The supralaryngeal vocal tract, consisting of the pharynx, the oral cavity, and the nasal cavity (Figure 1), can serve as a time-varying acoustic filter that suppresses the passage of sound energy at certain frequencies while allowing its passage at other frequencies. Formants are those frequencies at which local energy maxima are sustained by the supralaryngeal vocal tract and are determined, in part, by the overall shape, length and volume of the vocal tract. The detailed shape of the filter (transfer) function is determined by the entire vocal tract serving as an acoustically resonant system, combined with losses that include those due to radiation at the lips. An idealized filter function for the neutral vowel /ə/ is shown in the center panels of Figure 2 for a supralaryngeal vocal tract approximately 17 cm long, approximated by a uniform tube. The formant frequencies, corresponding to the peaks in the function, represent the center points of the main bands of energy that are passed by a particular shape of the vocal tract. In this idealized case they are 500, 1500, and 2500 Hz with bandwidths of 60 to 100 Hz, and are the same regardless of fundamental frequency (i.e., they are the same in both the top and bottom center of Figure 2).

The spectrum of the glottal air flow, which has energy at the fundamental frequency (100 Hz) and at the harmonics (200, 300 Hz, etc.), is plotted in the top left of Figure 2. The amplitude of the harmonics, which for the purposes of this figure combines the effects of both the source spectrum and radiation, decreases by approximately 6 dB per octave. The top right of the figure shows the spectrum that results from filtering the laryngeal source spectrum in the top left panel with the idealized filter function shown in the center of Figure 2. Note that the laryngeal source energy has been "shaped " by the filter function. Energy is present at all harmonics of the fundamental frequency of the glottal source function, but the amplitude of an individual harmonic is determined by both its source amplitudes and the filter function. The bottom half of Figure 2 shows the effect of using a different source function, while retaining the same filter function. In this case, the fundamental frequency of the glottal source is 200 Hz, with harmonics at integer multiples of the fundamental (400 Hz, 600 Hz, etc.). The spectrum that results from combining this glottal energy with the filter function for an idealized /ə/ has the same overall pattern as that shown above it. However, there are differences in the details. Note, for example, that the lowest formant for /ə/ has a center frequency of 500 Hz. A glottal waveform with a fundamental of 100 Hz has a harmonic at this frequency. A source function with a fundamental of 200 Hz has harmonics that straddle the lowest formant (i.e., at 400 and 600 Hz), as shown in the bottom right of Figure 2. Since the overall shapes are the same, these details do not change the perceived vowel quality,

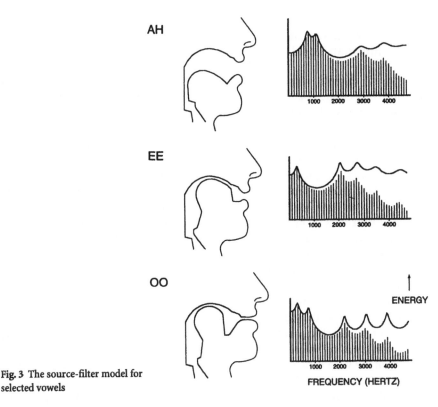

Fig. 3 The source-filter model for selected vowels

which would be that of an /ə/. However, the example shown at the top right of Figure 2 would be perceived to have lower pitch because of its lower fundamental frequency.

The flexibility of the human vocal tract, in which the articulators can easily adjust to form a variety of cavity shapes, results in the potential to produce a wide range of sounds. For example, the particular vowel quality of a sound is determined mainly by the shape of the supralaryngeal vocal tract, as reflected in the filter function. Figure 3 illustrates this effect. Three different vocal tract shapes are shown corresponding to the vowels "ah "(/a/), "ee" (/i/), and "oo" (/u/). In this example, we have used schematized vocal tract shapes from the Haskins Laboratories articulatory synthesizer (see Sect. 9). Plotted in the same graph for each tract shape is the smoothed transfer function that is computationally derived by the synthesizer, as well as the hypothetical energy spectrum that would result from using these functions to filter a glottal source spectrum with a fundamental frequency of 100 Hz. Note that although all three vowels have the same fundamental frequency, their spectra differ according to the filter characteristics of the various vocal tract shapes. Detailed accounts of the acoustic properties of the vocal tract can be found in a number of sources, including Fant (1960), Flanagan (1965), Fry (1979), and Lieberman and Blumstein (1988).

3
A Brief Comparison of Human
and Nonhuman Vocal Tract Characteristics

The acoustic theory of speech production (Fant 1960; Flanagan 1965; Lieberman 1975) described in Section 2 relates changes in the vocal tract shape to the resultant acoustics. In an application of this approach, Lieberman and colleagues (Lieberman 1969; Lieberman et al. 1969) studied animal vocalization using acoustic analysis procedures similar to those used for human speech. In particular, spectrographic analysis, waveform measurements, and computer modeling were used to study the vocalic repertoire of a rhesus monkey. Formant frequencies were calculated using area functions derived from plaster castings of the animal's oral cavities and the simulated acoustic properties of the monkey's vocal tract were compared with those of the human. A plot of the lowest two formant frequencies (F1 by F2) revealed that both simulated monkey formant frequencies and values for actual vocalizations lie in an extremely limited acoustic range compared with that of the human vowel space. This limitation was attributed to an anatomical difference between humans and other mammals: While the pharyngeal and oral cavities of the human vocal tract lie at right angles to one another, those of nonhuman vocal tracts are more nearly in a straight line (see Figure 4). Humans can position their tongues in a manner that changes the point of maximum constriction in the tract, allowing differential shaping of the entire structure into two (or three) cavities (tubes). This flexibility in the shaping of the vocal tract's cavity relationships permits the production of the wide variety of sounds observed in speech (Lieberman 1975, 1984). Although Lieberman's simulations are of interest, it should be pointed out that he did not empirically explore the range of sounds that rhesus monkeys are capable of producing (see below).

Comparison of the anatomy of the human larynx with those of other vertebrates suggests that in attaining the power of speech, some of the protective functions provided by more primitive larynges were relinquished (Lieberman 1984; Kirchner 1988). In nonhuman mammals, the larynx and epiglottis are higher in the pharynx than in adult humans. Indeed, for most mammals, especially the herbivores, the larynx is so high that it contacts the soft palate, thus causing a separation of breathing and eating functions. As shown in Figure 4, the epiglottis is in a relatively high position in these animals. It is raised during drinking or eating, sealing off the oral cavity from the nasopharyngeal passage and making it almost impossible for liquid or solid food to pass into the pharynx. This arrangement is also observed in human neonates, and prevents them from suffocating while nursing (Laitman et al. 1977; Sasaki et al. 1977). The adult human is not so fortunate, and is subject to choking while eating. In this case, the epiglottis and larynx are lower and the pharynx serves as a common pathway for air, liquids, and food — increasing the chance that objects may fall into the larynx and block the airway to the lungs (Lieberman 1984).

Recent work (reviewed by Owren and Linker 1995; Hauser 1996) questions the presumption that nonhuman primates lack both the laryngeal control and the flexibility in vocal tract movement needed to produce a variety of meaningful utterances. There

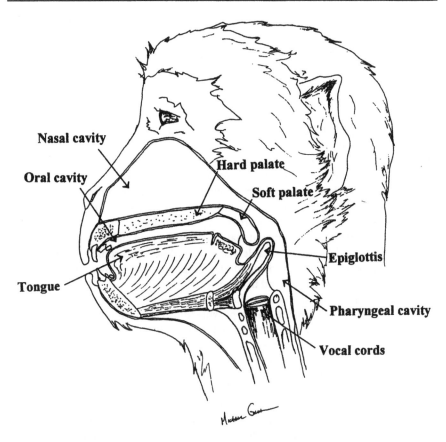

Fig. 4. Midsagittal view of a rhesus monkey vocal tract. (Drawing by Michael Graham)

is strong evidence of vocal tract filtering in the productions of a number of nonhuman primates, including baboon *grunts* (Andrew 1976; Owren et al. 1993, 1995), tonal *coo* calls and other vocalizations in macaques (Hauser et al. 1993), semantic alarm calls in vervet monkeys (Owren and Bernacki 1988; Owren 1990), and *double grunts* produced by wild mountain gorillas (Seyfarth et al. 1994). The work of Hauser and colleagues (Hauser 1991, 1992, 1996; Hauser and Fowler 1992; Hauser et al. 1993; Hauser and Schön Ybarra 1994) provides particularly compelling evidence for the ability of monkeys and apes to alter the shape and length of their vocal tracts using a variety of articulatory maneuvers, resulting in a range of resonant frequency patterns. Their findings suggest that spectral properties can be controlled independently of the glottal source. There is also substantial evidence for precise control of vocal fold vibration in monkeys and apes (Andrew 1976; Hauser and Fowler 1992; Hauser et al. 1993; Owren et al. 1995). Additionally, Nowicki and his colleagues (see Gaunt and Nowicki, this Volume) have demonstrated that the vocal tract above the syrinx modifies sounds produced by songbirds. These animals can also modify sounds both through "articulatory" movements of the beak and by elongating the neck, which changes the resonance of the vocal tract. Taken together, these results stand in strong opposition to earlier claims that the source-filter approach applies only in the case of human speech production.

4
Acoustic Analysis

A rich set of tools and techniques is available for the detailed study of speech acoustics (Rabiner and Schafer 1978; Witten 1982; Fallside and Woods 1985; O'Shaughnessy 1987, 1995; Kent and Read 1992). The ability to convert the acoustic signal to a digital record means that computers, with an arsenal of numerical and statistical methods, can be used to tease apart the fine structure of a sound both rapidly and reliably. Digital techniques have, in large part, replaced oscilloscopes and analog spectrographic devices. Digital analysis techniques generally provide a greater dynamic range, are more flexible, and use a representation of the signal that can be reanalyzed and reviewed in a variety of ways. The availability of affordable and powerful personal computers, with high-resolution graphical displays and sound digitizing interfaces, allows these tools to be used both in a desktop system, and in the field.

Techniques that examine the time-domain characteristics of the signal can be used to make duration measurements and to provide information about intensity and periodicity. Frequency – domain analysis is used to examine the spectral characteristics of the signal – the underlying formant structure, in particular the details of individual spectral cross-sections. A common approach uses analysis based upon the Fourier transform, which represents signals as sums of weighted sinusoids (Brigham 1974; Fallside 1985). Another common technique that provides information about frequency values is linear prediction coding (LPC) (Atal and Hanauer 1971; Markel and Gray 1976; Atal 1985), a time-domain coding method that is used in analysis, storage, and synthesis. LPC implements a source-filter model for separating the source characteristics of the signal (e.g., fundamental frequency) from the filter (vocal-tract) characteristics. Examples of the use of LPC analysis in the study of animal vocalization are described in more detail in the chapter by Owren and Bernacki (this Volume; see also Carterette et al. 1979, 1984; Owren and Bernacki 1988; Owren 1990; Shipley et al. 1991; Seyfarth et al. 1994).

Figure 5 shows the output of a system used at Haskins Laboratories, known as the Haskins Analysis Display and Experiment System, (HADES; Rubin 1995), that performs such analyses. The large window (on the left in Figure 5) contains three panels that provide both temporal and spectral information about a signal. The top panel is an acoustic waveform (energy vs. time) of the utterance, "The cow chewed its cud". The bottom panel is a spectrogram of the same utterance, calculated using fast Fourier transform (FFT) analysis — a computational method for efficiently calculating the discrete Fourier transform (Cochran et al. 1967). The spectrogram provides a representation of frequency (on the vertical axis) and amplitude (depicted by varying levels of darkness) over time (the horizontal axis). Note the formant structure in this broadband analysis of the signal, as shown by the darker bands of greater acoustic energy. The middle panel provides summary information from an LPC analysis of the signal. The short dark ver

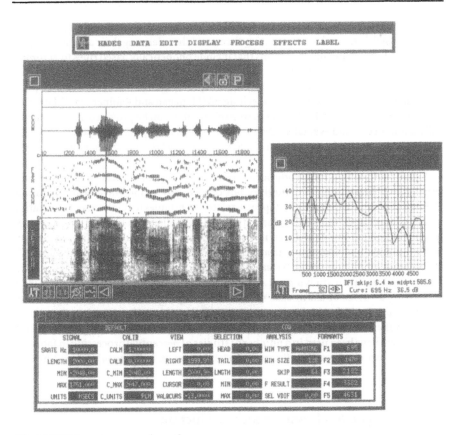

Fig. 5. HADES spectrogram and waveform

tical bars correspond to the peaks of the smoothed LPC spectra. An example of such a spectrum is shown in the window on the right Figure 5. This window shows a spectral cross-section for a portion of the vowel in the word "cow", at the point indicated by the vertical bar in the spectrogram. The "control panel" window at the bottom of the figure provides additional information about the signal, including the "formant" peaks (for the selected spectrum) automatically obtained from the LPC analysis (these are actually the peaks of the LPC spectrum for a selected frame). This sort of analysis is the standard approach for displaying and quantifying spectral and temporal aspects of both speech and non-speech signals. As mentioned above, tools for producing such analyses are now readily available on desktop and portable computers. Other chapters in this Volume provide additional details about speech analysis techniques.

Synthesis provides another tool for studying speech acoustics. In this technique, the acoustic waveform is created electronically, through the use of specialized circuitry or by combinations of computer hardware and software. A variety of methods can be used to provide an acoustic description. These include replaying stored samples of segments, words, and phrases; copying and regenerating aspects of the acoustic information from its spectrographic representation; specifying segmental descriptions (phonemes, diphones, syllables, etc.), concatenating them, and then deriving their acoustic properties; and using text-to-speech systems that include phonetically based rules to automatically convert strings of text into acoustic patterns. This methodology provides a practical means for experimentally testing the perceptual significance of particular aspects of the acoustic information in the speech signal (Flanagan and Rabiner 1973; Klatt 1980, 1987). For example, details of both the source and filter can be provided and systematically manipulated and evaluated. Overviews of speech perception, analysis, coding, and synthesis are available in Flanagan (1965), Witten (1982), and O'Shaughnessy (1987, 1995).

Signal synthesis has frequently been used to simulate animal vocalizations, although usually involving editing of natural sounds or relatively simple waveform generation. Perhaps the earliest application of speech-related techniques to nonhuman vocalization synthesis was Capranica's (1966) formative study of bullfrog mating calls. In this work, analog equipment was configured in a source-filter fashion with periodic pulse trains and white noise providing input to a parallel combination of resonant circuits. Digitally based linear-predictive synthesis was later used by Carterette and his colleagues (Carterette et al. 1984) in studying cat vocalizations (see also Shipley et al. 1991). Signal processing analyses of alarm calls of vervet monkeys by Owren and Bernacki (1988) provided the basis for evaluating species-typical perceptual processing using LPC-based synthetic stimuli (Owren 1990). Monkey calls generated specifically by speech synthesizers have been used in a number of studies of perceptual processing in Japanese macaques, including work by Petersen (1981), May et al. (1988, 1989), and Hopp et al. (1992).

5
Measuring and Analyzing Speech Production

The configuration of the human vocal tract, which "shapes" speech acoustics, depends on the position of the speech articulators (e.g., the tongue, lips, jaw, velum, and larynx). Furthermore, because the acoustics are continually changing during speech, it is the behavior of the speech articulators over time — the *changes* of articulatory configuration and their acoustic consequences — that must be analyzed. Recently, more adequate tools for observing speech production have been developed resulting in renewed interest in considering speech articulation and acoustics together. This Section provides a brief sketch of the history of acoustic and articulatory research and examines some major experimental and analytical techniques used to study speech articulation. Of particular relevance to research with animals, we suggest, are those techniques focusing on the dynamic behavior of individual articulators (e.g., jaw, tongue), the functional or task-specific spatiotemporal coordination among articulators (e.g., lip—jaw, tongue—jaw, lip—larynx), and the specification of an articulatory-acoustic relation from which anatomical and behavioral constraints may evolve.

6
Different Limitations of Human and Animal Research

Research using both human and animal subjects may endanger them. This has limited invasive experimentation, particularly in humans. Therefore, much of our detailed physiological and anatomical knowledge of sound production and perception systems has come from studies of other animals, often chosen because of presumed structural and/or functional similarities with humans — for example, hearing in chinchillas and cats (Neff 1964), laryngeal structure and innervation in dogs and monkeys (e.g., Larson 1988; Alipour-Haghighi and Titze 1991), and perceptual processing in nonhuman primates (Stebbins and Summers 1992). Direct knowledge of humans comes from *post mortem* anatomical and histological studies (Galaburda 1984), from studies of accidental trauma due to war, disease, injury, and surgery (e.g., Walsh 1957; Luria 1975; Weismer 1983), or from congenital abnormalities such as cleft-palate (Warren 1986) and deafness (Rubin 1983). The great advantage of using human subjects in behavioral studies is their ability to understand instructions and exert adequate control at various levels of perception and production. Behavioral research on other animals often must be conducted indirectly via discrimination tasks, which usually entail a considerable investment of resources for training (e.g., Norris and Møhl 1983; Petersen et al. 1984). However, at least until recently, research involving animals has been fairly free to combine systematically well controlled behavioral and physiological techniques – techniques which tend to be dangerously invasive (for overviews, see: Neff et al. 1975; Simmons and Grinnell 1988).

7
Overview of the Shifting Articulatory
and Acoustic Emphases in Speech Research

Before high-resolution data transduction and recording techniques became available in the second half of this century, speech was ephemeral and its record impressionistic. The best a transcriber could do was to write down what was said in a notation that would allow a rough reconstruction of what had been heard. By the 1880s, a standard orthography for phonetic transcription had emerged (the International Phonetic Alphabet, or IPA), which scholars hoped would be rich enough for precise transcription of utterances in any of the world's languages. This development was grounded in the use of symbols for minimally distinctive sound segments (e.g., the phonemes /d/ and /t/ in 'hid' vs. 'hit') augmented by diacritic marking of context-specific phonetic differences. However, the method was still susceptible to the biases or misperceptions of the listener/transcriber.

Early methods of in vivo investigation were restricted to what could be seen (e.g., movement of the lips and jaw), felt (e.g., vibration of the larynx, gross tongue position), or learned from practiced introspection of articulator position during production (e.g., Bell 1867). However, when the results of these methods were combined with those from anatomical and mechanical studies of cadavers, a great deal was correctly surmised about the relation between vocal tract shape and the resultant acoustics. This knowledge had practical applications such as teaching the deaf to speak, and provided the basic scheme for modeling articulation e.g., the vowel space, pitch dependence on pressure, elastic tension of the vocal folds, and height of the larynx.

By the end of the nineteenth century, the development of mechanical transduction techniques for slow moving events such as rhythmic motion of the jaw, thorax (respiration), and other rhythmically entrained structures (e.g., finger- and foot-tapping during speech) made fairly detailed kinematic analysis possible (Sears 1902; Stetson 1905). Physiological studies began as well, for instance, examining transduction of neuromuscular events (Stetson 1928). By modern standards, analysis of data from these studies was quite basic (e.g., measures of duration and observationally inferred estimates of articulator speed and impulse force). Nonethless, during this period many interesting claims and comprehensive hypotheses were advanced concerning the organization and control of learned voluntary behaviors (see Boring 1950 for a review). The culmination of this epoch, in which the basic research paradigm primarily entailed inference of articulatory events during production of minimally contrastive phonetic events, was the development of X-ray photography. Finally, the configuration of the entire vocal tract could be captured — first statically and later dynamically (cineradiography) — during speech production. Unfortunately, the characteristic events revealed by analysis of these data were very coarse-grained and difficult to quantify.

The rapid development of acoustic recording techniques and the sudden awareness of the dangers of X-ray exposure in the late 1920s helped drive the shift in interest from speech articulation to acoustics. During the 1930s and 1940s, recording, display, and analysis systems such as wire-recorders, the oscilloscope, and the sound spectrograph (Koenig et al. 1946), respectively, made it possible to study speech acoustic events in greater detail and revealed phoneme-specific information in the acoustic patterns. In particular, vowel formants and consonant-dependent formant transitions were recognized as key components to phoneme identity, and their patterning alone was shown to be sufficient for synthesis of acceptable and distinct syllables such as /ba/, /da/, /ga/ (Cooper et al. 1951; Liberman et al. 1959). Given the ability to synthesize speech from acoustic patterns, it seemed possible to conduct meaningful research in the acoustic domain alone, regardless of the underlying articulatory configurations.

Another contributor to the shift in focus from articulation to acoustics was the emergence of *distinctive-feature* theory. Heavily influenced by information theory (e.g., Fano 1949), this approach proposed that phonemes are composed of distinctive features whose binary values (+/-) enable minimal phonemic and, therefore, informational contrasts (Jakobson et al. 1963). Thus, consonant pairs such as /p,b/, /t,d/, and /k,g/ can be distinguished by the value of the voicing feature alone. Although feature detection was not restricted to the acoustic domain, the distinctive feature's role as the critical information-bearing element in the process of perceiving speaker intention required that the underlying message be encoded in the acoustic properties of the signal. This connection led quite naturally to the assumption that the medium of communicative interaction was strictly acoustic.

These developments were followed by an intense effort to decompose acoustic signals into minimally contrastive cues to "phonetic intent" (e.g., Lisker 1957; Abramson and Lisker 1965; Liberman et al. 1967; Liberman and Studdert-Kennedy 1978; Repp 1983, 1988; O'Shaughnessy 1987). However, despite demonstration of the perceptibility of a variety of acoustic cues, a number of problems became apparent. Taken together, these persistent difficulties suggest that consideration of acoustics alone may not reveal how listeners actually arrive at a given speech percept. One problem is that speech acoustics are highly variable both within and across speakers. Not only can the same phoneme have different acoustic properties when produced by different speakers, but also no two utterances by the same speaker are acoustically the same. Another problem is the difficulty in finding acoustic cues that persist across the range of phonetic contexts. Thus, the relatively long time between the release of a voiceless stop consonant (e.g., /p,t,k/) and the onset of voicing for a following vowel (as in /pa/) clearly distinguishes this kind of sound from its voiced counterpart (/b,d,g/, respectively) — if the syllable is stressed. In other contexts, however, this voice-onset time (VOT) dimension is not as strong a cue to voicing, or may be absent entirely (e.g., before a pause). The search for persistent cues led to a third problem, that multiple cues to the same phonetic feature may overlap or trade off, depending on the context, speaker, or other factors. For example, vowel quality and duration can both provide cues to consonant voicing (Lisker and Abramson 1967). The inability to identify unique cues for specific features suggested the possibility of multiple mappings between phonetic categories and acoustic distinctions. Finally, recent research has shown that both visual and acoustic information can be useful in speech perception (Benoît et al. 1992, 1994; Massaro et al. 1993; Sekiyama and Tohkura 1993). Indeed, the two modalities appear to complement each other in that some of the cues that are more unstable acoustically (for instance, cues to place of articulation that often occur very briefly in the fine structure of the spectrum) are visually the most consistently perceived, while the opposite is true for acoustically more stable cues such as nasality (for review see Summerfield 1987, 1991).

In articulation research, recognition of the mismatch between variable acoustic events and phonetic categories coincided with improvements in articulatory transduction techniques (e.g., greater subject safety, increased measurement accuracy, and reduced cost) and a growing corpus of physiological data from studies of other biological movement systems that suggested lawful constraints on the organization and production of biological behavior. Coupled with the notion that communicative intent must go through the production structure before any acoustic "encoding" can occur, these developments have led to an emphasis on considering phonetic abstractions in terms of underlying articulatory behavior.

As an example, consider production of the bilabial /b/ between two vowels (e.g., /aba/). Acoustically, no two productions are identical, even in the same speaker. Yet, the articulatory event is relatively simple. Following the first vowel, the lips come together, stopping the airflow from the vocal tract, and then are released for the next vowel. Vocal fold vibration, necessary for each of the two vocalic segments, may or may not stop during the closure period. Such stopping depends on whether or not the lips are closed long enough for the air pressure above the glottis to become the same as the pressure below the glottis, at which point the vocal folds will cease to vibrate. Thus, at least some of the acoustic variability can be explained by observing the timing of lip closure and release relative to aerodynamic factors such as air flow and intra-oral pressure.

The difference between an oral /d/ and a nasal /n/ provides another example of the articulatory perspective. Both of these sounds are produced by placing the tip of the tongue against the alveolar (maxillary) ridge and/or front teeth. Articulatorily, they are distinguished by whether or not the velar port is open. If it is, the result is a nasal sound. Acoustically, however, the difference is quite complex. In early articulatory synthesis, the degree of velar port opening was systematically varied in ordered steps — producing a stimulus continuum that ranged from acoustically perceptible /da/ to /na/ (Rubin et al. 1981) — simply by increasing velar opening from completely closed to the degree of opening necessary for an acceptable /na/. Producing a similar effect with acoustically based synthesis requires simultaneous control over a variety of parameters, including frequencies and bandwidths for three oral resonances, as well as the frequencies and bandwidths of nasal resonances and antiresonances. Another example of how simple variations in articulatory movement can result in complex changes in both acoustic and phonetic detail is provided in Section 8.

8
A Paradigm for Speech Research

A recurrent belief among speech researchers is that the listener extracts information about the production process itself from the speech signal — a process that, despite small differences in detail, is anatomically and physiologically constrained the same way for the entire species (Liberman et al. 1967; Mattingly and Liberman 1969; Fowler et al. 1980; Browman et al. 1984; Liberman and Mattingly 1985; Mattingly and Liberman 1988; Browman and Goldstein 1992; Fowler 1995). Thus, if the invariant aspects of production can be detected using analytical techniques, they can be used to determine the cognitive and neurophysiological underpinnings of speech behavior, as well as the mapping between the articulatory-to-acoustic domains. Furthermore, it seems increasingly likely that the anatomical and physiological constraints on speech production, while allowing extremely complex coordinated behaviors and acoustic output, may be quite similar in form to constraints affecting other biological behaviors. For example, the attempt to model the motion of the speech articulators may benefit from efforts to model other movements, such as locomotion, arm movement, and posture control.

Articulatory research of the past two decades has sought to describe the correspondence between phonetic units and the spatiotemporal behavior of various speech articulators, including attendant muscle activity, regulation of air supply, airflow through the larynx, and, to a lesser extent, airflow throughout the vocal tract. Whatever the actual subject of study, experiments have generally been designed to examine either the behavior of an articulator, or a set of articulators, over time and across different perturbing contexts (e.g., different vowel and consonant combinations, different speaking rates and intonation patterns, or experimentally controlled mechanical perturbations to the articulators). Measures in such studies have included characterizations of muscular activity, articulator motion and configuration, and resulting acoustic energy. The specific articulatory structures and combinations observed have been primarily determined by technological limitations on data acquisition and analysis. However, recent developments have facilitated simultaneous data collection from many sources, as outlined in Figure 6 (see also Borden and Harris 1984; Baken 1987; Fujimura 1988, 1990; Hardcastle and Marchal 1990; Kent et al. 1991; Bell-Berti and Raphael 1995).

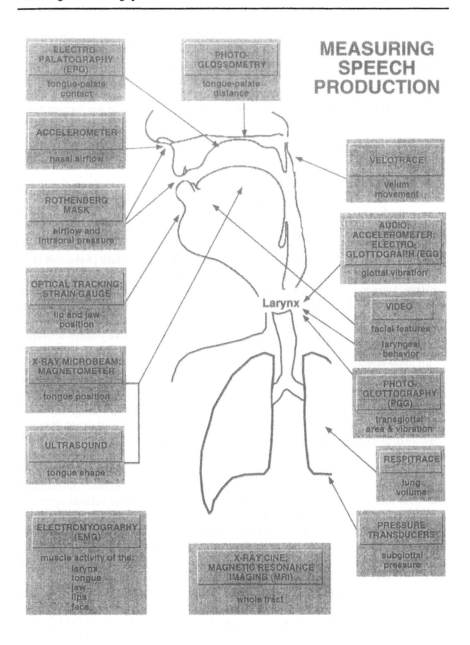

Fig. 6. Measuring speech production

Although by no means exhaustive, Figure 6 provides an overview of the various transduction devices and techniques that have been used to investigate vocal tract structures during speech production. Only a few of the structures involved in speech and nonspeech vocal production are readily accessible to noninvasive external view. Motions of the lips and jaw can be transduced optoelectronically (Sonoda and Wanishi 1982; Harrington et al. 1995; Vatikiotis-Bateson and Ostry 1995) or using strain gauges (Abbs and Gilbert 1973). Measurements of the lips and jaw, as well as other facial features, can also be made from video or film sources — examples in nonhuman vocal production can be found in the work of Hauser and colleagues (Hauser et al. 1993; Hauser and Schön Ybarra 1994), where frame-by-frame-video analysis was used to explore the role of mandibular position and lip configuration in rhesus monkey call production.

In human speech production, a number of other noninvasive techniques have been used. Air flow at the external boundaries of the nasal and oral cavities may be recorded using a Rothenberg mask (Rothenberg 1977). Glottal waveforms can be externally sensed using an electroglottograph, which measures impedance changes as the vocal folds open and close (Fourcin 1974, 1981; Kelman 1981; Rothenberg 1981), or an accelerometer, a transducer used to measure vibrations on the body surface (Askenfelt et al. 1980). In both cases, the device is strapped onto the neck in the vicinity of the thyroid cartilage. Lung volume can be measured using a spirometer (Beckett 1971) or a body plethysmograph (Hixon 1972) while the contributions of the ribcage and abdominal cavities to lung volume change can be evaluated using magnetometers (Mead et al. 1967), mercury strain gauges (Baken and Matz 1973), and inductive plethysmographs (Sackner 1980).

All other techniques included in the figure are to some extent invasive and require cooperation of the subject, particularly in the placement of transduction devices and sensors. Using a flexible fiber-optic endoscope, the larynx can be illuminated for video and photoglottographic recording of the laryngeal structures and transglottal areas (Sawashima et al. 1970; Fujimura 1977; Sawashima 1977; Fujimura et al. 1979). It is also possible to place miniature pressure transducers above and below the glottis for measurement of supra- and sub-glottal pressure (Cranen and Boves 1985; Gelfer et al. 1987).

Observation of the tongue, the most versatile and complex speech articulator, has been the most difficult of all. Optoelectronics, electromagnetic inductance, ultrasound, and X-ray imaging are currently available methods used to measure various aspects of tongue movement. For example, photoglossometry is an optoelectronic technique that uses reflection to measure the distance between the tongue surface and points on the hard palate (Chuang and Wang 1975). Electropalatography (Hardcastle 1972; Fletcher et al. 1975; Recasens 1984; Hardcastle et al. 1991) measures the pattern of contact between the tongue and the hard palate. The X-ray microbeam system tracks sagittal position of radio-opaque pellets on the surface of the tongue, lips, and jaw (Kiritani et al. 1975; Nadler and Abbs 1988; Westbury 1994). Electromagnetic techniques (e.g., magnetometers) can be used to recover similar information through transduction of field fluctuations at multiple points on the various articulator surfaces (Hixon 1971a,b; Schönle et al. 1987; Perkell et al. 1988, 1992; Tuller et al. 1990; Löfqvist et al. 1993; Löfqvist and Gracco 1994). Ultrasound has been used to acquire point-specific tongue data as well (Keller and Ostry 1983; Kaburagi and Honda 1994), but is used primarily to provide dynamic views of the tongue surface and other soft tissue structures (Morrish et al. 1985; Stone et al. 1988). While no system has yet surpassed the high resolution, sagittal view of the entire vocal apparatus provided by cineradiography (Perkell 1969; Subtelny et al. 1972; Wood 1979), the recently developed magnetic resonance imaging (MRI) technique is very promising (Baer et al. 1991; Moore 1992; Tiede 1993; Dang et al. 1994; Rubin et al. 1995). Although MRI applications have been limited to imaging the static vocal tract, improvements in scan rates and image-enhancement techniques may soon allow highly detailed, three-dimensional images of the vocal tract and surrounding structures during active speech production.

The ability to record various signal combinations in synchrony compensates substantially for the individual limitations of the available time-varying measurement techniques. There is a basic tradeoff between techniques that make rapid and accurate 'fleshpoint' measures of particular vocal tract structures (for instance, using tiny pellets placed on the tongue surface or mandible in the case of the X-ray microbeam, or markers placed on the lips and jaw in the case of optoelectronic position sensing devices), and those that provide more global views of the vocal tract (but with poorer spatiotemporal resolution), such as ultrasound or MRI. For example, while ultrasound has scan times that are short enough (approximately 35 ms) to allow tracking of the relevant motions of the tongue body, it cannot capture tongue-tip motion during production of /d,t,n/, where the tongue may contact the maxillary arch for as little as 15 ms. Ultrasound transduction of tongue-tip gestures is further complicated by the requirement that there be only one air—tissue boundary between the externally mounted transmitter/transducer and the articulator surface of interest. When raised to the maxillary arch, the tongue tip usually causes an additional air cavity to appear between the underside of the tongue and the mouth floor. However, when ultrasound is combined with fast transduction systems, such as the X-ray microbeam or magnetometer, a much better picture of the tongue's activity emerges (e.g., Stone 1990, 1991). Used together, then, the wide variety of transduction devices currently available makes it possible to assess the dynamic interaction of laryngeal and supralaryngeal structures at biomechanical, neurophysiological, and acoustic levels of analyses.

A multitude of articulatory studies have been conducted using these various techniques. Although many different issues have been addressed in this body of work (see Levelt 1989 for a review), the invariance issue has been fundamental. Many studies have attempted to identify the articulatory characteristics of different phonemes using experimental designs that manipulate stress and speaking rate (e.g., Kuehn and Moll 1976; Gay 1981), or phonetic context (e.g., Sussman et al. 1973). Stress-rate studies have been used to identify articulatory attributes of the phoneme that are independent of the changes induced when non-phonetic factors are varied. Experiments that vary the phonetic context (e.g., /aba, ibi, ubu/) are similar in that they allow the perturbing effects of the context to be distinguished from the inherent, and presumably stable, characteristics of the target phoneme (/b/ in this example). Another direction taken in the search for invariance has been to demonstrate that the articulators act in a flexible but highly coordinated fashion in achieving specific phonetic goals. These studies have used mechanically induced perturbations, either to reduce the number of articulators involved in a task (for instance, in bite-block and braking studies in which mandible position is fixed; e.g., Folkins and Abbs 1975; Lindblom et al. 1979), or to severely limit an articulator's contribution (for instance, through dynamic perturbation of the lips or jaw; e.g., Abbs and Gracco 1983; Kelso et al. 1984; Gracco and Abbs 1985; Saltzman et al. 1995).

Normal production processes always involve multiple articulators, even if some of the structures may not be moving. Thus, although the relative contributions of the two lips and the jaw may vary in bilabial productions (e.g., the final /b/ in baeb), the acoustic result is roughly the same in each case. Perturbing the system, for instance by removing the contribution of an articulator such as the jaw and examining kinematic and physiological effects, demonstrates that flexible and rapid compensatory effects act to prevent any loss of intelligibility in the speech signal. Because recording can be synchronized in the data channels of interest (including the perturbation delivery signal), the timing of articulatory events at both neuromuscular and kinematic levels of observation can be precisely examined (e.g., Tuller et al. 1982, 1983).

Although factors such as stress and speaking rate have fairly consistent articulatory laryngeal and supralaryngeal correlates, variability both within and across speakers is nonetheless quite high. One way to work around much of this variability is to focus on the relations among kinematic variables instead of on the individual measures themselves (e.g., Ostry et al. 1983; Kelso et al. 1985). For example, the relation between peak velocity and movement amplitude for a given articulator is quite stable and linear across most of its range of motion. At the same time, changes of stress and speaking rate are clearly marked by local changes in the relationship between these measures. Borrowing from the study of other biological movement systems, such as limb motion, researchers have begun to model such patterns of behavior as second-order mechanical systems. In this technique, the dynamic parameters of such systems (e.g., mass, stiffness, and viscosity) can be inferred from the relations among kinematic observables. Among other things, the approach promises the possibility of adducing stable (invariant) values of dynamic parameters from variable kinematic measures — values that can be compared across articulatory structures and many speaking contexts, including different languages (Vatikiotis-Bateson and Kelso 1993).

Figure 7 illustrates a qualitative method for discerning structure in the continuous kinematics of recurrent behavior. In this example, position and instantaneous velocity are plotted as trajectories in *phase space*, which provides a graphical description of the relation of the state (or phase) variables of dynamic systems. The result is a two-dimensional space where time is implicit in the continuity of consecutive data points but is not shown on a separate axis (see Abraham and Shaw 1982, 1987). Furthermore, such qualitative assessment can be used to direct subsequent quantitative analysis. Thus, even though gross effects of stress can be seen in the left panel of the figure as the roughly alternating sequence of larger and smaller movements, it is quite difficult to interpret the significance of these individual patterns of articulator position and instantaneous velocity, much less any relationships between them. But, when the two variables are plotted in phase space, as in the right panel of the figure, their continuous correlation can be seen in the stability of the trajectory shapes associated with the repetitive syllable sequence. Certain aspects of their variability also become readily apparent. For example, the phase portraits show motion of the articulators to be less variable during production of the consonant (top right) than of the vowel (bottom right). It can also be seen that the correlation between velocity and position is different for the two articulators, particularly during the closing phase of the movement cycle. The phase portraits reveal that there is greater tendency for covariation of movement amplitude and peak velocity for the jaw alone than for the lower lip — whose motion includes that of the jaw.

The phase-space representation also provides a means for considering distinctive differences in interarticulator timing. In Figure 8, for example, the relative timing of upper-lip movement toward closure for the second /b/ in bapab and the vowel—vowel movement cycle of the jaw is expressed in phase angle (Tuller and Kelso 1984; cf. Nittrouer et al. 1988). Across a range of available data (i.e., including changes in medial consonant identity, syllable stress, and overall speaking rate), the latency of upper-lip movement onset for the medial consonant has been found to be linearly proportional (Kelso et al. 1986a, 1986b), or nearly so (Nittrouer et al. 1988), to the period of the jaw-motion cycle associated primarily with production of the preceding vowel. Furthermore, phase-angle analysis of these data has demonstrated reliable differences for the medial consonants tested (Kelso et al. 1986a).

Thus, phase-angle analysis provides a precise means of transforming an articulatory database involving many dimensions of variability into a more functionally relevant form with fewer extraneous influences. Similar approaches are applicable across all articulator systems (Saltzman and Munhall 1989; Löfqvist 1990; Tuller and Kelso 1995). Examples include the coordination of laryngeal and supralaryngeal structures, such as the lips (Munhall et al. 1986; Munhall and Löfqvist 1992) or the tongue (Manuel and Vatikiotis-Bateson 1988), and functional coordination among various supralaryngeal articulators, such as coupling of the tongue and lip (Faber 1989) or the tongue and jaw (Stone and Vatikiotis-Bateson 1995).

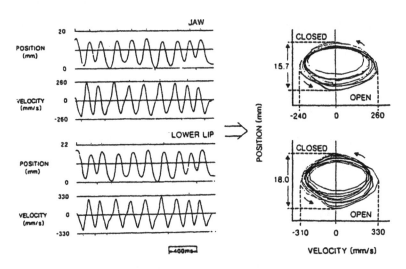

Fig. 7. *Left* Position and velocity over time of jaw and lower lip light-emitting diodes (LEDs) for sentences produced with reiterant /ba/ at a normal rate. *Right* Corresponding phase plane trajectories. *CLOSED* denotes the highest position achieved for /b/ and *OPEN* the lowest position for /a/. (From Kelso et al. 1985, ©1985 Acoustical Society of America)

9
A Model of the Human Vocal Tract

Although there is great interest in studying the speech production process, some of the methods discussed in the Section 8 place practical limits on the amount of data that can be gathered and analyzed. In addition, speakers cannot exercise the degree of control over their articulators needed for certain studies of the contributions of individual articulators. Paralleling the method for studying speech production and perception that uses speech synthesized from acoustic parameters as a fundamental tool, we use an articulatory synthesis (ASY) system at Haskins Laboratories that synthesizes speech through control of articulatory instead of acoustic variables (Mermelstein 1973; Rubin et al. 1981). ASY is designed for studying the linguistically and perceptually significant aspects of articulatory events. It allows quick modification of a limited set of key parameters that control the positions of the major articulators: the lips, jaw, tongue body, tongue tip, velum, and hyoid bone (whose position determines larynx height and pharynx width). Any particular set of parameter values provides a description of vocal tract shape that is adequate for research purposes, and that incorporates both individual

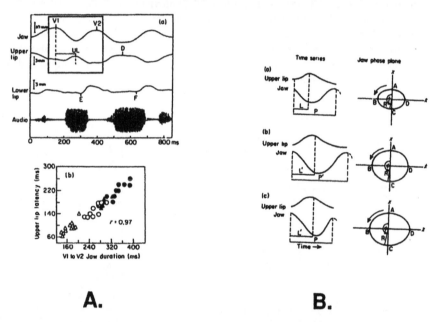

A. **B.**

Fig. 8. A. a Movements of the jaw, upper lip, and lower lip corrected for jaw movement, and the acoustical signal, for one token of /ba'pab/. Articulator position (*y axis*) is shown as a function of time. Onsets of jaw and lip movements are indicated (empirically determined from zero crossings in the velocity records). b Timing of upper lip associated with /p/ production as a function of the period between successive jaw lowerings for the flanking vowels for one subject's productions of /ba#pab/●, Slow rate, first syllable stressed; ○, slow rate, second syllable stressed; ▲, fast rate, first syllable stressed; △, fast rate, second syllable stressed B. Left: time series representations of idealized utterances. Right: corresponding jaw motions, characterized as a simple mass spring and displayed on the "functional" phase plane (i.e. position on the vertical axis and velocity on the horizontal axis). (a), (b) and (c) represent three tokens with vowel-to-vowel periods (*P and P'*) and consonan latencies (*L and L'*) that are not linearly related. Phase position of upper lip movement onset relative to the jaw cycle is indicated. (From Kelso et al. 1986a, used with permission)

articulatory control and links among articulators. Additional input parameters include excitation (the sound source) and movement-timing information. An important aspect of this model's design is that speech sounds used in perceptual tests can be generated through by varying the timing or position parameters. Another very important aspect of the system is that the synthesis procedure is fast enough to make interactive on-line research practicable.

Figure 9 shows a midsagittal view of the ASY vocal tract in which the six key articulators are labeled. These articulators can be grouped into two major categories: those whose movements are independent of the movements of other articulators (the jaw, velum, and hyoid bone), and those whose movements are dependent on the movements of other articulators (the tongue body, tongue tip, and lips). The articulators in the second group normally move when the jaw moves. In addition, the tongue tip can move relative to the tongue body. Individual gestures can thus be separated into components arising from the combined movement of several articulators. For example, the lip-closing gesture used in the production of the utterance /aba/ is a combined movement of the jaw and lips. Movements of the jaw and velum have one degree of freedom, while all others have two degrees of freedom. Movement of the velum has two effects. It alters the shape of the oral branch of the vocal tract and, in addition, changes the size of the coupling port to the nasal tract, which is fixed.

An overview of the steps involved in articulatory synthesis using our model is provided in Figure 10. Once the articulator positions have been specified (see below), the midsagittal outline is determined and can be displayed. Cross-sectional areas are cal-

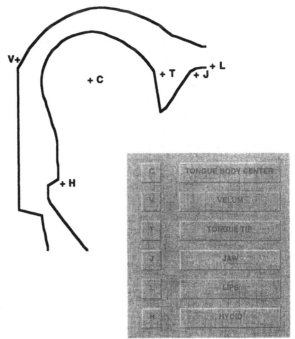

Fig. 9. Haskins Laboratories ASY vocal tract outline with key parameters labeled

culated by superimposing a grid structure, aligned with the maxilla in the outline, and computing the points of intersection of the outline and the grid lines. The resolution of this grid is variable, within certain limits. In general, parallel grid lines are set 0.25 cm apart and radial lines occur at 5° intervals. Sagittal cross-dimensions are calculated and converted to cross-sectional areas, using different formulas for estimating the shape in the pharyngeal, oral, and labial regions. These area values are then smoothed and approximated by a sequence of uniform tubes of fixed length (0.875 cm). The number of area values is variable because the overall length of the tract varies with both hyoid height and degree of lip protrusion. Improvements in the representation of the third dimension of vocal tract shape are eventually expected, and will be based on data from MRI of the vocal tract (Baer et al. 1991).

Once the area values have been obtained, the corresponding acoustic transfer function is calculated using a technique based on the model of Kelly and Lochbaum (1962), which specifies frequency-independent propagation losses within sections and reflections at section boundaries. Nonideal terminations at the glottis, lips, and nostrils are accurately modeled. However, the effects of other variables, such as tissue characteristics of the vocal-tract walls, are accounted for by introducing lumped-parameter elements at the glottis and within the nasal section.

In the interest of computational efficiency, and because the synthesizer was designed as a research tool to provide rapid feedback about changes in articulatory configuration, a number of compromises have been made in the details of the model. For example, acoustic excitation of the vocal tract transfer function is most commonly specified as an acoustic waveform, rather than through simulation of the physiological and aerodynamic factors of phonation. In this approach, control over the shape of individual glottal waveform pulses is limited to two parameters (Rosenberg 1971): the *open quotient* (i.e., the *duty cycle*, or relative durations of the open and closed portions of the glottal cycle) and the *speed quotient* (i.e., the ratio of rise-time to fall-time during the open portion). The fricative source is simulated by inserting shaped random noise anterior to the place of maximum constriction in the vocal tract. Acoustic output is obtained by supplying the glottal or fricative excitation as input to the appropriate acoustic transfer function, implemented as a digital filter.

Greater accuracy is achieved in the phonatory model through using a fully aerodynamic simulation of speech production that explicitly accounts for the propagation of sound along the tract (McGowan 1987, 1988). Such an approach provides a number of benefits, including more accurate simulation of voiced and fricative excitation, interaction of source and tract effects, and the effects of side branches and energy losses. However, it can result in slower overall calculation times. Because of such practical considerations, a choice of methods for calculating acoustic output has been implemented in the model.

Specification of the particular values for the key articulators can be provided in a number of ways. In the simplest approach, a display of the midsagittal outline of the vocal tract can be directly manipulated on-screen by moving one or more of the key articulators. The tract is then redrawn and areas and spectral values are calculated. This method of manual graphical modification continues until an appropriate shape is achieved. This shape can then be used for static synthesis of a vowel sound. Alterna-

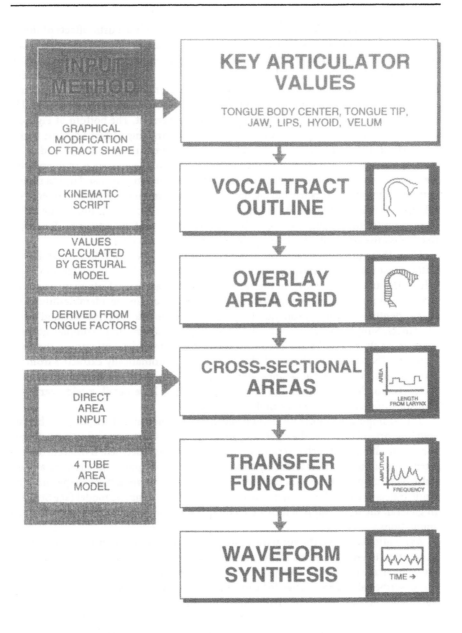

Fig. 10. The steps in articulatory synthesis

tively, individual shapes can be deposited in a table that will later be used as the basis for a "script" that specifies the kinematics of a particular utterance.

Using the method of kinematic specification, the complex acoustic effects of simple articulatory changes can be illustrated. The top of Figure 11 shows midsagittal vocal tract outlines for the major transition points (key frames) in the simulation of the articulations of four utterances: /bænænə/, /bændænə/, /bædnænə/, and /bædd ætə/ (i.e., "banana", "bandana", "bad nana", and "bad data", respectively). In this contrived example, the articulations of the four utterances are produced in very similar ways. The only parameter that varies is the timing of velar movement. The bottom of Figure 11 shows the degree of velar port size opening and the contrasting patterns of velar timing for the four different utterances. For utterance /bænnæə/ the velum is closed at the start, opens rapidly, and stays open throughout the rest of the utterance. In /bændænə/ the pattern of velar opening is similar, except that the velum closes and opens rapidly in the middle of the utterance, during the movement from /n/ to /d/. In /bædn ænə/ the velum stays relatively closed at the beginning of the utterance, opens during the movement from /d/ to /n/, and stays open throughout the rest of the utterance. Finally, in /bæddætə/ the velum stays relatively closed throughout the utterance.

All of these utterances have the same general form: C-V-CC-V-C-V, in which the initial consonant is /b/, and the vowel pattern is / æ,æ, ə /. With the exception of the velum, the articulators move in the same manner in each case. Note that the simple change in timing of velar opening in these four tokens results in considerable differences in the identities of the consonants that occur in the middle and near the end of the utterances. Simple changes in articulatory timing can also result in complex acoustic changes, as illustrated by the pseudo-spectrograms of the four utterances shown in Figure 11. These displays show only formant peaks, automatically extracted from the transfer functions of each utterance, where formant amplitude is indicated by height of the corresponding bar. Although these displays show little detail, a wide variety of acoustic differences can be seen in the four utterances.

This example illustrates one method, albeit a very schematized one, that can be used with the articulatory synthesis model to provide kinematic specifications. In real speech, production of such utterances is more variable and detailed. If desired, one can attempt to simulate these details by varying the model's input parameters on a pitch-pulse by pitch-pulse basis. Alternatively, specifications for vocal tract shape and kinematic trajectories can be calculated using an underlying dynamic model (see Sect. 10). In general, the latter approach is the technique most commonly used in our present simulations.

As mentioned above, the vocal tract model needs improvement in a number of ways. In addition to enhancements already achieved in its aeroacoustic simulation, changes are being made in its articulatory representation. The choice of particular key articulators described above has proven to be too limited. For example, additional tongue shape parameters are needed to simulate tongue bunching and for more accurate control of shape in the pharyngeal region. It would also be desirable to be able to fit the ASY model to a variety of head shapes, including those of females, infants, and, potentially, non-human primates. For this reason, a configurable version of ASY (CASY) is being developed that has increased flexibility in the model's internal linkages, the potential for adding new parameters, and a method for fitting vocal tract shape to actual X-ray or MRI data with concomitant adjustment of its internal fixed parameters.

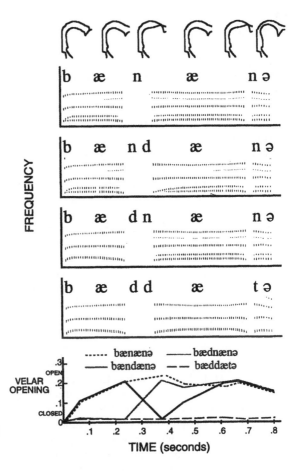

Fig. 11. Articulatory synthesis of
/bæn ænə/, /bænd ænə/,
/bædn ænə/, and /bædd ætə/

Finally, research is also underway to provide a more complete three-dimensional representation of vocal tract shape. A guiding principle of this approach is that actual physiological measurements and comparisons be used as the basis of improvements in both static and dynamic aspects of the simulation of speech production. In addition to our own interest in achieving more accurate physiological modeling, other researchers have focused on areas such as the control and dimensionality of jaw motion (Flanagan et al. 1990; Ostry and Munhall 1994; Vatikiotis-Bateson and Ostry 1995), as well as modeling soft-tissue structures, such as the tongue (Wilhelms-Tricarico 1995) and lips (Abry and Boë 1986; Badin et al. 1994; Benoît et al. 1994; Guiard-Marigny et al., in press).

10
Gestural Modeling

Modeling the speech production process requires a detailed consideration not only of the static anatomic and physiological aspects of the system, but also of how it changes over time. The speech articulators are continually in motion, producing a varying acoustic stream. The perceiver, in turn, is sensitive to both the local details of the resulting acoustic pattern and the global characteristics of change (Remez et al. 1981). In general, a greater emphasis has been placed on studying the static rather than the time-varying aspects of speech events. However, at Haskins Laboratories there has been a long-standing interest in the gestural basis of speech production and its relationship to perception (Liberman et al. 1967; Mattingly and Liberman 1969; Liberman and Mattingly 1985; Fowler 1995). Some of the techniques described in Section 8 are useful for examining the kinematics of the speech articulators. Theoretical approaches to studying action systems have also pointed out the necessity and desirability of examining the dynamic system that underlies these kinematic patterns (Bernstein 1967; Fowler 1977, 1984; Turvey 1977; Kelso et al. 1986a; Tuller and Kelso 1995).

Over the past several years, a computational model has been developed at Haskins Laboratories that combines these intersecting concerns in the form of a tool for representing and testing a variety of theoretical hypotheses about the dynamics of speech gestures and their coordination (Browman et al. 1984; Browman and Goldstein 1985). This approach merges a phonological model, based on gestural structures (Browman and Goldstein 1986, 1989, 1990, 1992), with an approach called *task dynamics* (see below) that characterizes speech gestures as coordinated patterns of goal-directed articulator movements. At the heart of both of these approaches is the notion of a gesture, which is considered in this context to be the formation of a constriction in the vocal tract by the organized activity of an articulator or set of articulators. The choice of gestural primitives is based upon observations of functional units in actual production. These models attempt to reconcile the linguistic hypothesis that speech involves an underlying sequence of abstract, context-independent units with "the empirical observation of context-dependent interleaving of articulatory movements" (Saltzman and Munhall 1989, p 333). The focus in this case is on discovering the regularities of gestural patterning and how they can be specified (see also Perrier et al. 1991; Shirai 1993 Kröger et al. 1995).

The computational model has three major components. First, a gesturally based phonological component (the linguistic-gestural model) provides, for a given utterance, a "gestural score" which consists of specifications for dynamic parameters for the set of speech gestures corresponding to the input phonetic string (Browman et al. 1986), and a temporal activation interval for each gesture, indicating its onset and offset times. These intervals are computed from the gesture's dynamic parameters in combination with a set of phasing principles that serves to specify the temporal patterning among the gestural set (Browman and Goldstein 1990). Second, the task-dynamic model is used to compute coordinated articulator movements from the gestural score in terms that are appropriate for the ASY vocal tract model. This model, in turn, allows computation of the speech waveform from these articulatory movements. An example of such a gestural score, for the utterance [pʰam], can be seen in Figure 12. This figure shows the periods of gestural activation (filled boxes) and trajectories generated during simulations (solid lines) for the four *tract variables* (see below) that are controlled in the production of this utterance: velic aperture, tongue body constriction degree, lip aperture, and glottal aperture.

The task-dynamic model used in this computational system has proved useful for describing the sensorimotor control and coordination of skilled activities of the limbs, as well as the speech articulators (Kelso et al. 1985; Saltzman 1986; Saltzman and Kelso 1987; Saltzman et al. 1987; Saltzman et al. 1988a, 1988b; Saltzman and Munhall 1989; Fowler and Saltzman 1993). For a particular given gesture, the goal is specified in terms of independent task dimensions, called *tract variables*. Each tract variable is associated with the specific set of articulators whose movements determine the value of that variable. For example, one such tract variable is lip aperture (LA), which corresponds to the vertical distance between the lips. Three articulators can contribute to changing LA: the jaw, the upper lip, and the lower lip. The standard set of tract variables in the computational model, and their associated articulators, can be seen in Figure 13. Recently, this set has been extended by incorporating aerodynamic and laryngeal components,

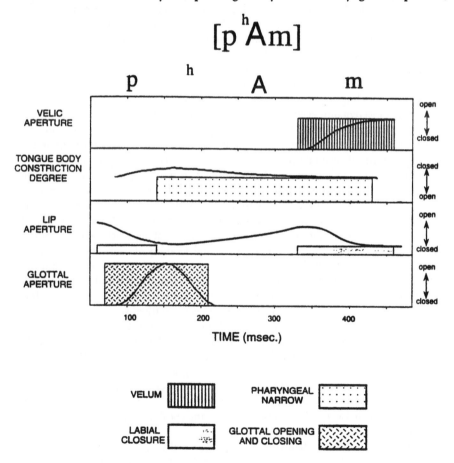

Fig. 12. Gestural score for the utterance [pʰam]

producing a more realistic model of source-related factors (McGowan and Saltzman 1995). Tract variables and articulators compose two sets of coordinates for gestural control in the model. In addition, each gesture is associated with its own *activation* coordinate, whose value reflects the strength with which the associated gesture "attempts" to shape vocal tract movements at any given point in time. Invariant gestural units are posited in the form of context-independent sets of dynamic parameters (e.g., protrusion target, stiffness, and damping coefficients, respectively, for the lips), and are associated with corresponding subsets of all three coordinate systems. Thus, the tract-variable and model articulator coordinates of each unit specify, respectively, the particular vocal tract constriction (e.g., bilabial) and the articulatory synergy that is affected directly by the associated unit's activation. Currently the model offers an intrinsically dynamic account of interarticulator coordination within the time span of single and temporally overlapping (coproduced) gestures, under normal conditions as well as in response to mechanical perturbations delivered to the articulators.

At the present stage of development, the task-dynamic model does not provide a dynamic account of intergestural timing patterns, even for simple speech sequences. Current simulations rely on explicit gestural scores to provide the timing patterns for gestural activation intervals in simulated utterances. While such explicitness facilitates research by enabling us to model and test current hypotheses of linguistically significant gestural coordination, an approach in which temporally ordered activation patterns are derived as implicit consequences of intrinsic *serial dynamics* would provide an important step in modeling processes of intergestural timing. Recent computational modeling of connectionist dynamic systems has investigated the control of sequences (e.g. Grossberg 1986; Tank and Hopfield 1987; Jordan 1989, 1990; Kawato 1989, 1991; Kawato et al. 1990). This serial-dynamics approach is well-suited for orchestrating the temporal activation patterns of gestural units in a dynamical model of speech production.

Connectionist models are also being applied to mapping relationships between acoustics and articulation. Investigators have long attempted to derive information about underlying articulation directly from the acoustic signal (Schroeder 1967; Atal et al. 1978; Wakita 1979; Levinson and Schmidt 1983; Kuc et al. 1985; Shirai and Kobayashi 1986; Sondhi and Schroeter 1987; Larar et al. 1988; Boë et al. 1992; Schroeter and Sondhi 1992; Badin et al. 1995; Beautemps et al. 1995), and a variety of connectionist or neural network methods are now being used for such mappings (e.g., Kawato 1989; Jordan 1989, 1990; Rahim and Goodyear 1990; Bailly et al. 1990,1991; Shirai and Kobayashi 1991; Papçun et al. 1992). In one example, Rahim and Goodyear (1990) trained a multilayer perceptron to map relationships between the power spectra of vowels and consonants and the parameters of a vocal tract model whose shape was specified by the areas of a fixed number of acoustic tube sections. For each member of the training set, input acoustic data consisted of 34 samples of the log power spectrum (from 100 to 4000 Hz). In an analysis-by-synthesis approach, a first-order gradient descent optimization procedure was used to minimize the spectral error between the target and synthesized spectra by adjusting the area values to reduce the acoustic mismatch. This mapping technique provided an efficient method for deriving vocal tract synthesis values directly from acoustic data.

In a related approach, Bailly and colleagues (Bailly et al. 1990;1991) proposed a method of control for an articulatory synthesis model (Maeda 1979) based on the opti-

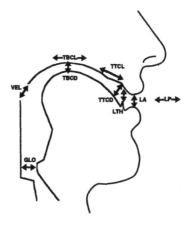

TRACT VARIABLE		ARTICULATORS INVOLVED
LP	lip protrusion	upper and lower lips, jaw
LA	lip aperture	upper and lower lips, jaw
LTH	lip-teeth height	jaw
TTCL	tongue tip constriction location	tongue tip, tongue body, jaw
TTCD	tongue tip constriction degree	tongue tip, tongue body, jaw
TBCL	tongue body constriction location	tongue body, jaw
TBCD	tongue body constriction degree	tongue body, jaw
VEL	velic aperture	velum
GLO	glottal aperture	glottis

Fig. 13. Task dynamic tract variables

mization approach for motor skill learning developed by Jordan (1988, 1989, 1990; see also Rahim et al. 1993). This modified version of Jordan's sequential network operated under certain constraints arising from the kinematic properties of the biological system being controlled and from the phonological task being simulated. Specifically, these constraints restricted the possible solutions that the feedforward multilayered perceptron could use to model the mapping between the production of vocalic gestures and the trajectories of the first three formants. An additional feature of the modeling approach was that it could generalize its movement pattern by interpolating new trajectories based on existing learned trajectories.

Similarly, a number of approaches are being used at Haskins Laboratories to model the development of the connection between articulation and acoustics. The interest is in studying how a neural network model comes to constrain the potential movements of a dynamically changing vocal tract as it "learns" the relationship between acoustics and vocal tract variables and/or gestural scores — a process that may be similar to the exploratory activities found in infants during speech development. Examples include work by McGowan (1994, 1995) and Hogden (Hogden et al. 1993;1996). McGowan has used genetic algorithms (which simulate crossover, mutation, and selection processes) in conjunction with the task-dynamic model to recover articulatory movments from formant frequency trajectories. The result is an analysis-by-synthesis optimization procedure in which the fitness of each gestural score is based on how well its corresponding formants match those of the original signal to recover articulatory movements from formant frequency trajectories. Hogden's approach uses continuity constraints in the process of recovering the relative positions of simulated articulators from speech signals generated through articulatory synthesis.

Finally, connectionist models may be particularly well-suited to directly examining the dynamic properties of the musculo-skeletal system, where previous efforts to characterize the mapping between motor commands to muscles and the resulting behavior of speech articulators have been severely hampered in several ways. For instance, the muscles associated with speech articulation are typically either small and highly interconnected (e.g., tongue muscles), or are hard to monitor safely (e.g., the masseter — the large jaw-raising muscle). Thus, it is difficult to ascertain the muscle sources of electromyographic (EMG) records, which are themselves very complex. Despite the use of signal conditioning and numerical techniques, such as signal rectification and integration, smoothing (low-pass filtering), and ensemble-averaging over multiple trials, identification of "key" events has been restricted to visually observable landmarks in the signal, such as the onset or peak of EMG activity. Interpretation of this restricted set of events has relied primarily on statistical analysis of highly variable mean values, which must then be reliably correlated with other arbitrarily chosen, discrete events in the articulator movement behavior.

In contrast, artifical neural networks, for example, have been used to obtain the forward mapping between muscle activity and resulting articulator motion. Such muscle-based models are inherently dynamic because they estimate the muscle forces required to move the articulators. They also enable the entire EMG to be used as the "motor command input", rather than just those events that stand out visually on a display screen. Hirayama and colleagues (Hirayama et al. 1992, 1993, 1994; Vatikiotis-Bateson et al. 1991) have used real physiological data — articulator movements and EMG from muscle activity — to develop a preliminary model of speech production based on the articulatory system's dynamic properties. Using these EMG data, a neural network learned the forward-dynamics model of the articulators , i.e., mapping from current input (EMG information) and current state (position and velocity) to the next state. After training, the acquired model was incorporated into a recurrent network (i.e, one with feedback loops) that was found to successfully predict continuous articulator trajectories using the EMG signals as the motor command input. Simulations of articulator perturbation were then used to assess the properties of the acquired model.

This kind of modeling implicitly assumes a causal link between muscle activity and movement, rather than taking the more traditional and difficult approach of attempting to reject the implausible null hypothesis of the absence of a connection. Because the goal of the network is to formalize or "learn" that link, any degree of correlation between muscle and articulator behavior is useful in determining the proper coefficients or "weights" of the model equation. In a different, but related, approach, Wada and colleagues (Wada and Kawato 1995; Wada et al. 1995) have demonstrated that a tight coupling exists between formation of movement patterns and recognition of such patterns.

Using examples from cursive handwriting and estimation of phonetic timing in natural speech, they have developed a computational theory of movement pattern recognition that is based on a theory for optimal movement pattern generation and that may be widely applicable across movement systems.

11
Summary

The recent period of rapid evolution in the methods for studying speech has seen a growing interest in a variety of areas, including modeling of the speech production process, methods for examining the kinematics of the speech articulators (often with reference to underlying dynamic models), and initial steps exploring the relationship between acoustics and articulation using both connectionist and other kinds of models. Historically, the use of the sound spectrograph established a research paradigm in which invariant cues to phonetic segments were sought in the acoustic signal. In the last decade, this form of analysis has been supplemented by computer-based analysis systems using both frequency- and time-domain techniques to segregate the signal source from subsequent filtering effects, and to provide detailed information about filter characteristics.

In spite of this significant emphasis on understanding the physical signal, however, invariant links between acoustic and phonetic aspects of speech have proved elusive. Due both to these difficulties and for a variety of theoretical reasons, researchers have increasingly turned to the sound production process itself in attempting to explain the stability and efficiency of speech. In addition, the momentary and punctate cue-based acoustic approach is expanding to include event-based analysis techniques. As these theoretical approaches evolve, computer simulations are increasingly being used as vehicles to explore models of production and coordination, while the availability and use of transduction equipment is providing a means for improving the inherent realism of these models. Overall, a variety of approaches suggest that the future holds promise for successful study of speech production. For instance, both new and existing technologies are becoming more accessible, including X-ray microbeam, alternating field magnetic tracking, and imaging based on magnetic resonance and ultrasound techniques. The utility of dynamic approaches, including serial dynamics, is increasing and will continue to be driven by the intense interest that exists in neural network modeling and other optimization techniques. Advances in these areas will in turn be occurring in a context of increasing power and ease in acoustic and statistical analyses of signals, using desktop workstations or personal computers that are directly accessible both in the laboratory and in the field, rather than being located in remote mainframe computing centers.

It has been proposed that speech perception is a highly specialized process that can be differentiated from other forms of auditory perception (Whalen and Liberman 1987; Mattingly and Liberman 1988). A major factor in that specialization is the degree to which the speech perception and production systems are structurally and behaviorally linked. For example, the fact that articulatory gestures encode information in parallel provides an efficient means for overcoming inherent temporal constraints on resolution in both the auditory and the articulatory systems. Thus, the gestural/articulatory models and analyses discussed here are intended to illuminate at least some of the organizational principles underlying the intimate connection between perception and production in human speech systems. While little information is available as to whether analogous links may be involved in the acoustic communication processes of other species, production and perception processes are also clearly biologically specialized in nonhumans. Thus, while the choice of analysis, description, and modeling methods must entail careful consideration of *what* is important to a given animal in a particular ecological niche, the speech-related approaches discussed here may prove useful in providing both applicable technological innovations and theoretical models that can be adapted to the perception and production processes of nonhumans as well.

Acknowledgments. Preparation of this chapter was supported, in part, by NIH grants HD-01994 and DC-00121 and NSF grant BNS-8820099 to Haskins Laboratories. We thank Catherine Browman, Carol Fowler, Louis Goldstein, Vincent Gracco, Katherine S. Harris, Ignatius Mattingly, Richard McGowan, Robert Remez, Elliot Saltzman, Michael Studdert-Kennedy, and D. H. Whalen for their comments.

References

Abbs JH, Gilbert BN (1973) A strain gage transduction system for lip and jaw motion in two dimensions: design criteria and calibration data. J Speech Hear Res 16: 248–256
Abbs JH, Gracco VL (1983) Sensorimotor actions in the control of multimovement speech gestures. Trends Neurosci 6: 391–395
Abraham R, Shaw C (1982) Dynamics – the geometry of behavior, part I. Periodic behavior. Aerial Press, Santa Cruz,
Abraham R, Shaw C (1987) Dynamics: a visual introduction. In: Yates FE (ed) Self-organizing systems: the emergence of order. Plenum, New York, p 543
Abramson AS, Lisker L (1965) Voice onset time in stop consonants: acoustic analysis and synthesis. Proc 5th Int Congr of Acoustics, Liege
Abry C, Boë LJ (1986) Laws for lips. Speech Commun 5: 97–193
Alipour-Haghighi F, Titze IR (1991) Elastic models of vocal fold tissues. J Acoust Soc Am 90: 1326–1331
Andrew RJ (1976) Use of formants in the grunts of baboons and other nonhuman primates. In: Harnad SR, Steklis HD, Lancaster J (eds) Origins and evolution of language and speech. Ann NY Acad Sci 280: 673–693
Askenfelt A, Gauffin J, Sundberg J, Kitzing P (1980) A comparison of contact microphone and electroglottograph for the measurement of vocal fundamental frequency. J Speech Hear Res 23: 258–273
Atal BS (1985) Linear predictive coding of speech. In: Fallside F, Woods WA (eds) Computer speech processing. Prentice-Hall, London, pp 81–124
Atal BS, Hanauer SL (1971) Speech analysis and synthesis by linear prediction of the acoustic wave. J Acoust Soc Am 50: 37–655
Atal BS, Chang JJ, Mathews MV, Tukey JW (1978) Inversion of articulatory-to-acoustic transformation in the vocal tract by a computer sorting technique. J Acoust Soc Am 63: 1535–1555
Badin P, Motoki K, Miki N, Ritterhaus D, Lallouache M-T (1994) Some geometric and acoustic properties of the lip horn. J Acoust Soc Jpn E15: 243–253
Badin P, Beautemps D, Laboissiere R, Schwartz JL(1995) Recovery of vocal tract geometry from formants for vowels and fricative consonants using a midsagittal-to-area function conversion model. J Phonet 23: 221–229
Baer T, Gore JC, Gracco LC, Nye P (1991) Analysis of vocal tract shape and dimensions using magnetic resonance imaging: vowels. J Acoust Soc Am 90: 799–828
Bailly G, Jordan M, Mantakas M, Schwartz JL, Bach M, Olesen M (1990) Simulation of vocalic gestures using an articulatory model driven by a sequential neural network. J Acoust Soc Am 87: S105
Bailly G, Laboissière R, Schwartz JL (1991) Formant trajectories as audible gestures: an alternative for speech synthesis. J Phonetics 19: 9–23

Baken RJ (1987) Clinical measurement of speech and voice. Little, Brown, Boston,

Baken RJ, Matz BJ (1973) A portable impedance pneumograph. Hum Commun 2: 28–35

Beautemps D, Badin P, Laboissière R (1995) Deriving vocal-tract area functions from midsagittal profiles and formant frequencies: a new model for vowels and fricative consonants based on experimental data. Speech Commun 16: 27–47

Beckett RL (1971) The respirometer as a diagnostic and clinical tool in the speech clinic. J Speech Hear Disord 36: 235–241

Bell AM (1867) Visible speech or self-interpreting physiological letters for the writing of all languages in one alphabet. Simpkin and Marshall, London

Bell-Berti F, Raphael LJ (eds) (1995) Producing speech: contemporary issues. For Katherine Safford Harris. AIP Press, New York

Benoît C, Lallouache T, Mohamadi T, Abry C (1992) A set of French visemes for visual speech synthesis. In: Bailly G, Benoît C (eds) Talking machines: theories, models and applications, Elsevier, Amsterdam, p 485

Benoît C, Mohamadi T, Kandel S (1994) Audio-visual intelligibility of French speech in noise. J Speech Hear Res 37: 1195–1203

Bernstein NA (1967) The coordination and regulation of movements. Pergamon, London

Boë LJ, Perrier P, Bailly G (1992) The geometric vocal tract variables controlled for vowel production: proposals for constraining acoustic-to-articulatory conversion. J Phonet 20:27–38

Borden GJ, Harris KS (1984) Speech science primer. Williams & Wilkins, Baltimore

Boring EG (1950) A history of experimental psychology, 2nd edn. Appleton-Century-Crofts, New York

Brigham EO (1974) The fast Fourier transform. Prentice Hall, Englewood Cliffs

Browman CP, Goldstein L (1985) Dynamic modeling of phonetic structure. In: Fromkin VA (ed) Phonetic linguistics. Essays in honor of Peter Ladefoged. Academic Press, New York, pp 35–53

Browman CP, Goldstein L (1986) Towards an articulatory phonology. Phonol Year 3: 219–252

Browman CP, Goldstein L (1989) Articulatory gestures as phonological units. Phonology 6: 201–251

Browman CP, Goldstein L (1990) Tiers in articulatory phonology, with some implications for casual speech. In: Kingston J, Beckman M (eds) Papers in laboratory phonology: I. Between the grammar and the physics of speech. Cambridge University Press, Cambridge, England, pp 341–376

Browman CP and Goldstein L (1992) Articulatory phonology: an overview. Phonetica 49: 222–234

Browman CP, Goldstein L, Kelso JAS, Rubin P, Saltzman E (1984) Articulatory synthesis from underlying dynamics. J Acoust Soc Am 75: S22–S23

Browman CP, Goldstein L, Saltzman E, Smith C (1986) GEST: a computational model for speech production using dynamically defined articulatory gestures. J Acoust Soc Am 80: S97

Capranica RR (1966) Vocal response of the bullfrog to natural and synthetic mating calls. J Acoust Soc Am 40: 1131–1139

Carterette E, Shipley C, Buchwald J (1979) Linear prediction theory of vocalization in cat and kitten. In: Lindblom B, Ohman S (eds) Frontiers of speech communication research, Academic Press, New York, pp 245–257

Carterette E, Shipley C, Buchwald J (1984) The speech of animals. In: Bristow G (ed) Electronic speech synthesis. Techniques, technology and applications. McGraw Hill, New York, pp 292–302

Chuang C, Wang W (1975) A distance-sensing device for tracking tongue configuration. J Acoust Soc Am Suppl 1 59: S11

Cochran WT, Cooley JW, Favin DL, Helms HD, Kaenel RA, Lang WW, Maling GC Jr, Nelson DE, Rader CM, Welch PD (1967) What is the fast Fourier transform? IEEE Trans Audio Electroacoust AU-15: 45–55

Cooper FS, Liberman AM, Borst JM (1951) The interconversion of audible and visible patterns as a basis for research in the perception of speech. Proc Natl Acad Sci USA 37: 318–325

Cranen B, Boves L (1985) Pressure measurements during speech production using semiconductor miniature pressure transducers: impact on models for speech production. J Acoust Soc Am 77: 1543–1551

Dang J, Honda K, Suzuki H (1994) Morphological and acoustical analysis of the nasal and the paranasal cavities. J Acoust Soc Am 96: 2088–2100

Faber A (1989) Lip protrusion in sibilant production. J Acoust Soc Am 86: S113

Fallside F (1985) Frequency-domain analysis of speech. In: Fallside F, Woods WA (eds) Computer speech processing. Prentice-Hall, London, pp 41–80

Fallside F, Woods WA (eds) (1985) Computer speech processing. Prentice-Hall, London

Fano RM (1949) The transmission of information. MIT RLE Tech Rep 65, Cambridge

Fant G (1960) Acoustic theory of speech production. Mouton, The Hague

Fay RR (1988) Hearing in vertebrates: a psychophysics databook. Hill-Fay, Winnetka, Illinois

Flanagan JL (1965) Speech analysis, synthesis, and perception. Springer Berlin Heidelberg New York

Flanagan J, Rabiner L (eds) (1973) Speech synthesis. Dowden, Hutchinson and Ross, Stroudsburg
Flanagan JR, Ostry DJ, Feldman AG (1990) Control of human jaw and multi-joint arm movements. In: Hammond GR (ed) Cerebral control of speech and limb movements. North-Holland, Amsterdam, pp 29–58
Fletcher S, McCutcheon M, Wolf M (1975) Dynamic palatometry. J Speech Hear Res 18: 812–819
Folkins JW, Abbs JH (1975) Lip and jaw motor control during speech: responses to resistive loading of the jaw. J Speech Hear Res 18: 207–220
Fourcin AJ (1974) Laryngographic examination of vocal fold vibration. In: Wyke B (ed) Ventilatory and phonatory control systems. Oxford University Press, New York, pp 315–333
Fourcin AJ (1981) Laryngographic assessment of phonatory function. In: Ludlow CL, Hart MO (eds) Proc Conf on the Assessment of vocal pathology, ASHA Rep 11, pp 116–127
Fowler CA (1977) Timing control in speech production. Indiana University Linguistics Club, Bloomington
Fowler CA (1984) Current perspectives on language and speech production: a critical overview. In: Daniloff R (ed) Recent advances in speech, hearing and language, vol. 4. College-Hill Press, Boston, pp 195–278
Fowler CA (1995) Speech production. In: Miller J, Eimas P (eds) Speech, language and communication. Academic Press, New York, pp 29–61
Fowler CA, Saltzman E (1993) Coordination and coarticulation in speech production. Lang Speech 36: 171–195
Fowler CA, Rubin P, Remez RE, Turvey MT (1980) Implications for speech production of a general theory of action. In: Butterworth B (ed) Language production. Academic Press, New York, pp 373–420
Fry DB (1979) The physics of speech. Cambridge University Press, Cambridge, England
Fujimura O (1977) Stereo-fiberscope. In: Sawashima M, Cooper, FS (eds) Dynamic aspects of speech production. University of Tokyo Press, Tokyo, pp 133–137
Fujimura O (1988) Vocal fold physiology vol 2. Vocal physiology: voice production, mechanisms and functions. Raven Press, New York
Fujimura O (1990) Methods and goals of speech production research. Lang Speech 33: 195–258
Fujimura O, Baer T, Niimi S (1979) A stereo-fiberscope with a magnetic interlens bridge for laryngeal observation. J Acoust Soc Am 65: 478–480
Galaburda AM (1984) Anatomical asymmetries. In: Geschwind N, Galaburda AM (eds) Cerebral dominance: the biological foundations. Harvard University Press, Cambridge, MA, pp 11–25
Gay TJ (1981) Mechanisms in the control of speech rate. Phonetica 38: 148-158
Gelfer CE, Harris KS, Baer T (1987) Controlled variables in sentence intonation. In: Baer T, Sasaki C, Harris, K (eds) Laryngeal function in phonation and respiration. College-Hill Press, Boston, pp 422–435
Gracco VL, Abbs JH (1985) Dynamic control of the perioral system during speech: kinematic analyses of autogenic and nonautogenic sensorimotor processes. J Neurophysiol (Bethesda)54: 418–432
Gracco VL, Abbs JH (1988) Central patterning of speech movements. Exp Brain Res 71: 515–526
Grossberg S (1986) The adaptive self-organization of serial order in behavior: speech, language, and motor control. In: Schwab EC, Nusbaum HC (eds) Pattern recognition by humans and machines, vol 1. Academic Press, Boston, p 187
Guiard-Marigny T, Adjoudani A, Benoît C (1996) A 3D model of the lips and of the jaw for visual speech synthesis. In: Progress in speech synthesis, Springer Berlin Heidelberg New York
Hardcastle WJ (1972) The use of electropalatography in phonetic research. Phonetica 25: 197–215
Hardcastle WJ, Marchal A (eds) (1990) Speech production and speech modelling. Kluwer, Dordrecht
Hardcastle WJ, Gibbon F, Nicolaidis K (1991) EPG data reduction methods and their implications for studies of lingual coarticulation. J Phonetics 19: 251–266
Harrington J, Fletcher J, Roberts C (1995) Coarticulation and the accented/unaccented distinction: evidence from jaw movement data. J Phonetics 23: 305–322
Hauser MD (1991) Sources of acoustic variation in rhesus macaque vocalizations. Ethology 89: 29–46
Hauser MD (1992) Articulatory and social factors influence the acoustic structure of rhesus monkey vocalizations: a learned mode of production? J Acoust Soc Am 91: 2175–2179
Hauser MD (1996) Nonhuman primate vocal communication. In: Cochran M (ed) Handbook of acoustics. John Wiley, New York
Hauser MD, Fowler C (1992) Declination in fundamental frequency is not unique to human speech: evidence from nonhuman primates. J Acoust Soc Am 91: 363–369.
Hauser MD, Evans CS, Marler P (1993) The role of articulation in the production of rhesus monkey (Macaca mulatta) vocalizations. Anim Behav 45: 423–433

Hauser MD, Schön Ybarra M (1994) The role of lip configuration in monkey vocalizations: experiments using xylocaine as a nerve block. Brain and Language 46: 423–433

Hirayama ME, Vatikiotis-Bateson E, Kawato M, Jordan MI (1992) Forward dynamics modeling of speech motor control using physiological data. In: Moody JE, Hanson SJ, Lippmann RP (eds) Advances in neural information processing systems 4. Morgan Kaufman, San Mateo, p 191

Hirayama M, Vatikiotis-Bateson E, Kawato M (1993) Physiologically based speech synthesis using neutral networks. IEICE Trans E76-A: 1898–1910

Hirayama M, Vatikiotis-Bateson E, Kawato M (1994) Inverse dynamics of speech motor control. Adv Neural Inf Proc Syst 6: 1043–1050

Hixon TJ (1971a) Magnetometer recording of jaw movements during speech. J Acoust Soc Am 49: 104

Hixon TJ (1971b) An electromagnetic method for transducing jaw movements during speech. J Acoust Soc Am 49: 603–606

Hixon TJ (1972) Some new techniques for measuring the biomechanical events of speech production: one laboratory's experiences. ASHA Rep 7: 68–103

Hogden J, Löfquist A, Gracco V, Oshima K, Rubin P, Saltzman E (1993) Inferring articulator positions from acoustics: an electromagnetic midsagittal articulometer experiment. J Acoust Soc Am 94: 1764

Hogden J, Rubin P, Saltzman E (1996) An unsupervised method for learning to track tongue position from an acoustic signal. Bull Commun Parl 3:101–116

Hopp SL, Sinnott JM, Owren MJ, Petersen MR (1992) Differential sensitivity of Japanese macaques (*Macaca fuscata*) and humans (*Homo sapiens*) to peak position along a synthetic coo call continuum. J Comp Psychol 106: 128–136

Jakobson R, Fant G, Halle M (1963) Preliminaries to speech analysis. MIT Press, Boston

Jordan MI (1988) Supervised learning and systems with excess degrees of freedom. COINS Tech Rep . University of Massachusetts, Computer and Information Sciences, Boston, pp 99–127

Jordan MI (1989) Serial order: a parallel, distributed processing approach. In: Elman JL, Rumelhart DE (eds) Advances in connectionist theory: speech. Erlbaum, Boston, pp 44–93

Jordan MI (1990) Motor learning and the degrees of freedom problem. In: Jeannerod M (ed) Attention and performance, XIII. Erlbaum, Boston

Kaburagi T, Honda M (1994) A trajectory formation model of articulatory movements based on the motor tasks of phoneme-specific vocal tract shapes. Proc Int Conf on spoken language processing II (ICLSP 94), Acoustical Society of Japan, Yokohama, Japan, pp 579–588

Kawato M (1989) Motor theory of speech perception revisited from minimum torque-change neural network model. In: Proc 8th Symp on Future electron devices, 30–31 October, 1989, Tokyo

Kawato M (1991) Optimization and learning in neural networks for formation and control of coordinated movement. In: Meyer D (ed) Attention and performance, XIV. Erlbaum, Hillsdale, New Jersey

Kawato M, Maeda Y, Uno Y, Suzuki R (1990) Trajectory formation of arm movement by cascade neural network model based on minimum torque-change criterion. Biol Cybern 62: 275–288

Keller E, Ostry DJ (1983) Computerized measurement of tongue dorsum movement with pulsed-echo ultrasound. J Acoust Soc Am 73: 1309–1315

Kelly JL Jr, Lochbaum C (1962) Speech synthesis. In: Proc Stockholm Speech Commun Seminar, RIT, Stockholm

Kelman AW (1981) Vibratory pattern of the vocal folds. Folia Phoniatr 33: 73–99

Kelso JAS, Tuller B, Vatikiotis-Bateson E, Fowler CA (1984) Functionally specific articulatory cooperation following jaw perturbations during speech: evidence for coordinative structures. J Exp Psychol: Hum Percept Perform 10: 812–832

Kelso JAS, Vatikiotis-Bateson E, Saltzman EL, Kay B (1985) A qualitative dynamic analysis of reiterant speech production: phase portraits, kinematics, and dynamic modeling. J Acoust Soc Am 77: 266–280

Kelso JAS, Saltzman EL, Tuller B (1986a) The dynamical perspective in speech production: data and theory. J Phonetics 14: 29–59

Kelso JAS, Saltzman EL, Tuller B (1986b) Intentional contents, communicative context, and task dynamics: a reply to the commentators. J Phonetics 14: 171–196

Kent RD, Read C (1992) The acoustic analysis of speech. Singular Publishing Group, San Diego, California

Kent RD, Atal BS, Miller JL (eds) (1991) Papers in speech communciation: speech production. Acoustical Society of America, Woodbury, New York

Kirchner JA (1988) Functional evolution of the human larynx: variations among the vertebrates. In: Fujimura O (ed) Vocal physiology: voice production mechanisms, and functions. Raven Press, New York

Kiritani S, Itoh K, Fujimura O (1975) Tongue-pellet tracking by a computer-controlled x-ray microbeam. J Acoust Soc Am 57: 1516–1520

Klatt DH (1980) Software for a cascade/parallel formant synthesizer. J Acoust Soc Am 67: 971–995

Klatt DH (1987) Review of text-to-speech conversion for English. J Acoust Soc Am 82: 737–793

Koenig W, Dunn HK, Lacey LY (1946) The sound spectrograph. J Acoust Soc Am 18: 19–49

Kröger BJ, Schröder G, Opgen-Rhein C (1995) A gesture-based dynamic model describing articulatory movement data. J Acoust Soc Am 98: 1878–1889

Kuc R, Tuteur F, Vaisnys JR (1985) Determining vocal tract shape by applying dynamic constraints. ICASSP 95. Proc of the Int Conf on Acoustics, speech and signal processing, Tampa IEEE, New York, pp 1101–1104

Kuehn DP, Moll K (1976) A cineradiographic study of VC and CV articulatory velocities. J Phonetics 4: 303–320

Ladefoged P (1975). A course in phonetics. Harcourt Brace Jovanovich, Inc, New York

Laitman JT, Crelin ES, Conlogue GJ (1977) The function of the epiglottis in monkey and man. Yale J Biol Med 50: 43–48

Larar JN, Schroeter J, Sondhi MM (1988) Vector quantization of the articulatory space. IEEE Trans Acoust, Speech Signal Process 36: 1812–1818

Larson CR (1988) Brain mechanisms involved in the control of vocalization. J Voice 2: 301–311

Levelt WJM (1989) Speaking: from intention to articulation. MIT Press, Cambridge

Levinson SE, Schmidt CE (1983) Adaptive computation of articulatory parameters from the speech signal. J Acoust Soc Am 74: 1145–1154

Liberman AM, Studdert-Kennedy M (1978) Phonetic perception. In: Held R, Leibowitz H, Teuber HL (eds) Handbook of sensory physiology, vol. 8 Perception. Springer, Berlin Heidelberg New York, p 143

Liberman AM, Mattingly IG (1985) The motor theory of speech perception revised. Cognition 21: 1–36

Liberman AM, Ingemann F, Lisker L, Delattre P, Cooper F (1959) Minimal rules for synthesizing speech. J Acoust Soc Am 31: 1490–1499

Liberman AM, Cooper FS, Shankweiler DP, Studdert-Kennedy M (1967) Perception of the speech code. Psychol Rev 74: 431–461

Lieberman P (1969) On the acoustic analysis of primate vocalizations. Behav Res Methods Instrum 5: 169–174

Lieberman P (1975) On the origins of language. MacMillan, New York

Lieberman P (1984) The biology and evolution of language. Harvard University Press, Cambridge

Lieberman P, Blumstein SE (1988) Speech physiology, speech perception, and acoustic phonetics. Cambridge University Press, New York

Lieberman PH, Klatt DH, Wilson WH (1969) Vocal tract limitations on the vowel repertoires of rhesus monkey and other nonhuman primates. Science 164: 1185–1187

Lindblom B, Lubker J, Gay T (1979) Formant frequencies of some fixed mandible vowels and a model of speech motor programming by predictive simulation. J Phonetics 7: 147–161

Lisker L (1957) Closure duration and the intervocalic voiced-voiceless distinction in English. Language 33: 42–49

Lisker L, Abramson AS (1967) Some effects of context on voice onset time in English stops. Lang Speech 10: 1–28

Löfqvist A (1990) Speech as audible gestures. In: Hardcastle W, Marchal A (eds) Speech production and speech modelling. Kluwer, Dordrecht, pp 289–322

Löfqvist A, Gracco VL (1994) Tongue body kinematics in velar stop production: influences of consonant voicing and vowel context. Phonetica 51: 52–67

Löfqvist A, Gracco VL, Nye PW (1993) Recording speech movements using magnetometry: one laboratory's experience. Proc ACCOR Worksh on Electromagnetic articulography in phonetic research. Forschungsberichte des Instituts fr Phonetik und Sprachliche Kommunikation der Universität München 31, pp 143–162

Luria AR (1975) The man with a shattered world. Penguin, London

Maeda S (1979) An articulatory model of the tongue based on a statistical analysis. J Acoust Soc Am 65: S22

Manuel S, Vatikiotis-Bateson E (1988) Oral and glottal gestures and acoustics of underlying /t/ in English. J Acoust Soc Am 84: S84

Markel JD, Gray AH (1976) Linear prediction of speech. Springer, Berlin Heidelberg New York

Massaro DW, Cohen MH, Gesi A, Heredia R, Tsuzaki M (1993) Bimodal speech perception: an examination across languages. J Phonet 21: 445–478

Mattingly IG, Liberman AM (1969) The speech code and the physiology of language. In: Leibovic KN (ed) Information processing in the nervous system. Springer, Berlin Heidelberg New York, p 97

Mattingly IG, Liberman AM (1988) Specialized perceiving systems for speech and other biologically significant sounds. In: Edelman GM, Gall WE, Cowan ME (eds) Auditory function: the neurobiological bases of hearing. Wiley, New York, pp 775–793

May B, Moody D, Stebbins W (1988) The significant features of Japanese macaque coo sounds: a psychophysical study. Anim Behav 36: 1432–1444

May B, Moody D, Stebbins W (1989) Categorical perception of conspecific communication sounds by Japanese macaques. J Acoust Soc Am 85: 837–847

McGowan R (1987) Articulatory synthesis: numerical solution of a hyperbolic differential equation. Haskins Laboratories Status Report on Speech Research SR-89/90, New Haven pp 69–79

McGowan R (1988) An aeroacoustic approach to phonation. J Acoust Soc Am 83: 696–704

McGowan R (1994) Recovering articulatory movement from formant frequency trajectories using task dynamics and a genetic algorithm: preliminary model test. Speech Commun 14: 19–48

McGowan R (1995) Recovering task dynamics from formant frequency trajectories: results using computer "babbling" to form an indexed data base. In: Bell-Berti F, Raphael LJ (eds) Producing speech: contemporary issues. For Katherine Safford Harris. AIP Press, New York, pp 489–504

McGowan R, Saltzman E (1995) Incorporating aerodynamic and laryngeal components into task dynamics. J Phonet 23: 255–269

Mead J, Peterson N, Grimby C, Mead J (1967) Pulmonary ventilation measured from body surface movements. Science 156: 1383–1384

Mermelstein P (1973) Articulatory model for the study of speech production. J Acoust Soc Am 53: 1070–1082

Moore CA (1992) The correspondence of vocal tract resonance with volumes obtained from magnetic resonance images. J Speech Hearing Res 35: 1009–1023

Morrish K, Stone M, Shawker T, Sonies B (1985) Distinguishability of tongue shape during vowel production. J Phonetics 13: 189–203

Müller J (1848) The physiology of the senses, voice and muscular motion with the mental faculties. (Translation) W. Baly, Walton and Maberly, London

Munhall K, Löfqvist A (1992) Gestural aggregation in speech: laryngeal gestures. J Phonet 20: 111–126

Munhall KG, Löfqvist A, Kelso JAS (1986) Laryngeal compensation following sudden oral perturbation. J Acoust Soc Am Suppl 1 80: S109

Nadler R, Abbs J (1988) Use of the x-ray microbeam system for the study of articulatory dynamics. J Acoust Soc Am Suppl 1 84: S124

Neff WD (1964) Temporal pattern discrimination in lower animals and its relation to language perception in man. In: de Reuck, O'Connor (eds) Ciba foundation symposium on disorders of language. Churchill, London, p 183

Neff WD, Diamond IT, Casseday JH (1975) Behavioral studies of auditory discrimination: central nervous system. In: Keidel Neff (eds) Handbook of sensory physiology, vol V/2. Springer, Berlin Heidelberg New York , p 307

Nittrouer S, Munhall K, Kelso JAS, Tuller B, Harris KS (1988) Patterns of interarticulator phasing and their relation to linguistic structure. J Acoust Soc Am 85: 1653–1661

Norris KS, Møhl B (1983) Can odontocetes debilitate prey with sound? Am Nat 122: 85–114

O'Shaughnessy D (1987) Speech communication: human and machine. Addison-Wesley, New York

O'Shaughnessy D (1995) Speech technology. In: Syrdal A, Bennett R, Greenspan S (eds) Applied speech technology. CRC Press, Boca Raton, pp 47–98

Ostry DJ, Munhall KG (1994) Control of jaw orientation and position in mastication and speech. J Neurophysiol 71: 1528–1545

Ostry DJ, Keller E, Parush A (1983) Similarities in the control of speech articulators and the limbs: kinematics of tongue dorsum movement in speech. J Exp Psychol Hum Percept Perform 9: 622–636

Owren MJ (1990) Acoustic classification of alarm calls by vervet monkeys (Cercopithecus aethiops) and humans: II. Synthetic calls. J Comp PsychOL 104: 29–40

Owren MJ, Bernacki RH (1988) The acoustic features of vervet monkey alarm calls. J Acoust Soc Am 83: 1927–1935

Owren MJ, Linker CD (1995) Some analysis techniques that may be useful to acoustic primatologists. In: Zimmermann E, Newman J, Jurgens (eds) Current topics in primate vocal communication, Plenum Press, New York, pp 1–27

Owren MJ, Linker CD, Rowe MP (1993) Acoustic features of tonal "grunt" calls in baboons. J Acoust Soc Am 94: 1823

Owren MJ, Seyfarth RM, Cheney DL (1995) Acoustic indices of production mechanisms underlying tonal "grunt" calls in baboons. J Acoust Soc Am 98: 2965

Papçun G, Hochberg J, Thomas TR, Laroche F, Zacks J, Levy S (1992) Inferring articulation and recognizing gestures from acoustics with a neural network trained on X-ray microbeam data. J Acoust Soc Am 92: 688–700

Parush A, Ostry D, Munhall K (1983) A kinematic study of lingual coarticulation in VCV sequences. J Acoust Soc Am 74: 1115–1125

Perkell JS (1969) Physiology of speech production: results and implications of a quantitative cineradiographic study. MIT Press, Cambridge

Perkell J, Cohen M, Garabieta I (1988) Techniques for transducing movements of points on articulatory structures. J Acoust Soc Am (Suppl) 1 84: S145

Perkell JS, Cohen MH, Svirsky MA, Matthies ML, Garabieta I, Jackson MTT (1992) Electromagnetic midsagittal articulometer systems for transducing speech articulatory movements. J Acoust Soc Am 92: 3078–3096

Perrier P, Laboissière R, Eck L (1991) Modelling of speech motor control and articulatory trajectories. Proc 12th Int Congr Phonet Sci 2 pp 62–65

Petersen MR (1981) The perception of species-specific vocalizations by animals: developmental perspectives and implications. In: Aslin R, Alberts J, Petersen M (eds) Development of perception: psychobiological perspectives, vol. 1 Auditory, chemosensory and somatosensory systems. Academic Press, New York , pp 67

Petersen M, Beecher M, Zoloth S, Marler P, Moody D, Stebbins W (1984) Neural lateralization of vocalizations by Japanese macaques: communicative significance is more important than acoustic structure. Behav Neurosci 98: 779–790

Rabiner LR, Schafer RW (1978) Digital processing of speech signals. Prentice Hall, Englewood Cliffs

Rahim MG, Goodyear CC (1990) Estimation of vocal tract filter parameters using a neural net. Speech Commun 9: 49–55

Rahim MG, Goodyear CC, Kleijn WB, Schroeter J, Sondhi MM (1993) On the use of neural networks in articulatory speech synthesis. J Acoust Soc Am 93: 1109–1121

Recasens D (1984) Timing constraints and coarticulation: alveolar-palatals and sequences of alveolar and /j/ in Catalan. Phonetica 41: 125–139

Remez RE, Rubin PE, Pisoni DB, Carrell TO (1981) Speech perception without traditional speech cues. Science 212: 947–950

Repp B (1983) Trading relations among acoustic cues in speech perception are largely a result of phonetic categorization. Speech Commun 2: 341–361

Repp B (1988) Integration and segregation in speech perception. Lang Speech 31: 239–271

Rosenberg A (1971) Effect of glottal pulse shape on the quality of natural vowels. J Acoust Soc Am 49: 583–590

Rothenberg M (1977) Measurement of airflow in speech. J Speech Hear Res 20: 155–176

Rothenberg M (1981) Some relations between glottal air flow and vocal fold contact area. In: Ludlow CL, Hart MOC (eds) Proc Con on the Assessment of vocal pathology, ASHA Rep 11, Am Speech-Language-Hearing Assoc, Rockville, Maryland, pp 88–96

Rubin JA (1983). Static and dynamic information in vowels produced by the hearing impaired. Doctoral Diss, City University of New York

Rubin PE (1995) HADES: a case study of the development of a signal analysis system. In: Syrdal A, Bennett R, Greenspan S (eds) Applied speech technology. CRC Press, Boca Raton, pp 501–520

Rubin PE, Baer T, Mermelstein P (1981) An articulatory synthesizer for perceptual research. J Acoust Soc Am 70: 321–328

Rubin PE, Tiede M, Vatikiotis-Bateson E, Goldstein L, Browman C, Levy S (1995) V-TV: the Haskins vocal tract visualizer CD-ROM. Paper presented at the ACCOR Workshop on Articulatory Databases, Munich, Germany, 25–26 May, 1995

Sackner MA (1980) Monitoring of ventilation without a physical connection to the airway. In: Sackner MA (ed) Diagnostic techniques in pulmonary disease, part I. Dekker, New York, pp 503–537

Saltzman E (1986) Task dynamic coordination of the speech articulators: a preliminary model. Exp Brain Res, Series 15: 129–144

Saltzman E, Kelso JAS (1987) Skilled actions: a task dynamic approach. Psychol Rev 94: 84–106

Saltzman E, Munhall KG (1989) A dynamical approach to gestural patterning in speech production. Ecol Psychol 1: 333–382

Saltzman E, Rubin P, Goldstein L, Browman CP (1987) Task-dynamic modeling of interarticulator coordination. J Acoust Soc Am 82: S15

Saltzman E, Goldstein L, Browman CP, Rubin P (1988a) Modeling speech production using dynamic gestural structures. J Acoust Soc Am 84: S146

Saltzman E, Goldstein L, Browman CP, Rubin P (1988b) Dynamics of gestural blending during speech production. Neural Networks 1: 316

Saltzman E, Löfqvist A, Kinsella-Shaw J, Kay B, Rubin P (1995) On the dynamics of temporal patterning in speech. In: Bell-Berti F, Raphael LJ (eds) Producing speech: contemporary issues. For Katherine Safford Harris. AIP Press, New York, pp 469–487

Sasaki CT, Levine PA, Laitman JT, Crelin ES Jr (1977) Postnatal descent of the epiglottis in man. Arch Otolaryngolica 103: 169–171

Sawashima M (1977) Current instrumentation and technique for observing speech organs. Technocrat 9-4 : 19–26

Sawashima M, Abramson AS, Cooper FS, Lisker L (1970) Observing laryngeal adjustments during running speech by use of a fiberoptics system. Phonetica 22: 193–201

Schönle P, Grabe K, Wenig P, Hohne J, Schrader J, Conrad B (1987) Electromagnetic articulography: use of alternating magnetic fields for tracking movements of multiple points inside and outside the vocal tract. Brain Lang 31: 26–35

Schroeder MR (1967) Determination of the geometry of the human vocal tract by acoustic measurements. J Acoust Soc Am 41: 1002–1010

Schroeter J, Sondhi MM (1992) Speech coding based on physiological models of speech production. In: Furui S, Sondhi MM (eds) Advances in speech signal processing. Dekker, New York, p 231–268

Sears CH (1902) A contribution to the psychology of rhythm. Am J Psychol 13: 28–61

Sekiyama K, Tohkura, Y (1993) Inter-language differences in the influence of visual cues in speech perception. J Phonet 21: 427–444

Seyfarth RM, Cheney DL, Harcourt AH, Stewart K (1994) The acoustic features of double-grunts by mountain gorillas and their relation to behavior. Am J Primatol 33: 31–50

Shipley C, Carterette EC, Buchwald JS (1991) The effects of articulation on the acoustical structure of feline vocalizations. J Acoust Soc Am 89: 902–909

Shirai K (1993) Estimation and generation of articulatory motion using neuronal networks. Speech Commun 13: 45–51

Shirai K, Kobayashi T (1986) Estimating articulatory motion from speech wave. Speech Commun 5: 159–170

Shirai K, Kobayashi T (1991) Estimation of articulatory motion using neural networks. J Phonetics 19: 379–385

Simmons JA, Grinnell AD (1988) The performance of echolocation: acoustic images perceived by echo-locating bats. In: Nachtigall PE, Moore PWB (eds) Animal sonar. Plenum, New York, p 353

Sondhi MM, Schroeter J (1987) A hybrid time-frequency domain articulatory speech synthesizer. IEEE Trans ASSP 35: 955–967

Sonoda Y, Wanishi S (1982) New optical method for recording lip and jaw movements. J Acoust Soc Am 72: 700–704

Stebbins WC, Sommers MS (1992) Evolution, perception and the comparative method. In: Webster DB, Fay RR, Popper AN (eds) The evolutionary biology of hearing. Springer, Berlin Heidelberg New York, p 211

Stetson RH (1905) A motor theory of rhythm and discrete succession. II. Psychol Rev 12: 293–350

Stetson RH (1928) Motor phonetics: a study of speech movements in action, 2nd edn. North Holland, Amsterdam (1951). (1st edn 1928 in Arch Neerl phonetique Exp 3)

Stone M (1990) A three-dimensional model of tongue movement based on ultrasound and x-ray micro-beam data. J Acoust Soc Am 87: 2207–2217

Stone M (1991) Toward a model of three-dimensional tongue movement. J Phonetics 19: 309–320

Stone M, Vatikiotis-Bateson E (1995) Coarticulatory effects on tongue, jaw, and palate beavior. J Phonet 23: 81–100

Stone M, Shawker TH, Talbot TL, Rich AH (1988) Cross-sectional tongue shape during the production of vowels. J Acoust Soc Am 83: 1586–1596

Subtelny J, Oya N, Subtelny JD (1972) Cineradiographic study of sibilants. Folia Phoniatrica 24: 30–50

Summerfield Q (1987) Some preliminaries to a comprehensive account of audio-visual speech perception. In: Dodd B, Campbell R (eds) Hearing by eye: the psychology of lip-reading. Erlbaum, Hillsdale, New Jersey, p 3

Summerfield, Q (1991) Visual perception of phonetic gestures. In: Mattingly IG, Studdert-Kennedy M (eds) Modularity and the motor theory of speech perception. Erlbaum, Hillsdale, New Jersey, pp 117–137

Sussman HM, MacNeilage PF, Hanson RJ (1973) Labial and mandibular dynamics during the production of bilabial consonants: preliminary observations. J Speech Hear Res 16: 397–420

Tank DW, Hopfield JJ (1987) Neural computation by concentrating information in time. Proc Natl Acad Sci USA 84: 1896–1900

Tiede MK (1993) An MRI-based study of pharyngeal volume contrasts in Akan. Haskins Laboratories Status Report on Speech Research SR-113, New Haven, pp 107–130

Tuller B, Kelso JAS (1984) The timing of articulatory gestures: evidence for relational invariants. J Acoust Soc Am 76: 1030–1036

Tuller B, Kelso JAS (1995) Speech dynamics. In: Bell-Berti F, Raphael LJ (eds) Producing speech: contemporary issues. For Katherine Safford Harris. AIP Press, New York, pp 505–519

Tuller B, Kelso JAS, Harris KS (1982) Interarticulator phasing as an index of temporal regularity in speech. J Exp Psychol: Hum Percept Perform 8: 460–472

Tuller B, Kelso JAS, Harris KS (1983) Converging evidence for the role of relative timing in speech. J Exp Psychol: Hum Percept Perform 9: 829–833

Tuller B, Shao S, Kelso JAS (1990) An evaluation of an alternating magnetic field device for monitoring tongue movements. J Acoust Soc Am 88: 674–679

Turvey MT (1977) Preliminaries to a theory of action with reference to vision. In: Shaw R, Bransford J (eds) Perceiving, acting, and knowing: toward an ecological psychology. Lawrence, Hillsdale, New Jersey, pp 211–265

Vatikiotis-Bateson E (1988) Linguistic structure and articulatory dynamics. Indiana University Linguistics Club, Bloomington

Vatikiotis-Bateson E, Kelso JAS (1984) Remote and autogenic articulatory adaptation to jaw perturbation during speech: more on functional synergies. J Acoust Soc Am 75: S23–24

Vatikiotis-Bateson E, Kelso JAS (1993) Rhythm type and articulatory dynamics in English, French, and Japanese. J Phonet 21: 231–265

Vatikiotis-Bateson E, Ostry J (1995) An analysis of the dimensionality of jaw motion in speech. J Phonet 23: 101–117

Vatikiotis-Bateson E, Stone M (1989) In search of lingual stability. J Acoust Soc Am 86: S115

Vatikiotis-Bateson E, Hirayama M, Kawato M (1991) Neural network modelling of speech motor control using physiological data. Phonetic Exp Res. Inst Linguistics, U Stockholm (PERILUS) 14: 63–68

Wada Y, Kawato M (1995) A theory for cursive handwriting based on the minimization principle. Biol Cybern 73: 3–13

Wada Y, Koike Y, Vatikiotis-Bateson E, Kawato M (1995) A computational theory for movement pattern recognition based on optimal movement pattern generation. Biol Cybern 73: 15–25

Wakita H (1979) Estimation of vocal-tract shapes from acoustical analysis of the speech wave: the state of the art. IEEE Trans on Acoustics, Speech, and Signal Proc, ASSP-27: 281–285

Walsh EG (1957) An investigation of sound localization in patients with neurological abnormalities. Brain 80: 222–250

Warren DW (1986) Compensatory speech behaviors in individuals with cleft palate: a regulation/control phenomenon? Cleft Palate J 23: 251–260

Weismer G (1983) Acoustic descriptions of dysarthric speech: perceptual correlates and physiological inferences. Speech Motor Control Laboratory, Preprints. University of Wisconsin, Madison

Westbury JR (1994) On coordinate systems and the representation of articulatory movements. J Acoust Soc Am 95: 2271–73

Whalen DH, Liberman AM (1987) Speech perception takes precedence over nonspeech perception. Science 237: 169–171

Wilhelms-Tricarico R (1995) Physiological modeling of speech production: methods for modeling soft-tissue articulators. J Acoust Soc Am 97: 3085–3098

Witten IH (1982) Principles of computer speech. Academic Press, Orlando

Wood S (1979) A radiographic analysis of constriction location for vowels. J Phonet 7: 25–43

CHAPTER 9

Sound Production in Birds:
Acoustics and Physiology Revisited

A. S. GAUNT AND S. NOWICKI

1
Introduction

In a volume concentrating on methods in the analysis of acoustic communication, two roles are served by a chapter on how one group of vocal virtuosos, namely birds, make sound. The first is to inform the reader of those innovations that have led to new insights into avian vocal physiology. There are several advances worth noting in this context, such as the use of miniature thermistors to monitor airflow and record acoustic signals in a bird's vocal organ during song (e.g., Suthers 1990), or the use of quantitative kinematic analysis to analyze motions of the vocal tract (e.g., Westneat et al. 1993).

Such innovations pale, however, by comparison to the technical advances in the analysis of sound that form the bulk of this Volume. A more important role for this chapter, then, is to lay the groundwork for the interface between advances in the analysis of vocal behavior and our understanding of how sounds are produced. Indeed, in his classic work, *Bird song: acoustics and physiology*, Crawford Greenewalt (1968) relied almost entirely on acoustic evidence to produce what still stands as the most comprehensive analysis of song production. His hypotheses were formulated largely by deducing how the available anatomy was likely to work based on detailed analysis of the sounds that were actually produced. Perhaps the best example of Greenewalt's success is his masterful argument that songbirds are capable of producing two sounds simultaneously as a kind of *internal duet*. Although this *two-voice theory* had been suggested earlier (Potter et al. 1947; Borror and Reese 1956), Greenewalt (1968) presented evidence in the form of oscillograms and hand-calculated spectrograms of electronically filtered sounds that convincingly argued for the existence of two independently modulated *voices* in the songs and calls of over two dozen species of birds.

The limitations of an acoustic-deductive approach to understanding vocal mechanisms are obvious, and experimental verification is necessary to elucidate the strengths and weaknesses of any theory. Nonetheless, it is clear that there is also limited value in trying to understand the mechanisms responsible for the production of a behavior such as birdsong without first understanding that behavior in some detail. In the case of vocal behavior, that understanding necessarily entails acoustic analysis.

The reverse is also true. Animal communication is best described as a system that includes the sender, signal, and receiver in all aspects of functional analysis (Shannon and Weaver 1949; Cherry 1966). Thus, a full appreciation of communication must consider the mechanism by which the signal is produced. From the point of view of a functional understanding of communication, the mechanism by which a signal is produced will restrict the problem by defining the possible "acoustic space" that a signal can occupy (e.g., Dabelsteen and Pedersen 1985; Nelson 1989; Nelson and Marler 1990; see also Nowicki et al. 1992). From an analytical point of view, the biomechanics of production will at least help to define the technologies necessary to approach the whole communication system, or more specifically, to analyze the relevant signal space (see, for example, Rubin and Vatikiotis-Bateson, this Volume).

In the following discussion, we first describe our current understanding of the relationship between syringeal structure and function. This section is brief, as the subject has been treated extensively in recent years (Brackenbury 1982, 1989; Gaunt and Gaunt 1985b; Gaunt 1987; Nowicki and Marler 1988; King 1989). We then consider recent modifications of available functional models, some of which have arisen from new data garnered with old techniques, some from new interpretations of older data, and some from the applications of newer techniques. Our goal is not to review this material exhaustively, but instead to highlight particular problems of recent interest and their solutions. We conclude with an overview of central unresolved issues and the technical approaches that might be taken to solve these problems.

1.1
Diversity and Uniformity of Syringeal Structure and Function

Birds are capable of producing a remarkable variety of sounds, especially if one considers the entire class Aves as a group. Even within a single suborder such as the Oscines (true songbirds), an astounding diversity of sounds is produced. Typical avian vocalizations include (but are hardly limited to) pure-tone whistles, tones with clearly defined harmonics, broad-band sounds with formant-like structures, coupled amplitude and frequency modulations, click-like sounds, and noise (Greenewalt 1968; Marler 1969; Gaunt 1987; Nowicki and Marler 1988). The simultaneous presence of two harmonically unrelated, often independently modulated tones is common, suggesting that the membranous portions of some syrinxes that serve as acoustic sources may be independently activated and controlled in some cases (Borror and Reese 1956; Greenewalt 1968). This diversity in the kinds of sounds produced by birds leads readily to the suggestion that avian vocal systems perhaps work (function) in more than one way. In other words, the same anatomical structures may, through differences in how they are activated and operated, produce qualitatively different kinds of sounds.

When comparing various orders of birds, such as chickens (*Galliformes*) and parrots (*Psittaciformes*), some differences in sounds produced are obviously attributable to profound differences in syringeal anatomy (Youngren et al. 1974; Nottebohm 1976; Gaunt et al. 1976 1982; Gaunt and Gaunt 1977). However, even anatomically similar syrinxes, e.g., those of the Oscines, can produce radically different sounds, with resulting diversity that is as great within orders or families as between them. Even a single species may include in its repertoire a vast array of qualitatively different sounds. Given this diversity of output, and the possibility that similar vocal structures might function in different ways in the production of different kinds of sounds, it is a futile exercise in typological thinking to consider "the function" of "the syrinx." Rather, it may be more productive to think in terms of a flexible system in which structure-function relationships may vary not only among taxa but also transiently within a species, or even within an individual.

1.2
Structural Considerations

The organ primarily responsible for generating sound in birds is the syrinx. Like feathers, syrinxes are unique to the class Aves and evolutionary antecedents are not known outside the class (Beddard 1898; King 1989). That is, all birds have some form of syrinx (except New World vultures, who have lost it secondarily), whereas no other animals do. Despite this commonality within the Aves, the shared characteristics of syrinxes among orders of birds are remarkably few. The anatomical complexity and taxonomic diversity of the syrinx has been well illustrated by Beddard (1898) and more recently by King (1989).

Syrinxes, like the larynx in other animals, occur as elaborations of structures in the vocal tract. A syrinx always occur in a more caudal position than the larynx. Most syrinxes lie at or near the point where the primary bronchi join to form the trachea (see Figure 1), and all occur within the interclavicular airsac[1]. In some cases where the syrinx occurs high in the trachea (cranially), such as in hummingbirds, the air sac itself extends between the furcula (the fused clavicles of birds) and up into the neck (Beddard 1898).

All syrinxes contain at least one pair of relatively membranous, flexible walls that can be distended into the lumen of the airway. All are also associated with extrinsic tracheal musculature that, by changing the length of the trachea, can tighten or relax membranous portions, thereby allowing them to extend into or out of the airstream (Figure 1). Many, but not all syrinxes also have denser pads of tissue that are flexible and may be moved relative to the airstream. Most complex syrinxes, especially those of oscine birds, have associated musculature that is presumed to be able to change the position and configuration of their flexible membranes and the tissues of the syrinx itself. Such changes in syringeal configuration are presumed, in turn, to facilitate sound production during vocalization, and are further thought to be responsible for actively modulating the acoustic features of the sound being produced. As suggested by the two-voice phenomenon, this musculature can presumably also change the configuration of membranes in different parts of the organ independently.

The extensive variation in syringeal structures observed in different groups of birds, coupled with the suggestion that similar structures may exhibit a diversity of functions, might lead one to despair of uncovering any general principles of syringeal structure and function. Certainly, there is no obvious correlation between syringeal complexity, especially in terms of the extent of intrinsic musculature, and the complexity of sounds that may be produced (Greenewalt 1968; Gaunt 1983; Baptista and Trail 1992). Some

---

1   This airsac is one of the many membranous outpocketings of the bird's lungs that extend throughout the body cavity and even ramify into the hollow centers of bones. Most airsacs are bilaterally paired, but the intraclavicular airsac is a fused midline cavity that extends from the middle of the thoracic region anteriorly up to the furcula and the neck.

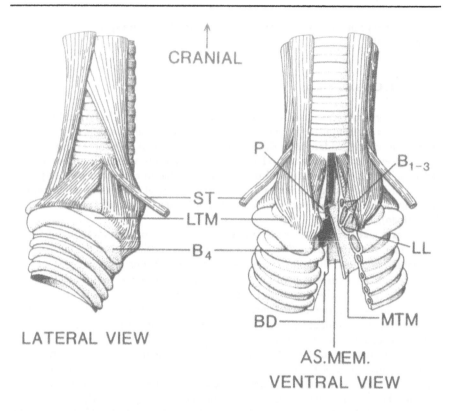

Fig. 1. Lateral and ventral views of a typical oscine syrinx. A portion of the left bronchus has been removed from the ventral view (*right*) to reveal internal structures. The bronchidesmus (*BD*), which has been cut in this view, extends between the medial surfaces of the two bronchi just caudal to the medial tympaniform membranes (*MTM*). The intraclavicular air sac (the dorsal wall of which is labeled *AS.MEM.*) surrounds the entire structure. Other abbreviations: B_{1-4} bronchial semi-rings; *LL*, lateral labium; *LTM* lateral tympaniform membrane; *ST* sternotrachealis muscle; *P* pessulus. (Reprinted with permission from Gaunt 1987, CRC Press, Boca Raton, Florida)

generalizations can be made, however. For example, the ability to learn vocal material through imitation is generally confined to species that possess intrinsic musculature (Gaunt 1983), even though the correlation between overall syringeal complexity and vocal learning is not strong (Baptista and Trail 1992). Another example is a general correlation between body size and emphasized frequencies in song, which likely results from allometry of syringeal morphology (Ryan and Brenowitz 1985). Unfortunately, such generalizations provide relatively little guidance in addressing more specific questions about the relationship between syringeal structure and vocal production.

1.3
Functional Considerations

It is generally agreed that sound is produced by the syrinx through the interaction of syringeal membranes with expiratory air flow. A variety of mechanisms for the induction and control of vibration in these membranes have been suggested (Greenewalt 1968; Stein 1968; Casey and Gaunt 1985; Gaunt and Gaunt 1985b; Fletcher 1988, 1989, 1992), but direct evidence is still lacking. Most models propose that air flowing past syringeal membranes induce the membranes to vibrate (see Sect. 3.1). One alternative model, dubbed the *whistle hypothesis*, suggests that syringeal membranes need not vibrate at all during sound production. Instead, the configuration of membranes and other tissues in the syrinx may enable the production of an aerodynamic *hole-tone whistle* (Chaunaud 1970) in which a stable pattern of vortices is produced (the vortices being the acoustic source) by the passage of air through two consecutive constrictions. This mechanism is found in human lip whistling and is also the way most tea kettles whistle. Available evidence (see below) shows that a true whistle is not likely to be involved in sound production in oscine birds. It might operate in doves (*Streptopelia*) and parrots, the two groups for which the mechanism was originally proposed (Nottebohm 1976; Gaunt et al. 1982), although recent work with the collared dove (*Streptopelia decaocto*) suggests that this species does not produce a true whistle (M. R. Ballintijn and C. ten Cate, pers. comm.).

The location of most types of syrinxes within the thoracic cavity and the high fundamental frequencies of vibration (often several thousand Hertz) make it difficult to measure the activity of syringeal membranes directly[2]. Indeed, the precise nature of vibration and its induction in the syrinx remains one of the least understood features of the system. Each of the current models does, however, incorporate the assumption that changes in the configuration of the syringeal membranes, mediated by the activity of the musculature associated with the syrinx, are responsible for actively modulating the acoustic characteristics (that is, time-varying frequency and amplitude variations) of the sounds produced. The most widely accepted model of syringeal function, developed independently by Greenewalt (1968) and Stein (1968), proposes that such modulations are affected by adjustments in the position and tension of the vibrating membranes. The anatomy of the syringeal musculature (e.g., Ames 1971), as well as electromyographic analyses (EMG; see for example Gaunt and Gaunt 1977, 1985; Gaunt 1987a; Vicario 1991), both support this general view.

A related supposition of the Greenewalt-Stein model is that *all* modulations are source-generated. In contrast, the *source-filter* model of human speech production (Fant 1960; see also Rubin and Vatikiotis-Bateson and Owren and Bernacki, this Volume) proposes that the supralaryngeal vocal tract makes a major contribution to producing the time-varying changes in acoustic structure that are characteristic of speech. The idea that the source alone must be responsible for all acoustic modulations in bird song arose, in part, from the two-voice phenomenon. How could a single resonating system independently modulate two simultaneous tones? Imagine two larynges pro-

---

2   In contrast, the relatively low fundamental frequency of vocal fold vibration (about 120 Hz in males) and the relatively accessible position of the larynx at the anterior end of the trachea in humans has allowed the development of many simple and precise techniques for measuring vocal fold activity.

ducing speech signals that passed through and were modified simultaneously by the same vocal tract. It is difficult, if not impossible, to conceive of how such a vocal system could independently modulate both source signals! In spite of this argument, recent empirical work (see Sect. 3.3) has demonstrated that the suprasyringeal vocal tract does play a critical role in sound production, at least in songbirds (Nowicki and Marler 1988). Although the acoustic effects of the avian vocal tract on the emitted signal are different from those observed in speech, some of the same physical and physiological principles appear to be at work in both systems.

Certainly, one of the most fascinating features of avian vocal production is the possibility of an internal duet being produced by birds that have syrinxes with two membranes that are capable of vibrating independently, such as oscines. This possibility was initially suggested as an afterthought in a book describing the use of the sound spectrogram in speech analysis (Potter et al. 1947). Twenty years later, Greenewalt (1968) used essentially the same acoustic techniques to marshall an impressive array of examples that seemed to require the production of two separate voices for their explanation (see Figure 2). Although few doubted the two-voice theory following Greenewalt's treatment, it was more than another twenty years before Suthers (1990) developed a technique that experimentally demonstrated the action of two separate sound sources in song production and provided a means for exploring the mechanisms underlying this phenomenon in more detail (see also Sect. 2.2).

In summary, birds have vocal systems of considerable capabilities. The mechanics of sound production almost certainly differ among different taxa, and may differ even during the production of various kinds of sounds within the same species. Moreover, the syrinx has proven to be difficult to observe directly, making it difficult to reach

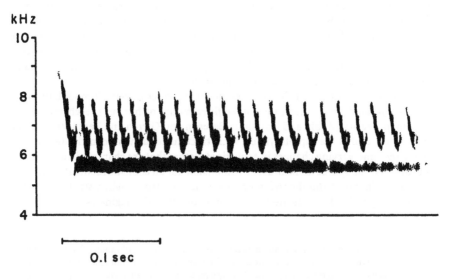

Fig. 2. Spectrogram of terminal trill of wood thrush song, illustrating the simultaneous presence of two voices that are independently modulated and harmonically unrelated (Kay Digital Sona-Graph model 7800, 16-kHz analysis range, 300-Hz frequency resolution) (from Nowicki and Marler 1988, Regents of the University of Californa, used with permission)

unequivocal and broadly applicable conclusions based on empirical data. Comparisons of data gathered with different techniques or from different species, not to mention different orders, must be made with extreme caution and be confined to those issues directly tested.

2
New and Old Insights from New and Old Techniques

2.1
Air Flow and Phonation

One topic of long-standing interest has been the mechanism by which birds are capable of producing loud, prolonged songs without apparently pausing for breath. With the possible exception of nightjars (Hunter 1980), few birds seem to vocalize on inhalation. Yet many birds, including some notably small species, emit continuous vocalizations for seemingly improbable durations (e.g., 41 s for a winter wren, *Troglodytes troglodytes*, Clark 1949; at least 60 s and possibly up to 117 s for a grasshopper warbler, *Locustella naevia*, Brackenbury 1978a; Schild 1986). Estimates of tidal volume suggest that birds should run out of breath long before the end of such sustained vocalizations (Brackenbury 1978a).

Calder (1970) proposed that songbirds might partially replenish their air supply during song by using *minibreaths*, which he defined as shallow inhalations between successive notes. He made this proposal based on data obtained from singing canaries (whose prolonged songs can continue for up to 45 s) by using an *impedance pneumograph*. This device uses skin electrodes attached to the front and back of the bird. The impedance between these two electrodes covaries with the size of the thoracic cavity, which presumably correlates with patterns of inhalation and exhalation. Calder (1970) discovered that there was not a sustained expiratory motion during production of rapidly trilled notes. Instead, the birds he tested appeared to inhale slightly before each note, even when the bird was singing a rapid trill.

Gaunt et al. (1976) objected to Calder's interpretation, suggesting that impedance measurements could not be used to distinguish among several competing hypotheses for the production of prolonged vocalizations. Specifically, there was no evidence that airflow was being reversed during song, as would occur during a small inhalation. Gaunt et al. (1976) proposed four models: (1) oscillating valves in the airstream, (2) pulsatile output of airflow, (3) reciprocal oscillating air chambers, and (4) true minibreaths, all of which might explain Calder's data equally well. Only the last two of these hypotheses would effectively extend the duration of vocalization beyond that predicted by the characteristic tidal volume for a given species.

Various studies using implanted devices for measuring flow or pressure, sometimes combined with EMG, have demonstrated the presence of pulsatile output in evening grosbeaks, *Coccothraustes vespertinus* (Berger and Hart 1968), starlings, *Sturnus vulgaris* (Gaunt et al. 1973), chickens (Gaunt and Gaunt 1977; Brackenbury 1978b) and doves (Gaunt et al. 1982). Similar data from parrots support the existence of an oscillating valve in this group (Gaunt and Gaunt 1985a). In each of these cases, however, the duration of vocalization is not so extreme as to require a mechanism that extends the tidal volume of the bird.

Calder's (1970) hypothesis was again addressed by Hartley and Suthers (1989), who used miniaturized flow transducers to directly monitor air flow in the trachea of singing canaries. Interestingly, Calder's main reason for using an indirect measure of respiratory activity, as stated in his paper, was the infeasibility at that time of obtaining direct flow measurements in a small bird. Hartley and Suthers (1989) overcame this difficulty by improving on a method used earlier by Suthers and Hector (1982, grey swiftlets, *Collocalia spodiopygia*; 1985, oilbirds, *Steatornis carpiensis*) in which the cooling of a microbead thermistor is calibrated to provide a very accurate measure of air flow. Suthers and his coworkers also implanted small pressure transducers in the subjects' air sacs. Pressure data from these instruments indicate the direction of the flow detected by the thermistors, and thus help to provide a complete picture of respiratory mechanics.

Hartley and Suthers' (1989) data unequivocally demonstrated reversal of flow between song notes at repetition rates of up to about 30 Hz. As the note repetition rate exceeded 30 Hz, the canaries switched to a pulsatile output. Thus, canaries do use minibreaths as Calder (1970) originally proposed. At the same time, short repeated syllables are produced by canaries in two different ways, by minibreaths or pulsatile output, depending on the repetition rate of these notes. Evidently there is some upper rate beyond which the use of mini-breaths is constrained by the functional morphology of respiration.

The resolution of the minibreath controversy is tied directly to a specific technological improvement — the miniaturization of flow transducers that can be implanted in a bird's vocal tract. The unequivocal evidence obtained this way that canaries *do* use minibreaths to extend the duration of their phonations shows that three of the four mechanisms proposed by Gaunt et al. (1976) have now been demonstrated. The fourth, reciprocally oscillating chambers, has been described only in anurans (Martin and Gans 1972). In birds, both among species and within a single species, we see evidence for multiple physiological mechanisms leading to similar acoustic results.

2.2
Two-Voicing and Peripheral Lateralization of Function

The two-voice theory suggests that birds possessing a tracheobronchial syrinx with separate syringeal membranes in both bronchi should be capable of simultaneously producing two independent sounds. This theory was finally verified by Suthers (1990), although it had been generally accepted for over two decades in spite of the lack of direct experimental evidence. Recently, it also has become apparent that the two sides of a syrinx, in addition to being able to operate in isolation, might interact in the production of sound in a variety of ways. The possible interactions between the two sides of a syrinx, as well as the implications of two potential sound sources for control of song production, has been the subject of considerable work and discussion. Much of this discussion involves the relationship between neural control and the action of the two sound sources in songbirds.

Following Greenewalt's (1968) acoustic analysis in support of the two-voice theory, Nottebohm (1971) asked whether both sides of a songbird's syrinx contribute equally to song production. He did so by selectively denervating the left or right syringeal musculature of chaffinches (*Fringilla coelebs*) and observing the resulting deficits in song production. This technique involved severing the roots of the XIIth cranial nerve (the hypoglossus nerve) proximal to the point where this nerve forms an anastomosis with the Xth cranial nerve (the vagus); later work showed that severing the descending branch of the hypoglossus (the tracheosyringealis) achieves very similar results (Nottebohm and Nottebohm 1976). Nottebohm (1971) found that denervation of the right side of a chaffinch's syrinx resulted in only minimal degradation in song production, whereas denervation of the left side almost totally disrupted a bird's ability to sing. This apparent left side dominance was also demonstrated in several other species, most notably the canary (Nottebohm and Nottebohm 1976; see Nottebohm 1980, for a review).

The suggestion that one side of a bird's syrinx appears to play a dominant role in song production suggested an intriguing parallel to the well-known phenomenon of cerebral lateralization in speech (Marler 1970; Nottebohm 1971, 1980). Nottebohm (1977) pursued this idea by performing unilateral lesions of regions in the central nervous system known to be involved in the motor control of song production (HVC and RA; Nottebohm et al. 1982) and comparing differences in the effects of left versus right lesions. Lesions of the left HVC and RA had profoundly disruptive effects on song production in canaries (similar to those observed following unilateral lesion of the motor nerve leading to the syrinx), whereas those on the right side had a minimal effect. Thus, Nottebohm's work both supported the idea that the two sides of a bird's syrinx work independently and further demonstrated, for at least some species, that one side plays a dominant role in song production (although see below). The latter finding, along with the attendant description of a discrete set of brain regions involved in song control, contributed to an explosion of interest in the neurobiological underpinnings of song production, perception, and development. An overview of work on the *song system*, as it is commonly called, is beyond the scope of this chapter, but reviews may be found in Nottebohm (1980, 1991), and Konishi (1989).

The two sides of a bird's syrinx *are capable of* independently producing two distinct sounds, but do they *always* do so during production? Using Nottebohm's denervation technique, Nowicki and Capranica (1986a,b) demonstrated a counterexample in which both halves of the syrinx work together when the black-capped chickadee (*Parus atricapillus*) produces its complex *dee* syllable. This note (which occurs at the end of the familiar *chick-a-dee* scold call of this species) is highly modulated, resembling a harmonic series in which the fundamental frequency and the next one or two harmonics are missing (shown in Figure 3). Based on Nottebohm's earlier work, one might expect that this complex sound is produced entirely by one side of the syrinx, or perhaps that some of its frequency components are produced by one side of the syrinx while others are produced simultaneously by the other. In either case, one would expect the contributions from the two syringeal halves to add in a linear fashion. Contrary to this expectation, unilateral denervation of the left and right halves of the syrinx resulted in comparable postoperative deficits (Figure 4). Furthermore, a simple summation of the frequency components that remained following unilateral left denervation with those that remained intact following unilateral right denervation did not approximate the normal

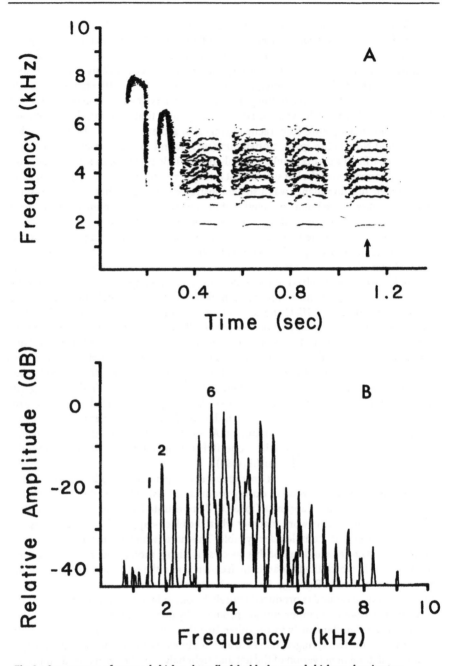

Fig. 3.a Spectrogram of a normal chick-a-dee call of the black-capped chickqee showing two introductory pure-tone notes and four dee notes (Kay Digital Sona-Graph model 7800, 8-kHz analysis range, 45-Hz frequency resolution). b Amplitude spectrum of a dee note, calculated from a 40 msec section of the last note as marked by an *arrow* in A (Nicolet Mini-Ubiquitous FFT analyzer, 25-Hz frequency resolution) (Reprinted with permission from Nowicki and Capranica 1986b, American Association for the Advancement of Science)

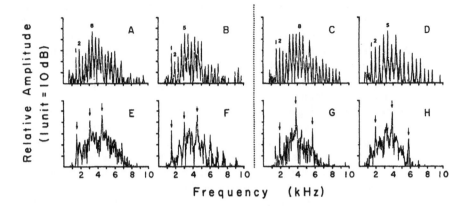

Fig. 4. Comparison of pre- and postoperative amplitude spectra of dee notes produced by four different chickadees. A-D Preoperative spectra of dee syllables. The first two frequency components with significant energy and the component of maximum amplitude are numbered. E,F Postoperative spectra of dee notes produced by the same birds as in A and B, respectively, following section of the right tracheosyringealis (TS) nerve. (G,H) Postoperative spectra of dee notes produced by the same birds as in C and D, respectively, following left TS nerve section. *Arrows* indicate postoperative harmonic components. Note the relationship between preoperative component 1 and the postoperative fundamental frequency in the case of a right nerve section, and between preoperative component 2 and the postoperative fundamental in the case of a left nerve section (Reprinted with permission from Nowicki and Capranica 1986b, American Association for the Advancement of Science)

sound. This result indicates that both sides are indeed independently innervated and are capable of operating independently, but that adding the signals produced by the two sides when working alone does not recreate the normal, intact signal.

Nowicki and Capranica (1986b) solved this apparent puzzle by mathematically modeling possible nonlinear interactions between the signals presumed to be generated by the two syringeal sources, as inferred from signals produced following unilateral denervation. They found that multiplying the two source signals, equivalent to an amplitude modulation of one by the other, produced a pattern of modulation *sidebands* (summation and difference frequencies) that closely matched the pattern observed in normal production. Thus, they argued that the *dee* syllable is normally produced by a coupling of the two syringeal sources, i.e., that the two sources do not act independently. Interestingly, other syllables produced by the chickadee in the same call (*chick-a*) have a simple, pure-tone structure and they are affected by unilateral nerve section in a fashion that is more similar to that observed by Nottebohm in the chaffinch or canary. This outcome suggests that the nonlinear interaction between the two syringeal sound sources can be "turned on and off," even in the context of a single call produced by the same bird.

Nowicki and Capranica's (1986a,b) demonstration of bilateral syringeal coupling did not rely on a new technique, but instead sprang from the combination of some mathematical intuition with the serendipitous application of an older technique to a different species. Their work did not suggest that the two-voice theory was incorrect, only that it was incomplete. Birds are able to sing an internal duet, with the two vocal sources operating independently from each other, but they also can produce more complex modulations in some cases when the two sources are coupled. The precise physical nature of this coupling is unclear, which is not surprising given how little is known about vibration even in a single syringeal membrane. Moreover, it is possible that the bilateral interaction discussed by Nowicki and Capranica is only the tip of an iceberg of possible ways in which the two sides of a bipartite syrinx may conspire to produce intricate (and beautiful) sounds.

McCasland (1987), working with canaries, also reexamined the issue of how the left and right sides of a syrinx contribute to song, but he did so using still another technique. McCasland suggested that a more direct way to assess the contribution of one side of the syrinx to song production is to plug physically the bronchus below the contralateral side. Using this bronchus-plug technique, he found much less profound differences between the left and right than Nottebohm and Nottebohm (1976) had, when using unilateral denervation in the same species. Based on these data, McCasland questioned whether the two sides of a canary's syrinx ever act independently. He suggested instead that Nottebohm's earlier findings might be explained if the two syringeal halves normally produce qualitatively similar signals, but with the larger left side normally producing sound at greater amplitudes. McCasland reasoned that when the left side is turned off by depriving it of muscular control, air still flows through, and the weaker activity of the right side is masked. Plugging the left side, in contrast, not only prevents it from producing sound, but also may increase airflow through the contralateral bronchus, causing the right side to produce its signal (which McCasland suggested was the same as that produced on the left) at higher amplitude.

Hartley and Suthers (1990) replicated McCasland's study, also in canaries, but failed to confirm his result. Instead, they found the effects of bronchus plugging in canaries to be very similar to those produced by unilateral denervation, with the left side being being clearly dominant in production. Differences in the results of these two studies; remain unexplained. The method used to physically plug the bronchi differed in the two studies; McCasland inserted a filled piece of tubing with an outer diameter that matched the inner diameter of the bronchus; Hartley and Suthers filled the bronchus with an injection of elastic impression medium. The plugs in the latter case were probably longer and extended further towards the lungs, but it is not obvious how this difference would contribute to the observed differences in the data.

After obtaining post-operative recordings from birds with a bronchus plug in place, Hartley and Suthers (1990) then denervated the syringeal musculature on the same side to determine what further effect, if any, this procedure had on the bird's ability to sing. They found that, whereas denervation of the right side did not produce additional effects on singing in a bird with a plug on the right, analogous denervation of the left side caused a further loss in the vocal repertoire. This outcome suggests that neuromuscular activity on the left side of the syrinx can influence the behavior of the contralateral side. Given that no peripheral neural "crossovers" have been identified in oscine birds (i.e., the left tracheosyringeal nerve does not appear to innervate the right side, and viceversa), Hartley and Suthers (1990) argued that the effect may be due either to an anatomical coupling (perhaps similar to that postulated by Nowicki and Capranica to account for bilateral interactions in the dee syllable), or to the loss of sensory feedback.

Another result using the bronchus-plugging technique was obtained by Nowicki (1989, unpubl.data). Working with swamp sparrows (*Melospiza georgiana*), Nowicki found that birds with either their left or their right bronchus plugged were capable of singing entirely normal songs. Unlike earlier studies, Nowicki measured acoustic similarities quantitatively, using the spectrogram cross-correlation technique of Clark et al. (1987). This quantitative approach revealed that post-operative birds sometimes produced normal song notes, but might also produce the same notes with measurable deficits. Ipsilateral denervation of a plugged bird always resulted in deficits that were similar to those seen after unilateral denervation alone. Finally, Nowicki added a third manipulation, immobilizing the syringeal membrane on the plugged side by physically anchoring the posteriormost one-third of the membrane with glue. This manipulation also resulted in deficits comparable to those seen after unilateral denervation alone.

The fact that swamp sparrows with bronchus plugs can sometimes produce normal songs seems at first to support McCasland's (1987) original finding, but this interpretation is misleading for several reasons. First, evidence from unilateral denervation alone shows that swamp sparrows do not exhibit lateral dominance (i.e., denervation of the left side does not result in more severe deficits than does denervation of the right, or vice versa). Many individual notes are partially affected by denervation of either side, suggesting that bilateral innervation, at least, is required for their production. Furthermore, McCasland's interpretation implies that denervation of the plugged syringeal half would not have any additional effect on production (although he did not himself perform this second manipulation), but Nowicki found just the opposite. In fact, Nowicki's findings are in this sense more similar to those of Hartley and Suthers (1990), who also found an effect of ipsilateral denervation following plugging on the left side.

Nowicki's (1989) results, then, are compatible with a two-voice model of production, in the context of a species that does not show functional dominance of one side in production overall. The finding that immobilization of the syringeal membrane ipsilateral to a plug has an effect similar to ipsilateral denervation is the most puzzling in that it suggests that vibration in the membrane on the side of the plug contributes to production. One interpretation is that left and right syringeal membranes are excited as a single unit in swamp sparrows, even by unilateral airflow, and that normal syringeal function requires both bilateral innervation *and* unimpeded mechanical activity of both syringeal membranes.

The bronchus-plugging technique seemed at first to offer an important new source of insight into the two-voice theory. However, the contradictory results obtained by McCasland (1987) and Hartley and Suthers (1990) when testing the same species, along with the additional differences seen by Nowicki (1989) in another species, have not clarified issues of peripheral lateralization of syringeal function. Certainly, comparisons between species must be considered with caution. Imperfect understanding of syringeal mechanics hampers interpretation of these data, as does the fact that bronchus plugging, like denervation, is an indirect method of understanding syringeal function.

The most direct technique for assessing the relative contributions of the left and right halves of the syrinx is that developed recently by Suthers (1990), in which miniature microbead thermistors are implanted in both the left and right bronchi just below the syringeal membranes. This technique is the same as that used by Hartley and Suthers (1989) to explore the mini-breath hypothesis (as described in Sect. 2.1), except that two transducers are used. Each transducer is in a separate airway (one in each bronchus) upstream from the syrinx, as opposed to a single transducer in the common airway (the trachea) downstream from the syrinx. Furthermore, the thermistors used by Suthers (1990) were sufficiently sensitive so as to detect the air oscillations corresponding to the near field of the acoustic signal generated by the syringeal membrane up to about 3 kHz. Thus, using this technique, Suthers could simultaneously measure respiratory flow and record the acoustic activity on each side of the syrinx independently in a singing bird.

Recordings made in this way from grey catbirds (*Dumatella carolinensis*) and brown thrashers (*Toxostoma rufum*) yielded several interesting and important findings (Suthers 1990; Suthers et al. 1994). First, of course, was the unequivocal verification of the two-voice theory as it was first proposed over three decades ago. In both species, 10-23 % of all syllables produced (among the six individuals studied) were produced by either the left or the right side of the syrinx acting alone. Syringeal activity sometimes switched from one side to the other for different syllables within a song, as expected from Greenewalt's (1968) model (see Figure 5). In 21-67 % of the syllables recorded from these birds, both sides of the syrinx were simultaneously active, also as predicted by Greenewalt. In some of these cases, both sides appeared to generate identical sounds, a result that is reminiscent of the argument advanced by McCasland (1987). Most importantly, however, in other such cases the two syringeal halves produced separate sounds with independent frequency modulations that were not harmonically related. These observations provide incontrovertible evidence in support of the long-standing two-voice hypothesis.

These data do not provide evidence, however, for lateral dominance in song production. None of the species Suthers has examined show a consistent pattern of favoring one side over the other (Suthers 1990; Suthers et al. 1994). Some individuals do favor one side in the production of syllables involving only that single side, although never as strongly as the effect observed in chaffinches by Nottebohm (1971) or in canaries by Nottebohm and Nottebohm (1976), but individuals vary as to which side is emphasized. This finding is surprising given the number of species in which lateral dominance was reported using Nottebohm's denervation technique (reviewed in Nottebohm 1980). It is unclear whether catbirds and thrashers are exceptional in showing a lack of lateral dominance in production, or whether canaries and chaffinches are more unusual in showing it strongly. Most of the earlier work using unilateral denervation relied on subjective visual analysis of spectrograms to assess post-operative effects. Studies that have quantified effects (e.g., Nowicki and Capranica 1986b; Nowicki 1989) have failed to find evidence for strong lateral dominance in several species, including black-capped chickadees, swamp sparrows, and song sparrows (*Melospiza melodia*). Taken together, this fact, the more direct nature of Suthers' thermistor technique, and the difficulties in interpreting bronchus-plugging results all suggest that the patterns observed by Suthers may be the most typical of how songbirds sing.

Studies using flow meters implanted in the bronchi show that different species of passerines exploit the potentials of a bilateral syrinx in various ways. Some, such as mimids (Suthers et al. 1994), appear to be able to use either side with equal facility. Even among mimids, however, there is a tendency to produce certain note types or certain portions of the frequency range using one side. In species such as cowbirds, *Molothrus*

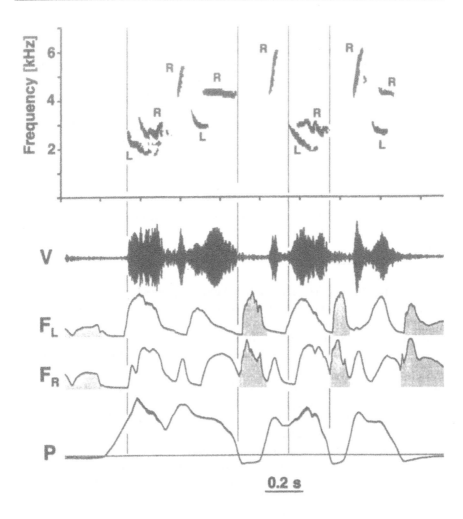

Fig.5. A segment of brown thrasher song including syllables with two-voice and single voice components. The *top panel* is a spectrogram of the song segment, with contributions from the left side *L* and right side *R* marked. The *lower traces* show the acoustic waveform of the vocalization (*V*), air flow through the left (*F_L*) and right (*F_R*) sides of the syrinx (measured by a heated thermistor in each primary bronchus), and respiratory pressure (*P*) recorded in the thoracic air sac. *Shaded areas* under the air flow traces indicate inspiration, *unshaded areas* indicate expiration (From Suthers et al. 1994 ©, reprinted by permission of John Wiley and Sons, Inc.)

ater (Allen and Suthers 1994) or northern cardinals, *Cardinalis cardinalis* (R.A. Suthers, pers. comm.), one or more of these tendencies may become fixed, so that the two sides are each dedicated to the production of only a portion of the species' sounds. Extreme dominance, with all or most of the repertoire being produced by one side, appears to be an extension of such specialization.

Suthers' work (Suthers 1990; Allen and Suthers 1994; Suthers et al. 1994) has begun
to reveal other, more complex patterns of how the two syringeal halves work together
in sound production. He has documented, for example, cases in which the two sides
alternate in the production of a sound that otherwise appears continuous. Responsibil-
ity for producing the sound appears to be passed back and forth between the two sides,
sometimes several times in a single syllable (illustrated in Figure 6). In other cases, one
side is active first, with the other side then joining in, again in the context of a single
continuous acoustic element. Perhaps most interesting are cases in which both sides
are active and the emitted signal is amplitude-modulated at a periodicity that is equal
to the difference in frequencies between the two sides. This phenomenon (referred to
as *beating*) is a linear interaction between the two syringeal sides that approaches the
complexity of the nonlinear cross-modulation proposed by Nowicki and Capranica
(1986a,b). We expect that work using microbead thermistors to measure airflow will
continue to reveal new and interesting facts about the functioning of the syrinx.

2.3
Tracheal Modulation

Until recently, it was widely assumed that the suprasyringeal vocal tract plays little or
no role in sound production in birds (see reviews in Brackenbury 1982, 1989; Gaunt and
Gaunt 1985b; Gaunt 1987). One reason for this assumption is that many bird songs include predominantly pure-tone sounds whose simple spectral characteristics do not
appear to require any involvement of vocal tract filtering (Marler 1969; Nowicki and
Marler 1988). In contrast, such effects are critical to the source-filter theory of speech
production (Fant 1960; see also Rubin and Vatikiotis-Bateson and Owren and Bernacki,
this Volume). This model suggests that the human supralaryngeal vocal tract acts as a
complex acoustic filter that modifies the amplitude spectrum of the laryngeal source
signal through its resonances and anti-resonances. Because the larynx produces a
broadband harmonic signal, the acoustic consequences of vocal tract filtering are observed as time-varying changes in the relative amplitudes of different frequency components in the signal. In pure-tone birdsongs, whose acoustic energy is concentrated at
a single frequency, there appears at first glance to be no functional role for an acoustic
filter that selectively attenuates different frequencies.

Sound production in birds can also be compared to characteristic production processes of musical wind instruments, such as trumpets and clarinets. For the latter, the
source energy (i.e., buzzing lips in the case of trumpets and other brass instruments,
vibrating wooden reeds in the case of clarinets and other woodwinds) is tightly coupled
to the acoustic resonances of the body of the instrument. Specifically, due to this coupling effect, the fundamental frequency of vibration (or its overtones) in the source is
constrained to match the resonance frequencies determined by the overall length of the
instrument's air flow tube, an aspect that the player may continuously change through
adopting various fingering positions (Benade 1976). A significant implication of this
model, then, is that there is a necessary relationship between the effective length of the
tube and the frequency of sound produced. In the case of birds, the relevant tube is most
likely to be the trachea. As the tonal nature of many birdsongs is more reminiscent of

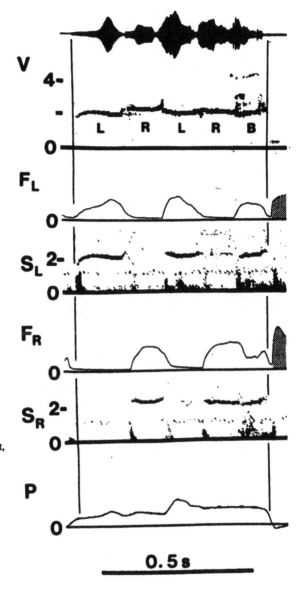

Fig. 6. Syllable from a catbird song showing alternating contributions from the left, right, and both sides of the syrinx. *Abbreviations* as in Fig. 5. Also shown are spectrograms of the isolated acoustic signals, as recorded directly from the two thermistors in the bronchi, produced by the left (S_L) and right (S_R) sides of the syrinx (From Suthers 1990, used with permission).

the typical sounds of musical instruments than of the broadband sounds of speech (Nowicki and Marler 1988), the analogy seems applicable, at least on the surface.

Greenewalt (1968) addressed this possible similarity by looking for a correlation between tracheal length and dominant song frequencies across many species of birds.

He failed to find such a correlation, however, and therefore argued that tracheal resonances are not important in the production of birdsong. Greenewalt further proposed
that the changes in tracheal length that would be necessary to account for the range of
acoustic frequencies often observed within a single species' vocal repertoire were too
great to provide a plausible mechanism for frequency modulation. Finally, Greenewalt
argued that if tracheal resonances influenced the acoustic properties of a bird's song in
any way, one should find evidence of resonances and anti-resonances in song notes
exhibiting large continuous changes in the fundamental frequency (*glissandi*). These
resonances and anti-resonances should be observed as amplitude increases and decreases, respectively, at frequencies predicted by the length of the birds vocal tract.
Greenewalt failed to find such evidence in his analysis of the songs of dozens of species.

A related reason why vocal resonances were discounted as an important factor in
song production was that the widely accepted Greenewalt-Stein model of song production appeared to adequately account for the modulations typically observed in bird
sounds based on the mechanics of the syringeal sound source alone. Changes in the
tension of the vibrating membranes and in the position of other tissues (such as the
external labia) were held to account for the characteristic patterns of frequency and
amplitude modulation that characterize most bird sounds. In other words, the Greenewalt-Stein model did not need to evoke extrasyringeal influences on sound production,
either in terms of frequency modulation occurring at the source (the wind instrument
analogy) or vocal-tract-based spectral shaping (the speech analogy).

However, these particular analogies represent only the extreme cases in a broad array
of ways in which the acoustic properties of a bird's vocal tract might affect sound production. The first question to ask, whether the vocal tract has any influence at all, was
readdressed by Nowicki (1987) using a simple technique for manipulating vocal tract
acoustic resonances. Acoustic resonance is dependent on the speed of sound, which is
in turn dependent on atmospheric density (Rayleigh 1896). If nitrogen, which comprises
about 80 % of normal air, is replaced by helium, then the velocity of sound increases
from 331 to 578 m/s, which produces a corresponding upward shift of roughly 75 % in
the acoustic resonance frequencies of air-filled chambers or tubes. Thus, if the vocal
output of a bird is affected by singing in a helium-oxygen atmosphere (i.e., *heliox*), it
is safe to conclude that acoustic properties of the vocal tract do play some role in production.

Nowicki (1987) observed striking changes in the acoustic structure of the songs of
ten different songbird species recorded in heliox. The most obvious effect of heliox on
song production was the appearance of harmonic overtones for song elements appearing as pure tones in normal atmosphere (see Figure 7). Another effect was an overall
reduction of as much as 10 dB in the amplitude of the sounds produced. Unlike the effect
that would be predicted for an analogous experiment using a wind instrument, however,
there was *not* an appreciable shift in the fundamental frequencies of various song elements. Such an effect would be predicted if coupling between acoustic resonances and
the frequency of vibration of the sound source was occurring. A shift of 3 - 5 % was
observed, but this outcome was an order of magnitude less than would be expected if
coupling were occurring. However, the effects that Nowicki (1987) observed were also
not obviously consistent with a source-filter model, such as that proposed for human
speech. The dominant effect of heliox on speech production is to shift the emphasized

frequencies (formants) in a broadband, harmonic sound to correspondingly higher values. The appearance of harmonics above a previously pure-tone fundamental frequency appears to be quite different. Thus, although the results of the heliox experiment suggested that acoustic resonances do have a role in birdsong production, it is not immediately apparent what this role might be.

In spite of the apparent differences from typical speech production, Nowicki (1987) argued that the effects of heliox on birdsong could be explained by modeling the songbird vocal tract as an acoustic filter. For instance, the appearance of harmonics in heliox might have been due to the upward shift of the filter, allowing acoustic energy that was previously heavily attenuated to pass more readily (see Figure 8). Consistent with this view, broadband bird sounds recorded in this study showed a change in the relative amplitude of overtones at different frequencies. An overall decrease in the amplitude of pure-tone sounds recorded in heliox would result from damping of the fundamental frequency, which would no longer be in the center of a resonance peak. In this view, then, the source-filter model of speech is an appropriate analogy for the role of the vocal tract in birdsong production.

Although the heliox experiment demonstrated that acoustic resonances can influence song production, at least two problems remained in interpreting the results. First, it could be argued that both the magnitude of the upward shift in acoustic resonance frequencies (more than 70 %) and the striking nature of the observed effects (the appearance of strong harmonics in a normally pure-tone sound) were too large to be biologically relevant in normal song production. Furthermore, an important implication of the heliox experiment result is that birds should be capable of dynamically modifying the acoustic properties of their vocal tracts while singing (Nowicki and Marler 1988). It remained unclear if and how birds could accomplish such modification.

Both of these criticisms were addressed by Westneat et al. (1993), who used kinematic analysis to quantify motions of the head, throat, and beak in singing birds. If acoustic properties of the vocal tract are important in production, then changes in resonances should correspond in turn to changes in the physical configuration of the tract. If the vocal tract of a bird is modeled as a tube that is open at one end, there are three simple ways in which its acoustic resonances might be altered (Nowicki and Marler 1988). First, the tube itself could be lengthened or shortened. Second, the open end of the tube could be occluded to varying degrees. Finally, the open end of the tube could be flared to varying degrees.

Greenewalt (1968) argued that it is unlikely that a bird can change the length of its vocal tract to any great degree, even though a bird's trachea may be extensible. He did not, however, address the last two possibilities. The beak is especially well-positioned to modify the acoustic properties of the vocal tract, either by changing the actual length of the tract through the degree to which it is flared (analogous to lip-rounding in human speech production (see Lieberman 1977) or by changing the impedance at the open end and thereby altering the effective length of the tract (a bird might also occlude the open end of its vocal tract, and be able change impedance by varying the opening of the glottis, i.e., the tracheal opening just posterior to the oral cavity). For instance, a more closed beak should correspond to lower-frequency vocal tract resonances, whereas a more open beak should correspond to higher-frequency vocal tract resonances. This is precisely the relationship Westneat et al. (1993) found in the songs of two species, the

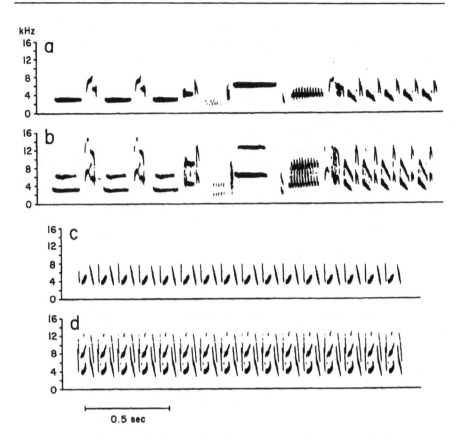

Fig. 7. Spectrograms of sparrow songs recorded in normal air and in heliox. A Song sparrow song in normal air. B The same song-type produced by the same bird as in A, but sung in heliox. C Swamp sparrow song in normal air. D The same song type produced by the same bird as in C, but sung in heliox (from Nowicki and Marler 1988, Regents of the University of California, used with permission)

white-throated sparrow (*Zonotrichia albicollis*) and the swamp sparrow. In white-throated sparrows, species-typical song comprises a series of unmodulated, pure-tone notes, each lasting up to 800 ms, with discrete frequency shifts occurring between adjacent notes. During song production, the beak opening was found to be relatively constant over the course of a given note, but changed abruptly between notes of different frequencies. Data from the swamp sparrow were even more compelling. This species' songs are composed of short, frequency-modulated notes that sweep through several thousand Hertz in as few as 10 to 20 ms. Even at this rapid modulation rate, each bird's beak was observed to open and close in accordance with note frequency (illustrated in Figure 9). In both species, other acoustic properties of the song, such as amplitude, were not correlated with beak opening. Other kinematic variables, such the angle of the head,

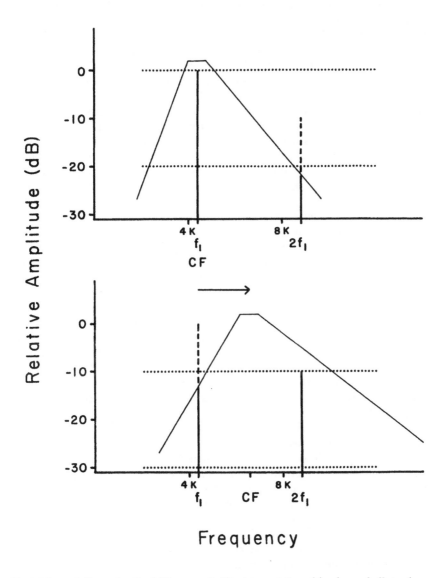

Fig. 8. Schematic illustrating the shifting acoustic filter interpretation of the observed effects of heliox on birdsong production. In normal air (*above*), the filter is centered over the fundamental frequency produced by the syringeal source, resulting in the attenuation of the second harmonic. In heliox (*below*), the vocal tract filter's center frequency is shifted upward by almost 70 %, attenuating the fundamental and revealing the second harmonic. The hypothetical filter roll-off is 24 dB/octave; the asymmetry is the result of representing the filter function on a linear frequency axis. *Horizontal dotted lines* represent a 20-dB dynamic range, characteristic of older analog spectrographs (From Nowicki and Marler 1988, Regents of the University of California, used with permission)

were also not related to acoustic properties of the song. Thus, Westneat et al. (1993) concluded that beak opening and closing could provide a mechanism by which a bird dynamically changes the acoustic properties of its vocal tract, consistent with the interpretation that such dynamic changes are involved in normal sound production.

Beak motions are not necessarily the only means by which birds might vary the acoustic properties of the vocal tract. For instance, postural changes involving the opening and closing of the glottis, as well as changes in the length and position of the trachea, might contribute to the acoustic properties of a bird's vocal tract just as the positioning of various articulators in the human vocal tract affect its resonances (Lieberman 1977). However, Westneat et al.'s (1993) kinematic analysis did not conclusively demonstrate that beak motions are necessary for normal production of birdsong. Such a demonstration would require that experimental manipulation of the putative mechanism controlling tract resonances (e.g., the beak) be found to produce a predictable change in vocal output.

Nowicki et al. (unpubl. data) provided just such a demonstration by comparing songs produced under normal conditions with those produced when a subject's beak had been immobilized. They argued that this manipulation would not prevent a bird from being able to produce the full range of frequencies observed in normal song, given that the heliox experiment had demonstrated little or no direct coupling between the activity of a bird's vocal source and suprasyringeal vocal tract resonances (i.e., changes in tract resonances, such as that observed in the heliox atmosphere, did not result in a shift in the fundamental frequency). However, limiting beak motion would restrict a bird's ability to modify vocal tract resonances if such motions do in fact represent a mechanism by which dynamic changes can be effected. In this case, the predicted effect of the manipulation is a frequency-dependent change in the amplitudes of sounds produced, with frequencies that are farther from the experimentally fixed vocal tract resonance being more attenuated than those falling closer to this fixed resonance. Nowicki et al. (unpubl.) recorded three white-throated sparrows and three swamp sparrows before, during, and after their beaks were immobilized. In all cases, the relative amplitudes of sounds at different frequencies were affected as predicted.

This last study conclusively demonstrates that changes in vocal tract acoustic properties, mediated by beak motions (and probably by other postural changes of the vocal tract), are a necessary feature of normal birdsong production. This finding is an important and quite unexpected departure from earlier thinking about this topic. The technical approaches used, however, were not particularly sophisticated or advanced. Instead, a suite of straightforward techniques were combined with simple physical insights to develop the necessary evidence. Not all advances await new technology; sometimes they simply await new ideas.

3
Quo Vadis?

Despite notable progress on several fronts, some of the oldest questions in the field of syringeal function still remain. Many of these questions concern the syringeal membranes and how they produce sound. Questions also remain about the two-voice theory.

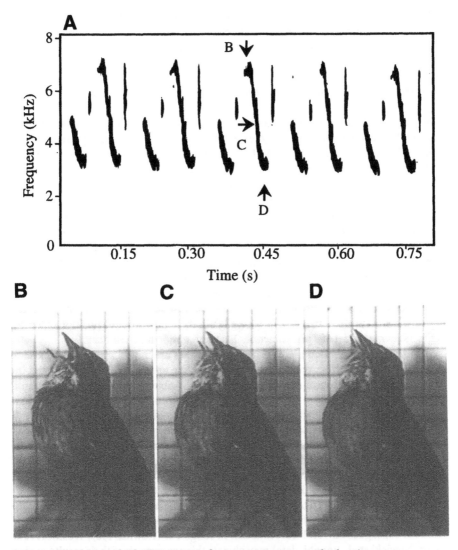

Fig. 9. A Spectrogram and video B-D images of a swamp sparrow song. The three images are adjacent video fields showing the progressive closure of the beak as the bird makes a downward frequency sweep. The *arrows* in A identify the points in time over the duration of the *type-V* note from which the video fields in B, C, and D were taken (From Westneat et al. 1993, used with permission)

Although it has now been demonstrated that birds are capable of singing a true internal duet (Suthers 1990), it also seems clear that multiple acoustic sources may interact in ways that we are only beginning to understand (e.g., Nowicki and Capranica 1986a). Finally, the role of the vocal tract in production remains unclear. Although the importance of the vocal tract in vocal production in songbirds is now firmly established (Westneat et al. 1993), the precise acoustic nature of this action, as well as the possible diverse modes of action it might take, are still to be explored. The role of vocal tract acoustics in other vocal production in other groups may well be different than in songbirds (e.g., Gaunt et al. 1987) and further study is clearly needed.

3.1
Membrane Physiology and Vibration

The Greenewalt-Stein model suggests that stable vibration in a syringeal membrane arises when airflow from the lungs creates a Bernoulli effect, lessening pressure on the side of the membrane exposed to the flow and thus drawing the membrane further into the lumen of the bronchial tube. As the membrane extends further into the tube, the Bernoulli force and the counteracting forces of the membrane's own elastic tension eventually become equal, and the former decreases through viscous flow losses. The membrane then retracts until the two forces again reach equilibrium and another cycle begins. Thus, the forces responsible for driving oscillation in the syringeal membranes are thought to be similar to those acting on the vocal folds in human speech production (see Rubin and Vatikiotis-Bateson, this Volume). One major problem with this model, however, is the possibility that the Bernoulli effect is not applicable given the particular pressures, flow rates, and membrane diameters characteristic of songbird syrinxes (Fletcher 1989; see also Vogel 1994). Because no one has ever directly observed the vibration of a syringeal membrane, we are left in the uncomfortable position of having an inadequate model of a phenomenon that may or may not even occur!

Granting the likelihood that syringeal membranes are indeed somehow induced to vibrate, there is a further controversy concerning the nature of that vibration and the kinds of sound a syringeal membrane is likely to be able to produce. The question is whether a freely oscillating syringeal membrane will produce pure-tone sounds, harmonic sounds, or sounds with some more complex set of overtones (Goldspink 1983). Casey and Gaunt (1985) raised objections to the possibility that an edge-clamped membrane, vibrating at right angles to the direction of air flow as the Greenewalt-Stein model suggests, could produce either pure tones or even a simple harmonic series. Fletcher (1988, 1992) presented a more sophisticated mathematical model that showed that, in a nonlinear system, the objections of Casey and Gaunt do not hold. Nonetheless, although Fletcher's model worked well for predicting the harmonic tones of a corvid, he also was unable to account for pure tone, or whistled, phonations. The work of Nowicki and his colleagues on vocal tract function in songbirds (Nowicki 1987; Westneat et al. 1993; Nowicki et al., unpubl. data) may obviate this concern, since their results suggest that the primary functional role of the vocal tract is to filter out or otherwise suppress overtones that are present in the source energy. Nonetheless, considerably more work with mathematical models is needed. As Fletcher (1988, p. 457) states: "Only when we can construct a quantitative model with reasonable success can we claim to understand the sound production mechanism".

Interestingly, almost all attempts at modeling (Greenewalt 1968; Gaunt and Wells 1973; Brackenbury 1979, 1982; Casey and Gaunt 1985; Fletcher 1988) have assumed a membrane that enters the lumen as a uniform, semi-hemispherical bulge. Yet some direct observations have suggested that the membrane folds into the lumen rather than distending, projecting an edge into the airstream rather than a dome-like surface with approximately uniform tension along all axes (Beebe 1925; Paulsen 1967; Suthers and Hector 1982, 1985). In fact, the membranous portions are supported by C-shaped half rings that are either bony and rigid, or cartilaginous and elastic. These rings are positioned so as to maintain tension on the membrane along either a dorso-ventral or medio-lateral axis, depending on its position. If movements of the trachea or syrinx relax the membrane along the antero-posterior axis, tension along the orthogonal axis is maintained. Moreover, most intrinsic syringeal muscles are so arranged that, regardless of whatever else they may do, they will apply an "opening out" vector to the ends of the C-shaped rings. Thus, their activity should tend to increase the uniaxial tension on the membrane.

Clearly, our models will be much improved when we understand the various configurations the membranes may adopt under different conditions. Two kinds of information will be useful. The first will come from more detailed studies of the anatomy and material properties of the syrinx and its membranes. For example, questions about how a membrane distends will be informed by understanding whether the membrane stretches uniformly in all dimensions, which in turn can be predicted from knowledge of the types and orientation of fibers in the membrane (see e.g., Wainwright 1988) and measured in vitro. Direct in vivo imaging of membrane actions during song production would be more informative still. Such imaging is technologically challenging due to both the inaccessible position of the syrinx in the thoracic cavity in most species and the extremely high vibration rates involved. Although the requisite laser interferometry and small-fiber optic cable might be technology that is available, its application to birdsong physiology requires the use of equipment that can detect a signal under the adverse conditions likely to be associated with the small size of the optic fiber needed to be implanted, as well as development of a surgical procedure for implanting, fixing, and aiming this cable in a bird in a way that still allows singing to occur.

3.2
Syringeal Physiology and Modulation

It appears likely that the characteristic time-varying changes in the frequency and amplitude of a typical birdsong fundamental (i.e., its *contour*), as well as more rapid frequency and amplitude modulations, are traceable to the syrinx and are controlled by the syringeal musculature (Gaunt and Gaunt 1985b; Vicario 1991). Further, we now have direct evidence that songbirds, at least, can use two sources independently in sound production (Suthers 1990; Suthers et al. 1994). Specifically how modulations are controlled at the source, however, is less clear. There is as yet little direct information about how the syringeal musculature affects the membranes and other aspects of syringeal configuration, let alone the specific acoustic features of a bird's song. The range of possible interactions and combinations of the activity of two sound sources has only just begun to be explored. In some cases, the two sources appear to work in complete independence (Suthers 1990). In other cases, interactions between them produce complex, nonlinear modulations (Nowicki and Capranica 1986a,b). These cases probably represent only two extremes in a continuum of possible interactions.

Studies that combine the use of miniature flow and pressure transducers with EMG have begun to yield important new information on syringeal function. For example, Suthers and Hector (1982, 1985) demonstrated that the gating of sound (passage of air pulses) is mediated by actions of extrinsic syringeal musculature in grey swiftlets and oilbirds. More recently, Goller and Suthers (pers. comm.) have obtained EMG, flow, and pressure data that suggest specific roles for individual intrinsic syringeal muscles. However, these preliminary results are not entirely consistent with earlier models of syringeal function. The intrinsic musculature seems to function as three groups: left-dorsal, right-dorsal, and ventral (Vicario 1991; Goller and Suthers 1995), at least as far as can be determined from present sampling methods. The dorsal muscles appear to be active primarily in gating, possibly by inserting the lateral labium into the syringeal lumen. Gating is independent for the two sides (as would be predicted by the two-voice theory). The ventral muscles appear to moderate the tension on the typaniform membranes. Oddly, both the phonatory and silent sides appear to be active during all song elements. Thus, the oscine syrinx may function in a less complex fashion than its anatomy would suggest. Coupling of flow measurements with precision electromyography will continue to elucidate this problem, but the exceedingly small size of the syrinx, the difficulty of identifying particular muscles and other functionally important elements involved in phonation, and the acoustic complexity of all but the most simple birdsongs will continue to present a challenge to the most adroit experimentalists.

Note added in proof. Progress in the analysis of syringeal function has outstripped the speed of the publishing process. Recent papers, especially from Suthers and his co-workers, have shown that 1) dominance of one side, ususally the left, is common; 2) the silent side is closed; 3) the left side is ususally responsible for low frequencies; and 4) some birds are capable of a seamless transition from one side to the other during a frequency sweep. Much of this work is reviewed in R.A. Suthers and F. Goller, 1997. Motor correlates of vocal diversity in songbirds. In: Current Ornithology (Val Nolan, Jr. and Ellen Ketterson, eds.) Vol. 14. Plenum Press, NY. (In press).

3.3
Vocal Tract Physiology and Tonal Quality

As it has now been demonstrated that the vocal tract is an important element in sound production in birds, a whole new can of worms has been opened for physiologists interested in vocalization. Major questions whose importance was not even recognized previously must now be addressed. One of these concerns how vocal tract properties physically influence sound production. An obvious possibility is that the avian vocal tract acts as an acoustic filter in a manner similar to that proposed for the human vocal tract in speech production (Fant 1960; Lieberman 1977). If so, pure-tone whistles observed in birdsong are in fact likely to be based on harmonically rich source signals, with the vocal tract acting to filter out all but a single frequency (such as the fundamental or second harmonic). Although this model obviates the need to account for production of sinusoidal sounds by the syringeal membranes, it does require a vocal tract filter function with sufficiently sharp roll-off characteristics that all frequency components but one are strongly attenuated.

An alternative model also derives from comparisons to human vocal production, but one that involves soprano singing rather than speech. In the source-filter model of speech, source and filter act more or less independently. In soprano singing, by contrast, recent work suggests that an overlap between the fundamental frequency of vocal fold vibration and the lowest-frequency formant results in nonlinear feedback from the vocal tract to the glottis that effectively suppresses production of harmonics at the source (Rothenberg 1981, 1987a,b). A similar overlap between the fundamental (or second harmonic) of the syringeal source energy and a dominant resonance of the vocal tract is probably very common in birdsong production (Nowicki and Marler 1988). In both the speech-based and *soprano* models, a bird must dynamically coordinate the acoustic properties of its vocal tract with the output of its syrinx during song. In the latter case, however, the effect of this coordination is to suppress the production of overtones at the source, potentially providing a more parsimonious explanation for a bird's ability to produce virtually pure-tone song.

Both models of vocal tract function are compatible with the experimental evidence currently available. For example, one result of the heliox experiment — the appearance of harmonics above sounds that are normally produced as pure tones (Nowicki 1987), could equally well be explained by the shifting of a resonance filter, allowing higher frequency overtones to pass through the filter with less attenuation, or by a decoupling of vocal tract acoustic properties from the source, enabling the production of overtones that are normally suppressed at the source. Two approaches may provide data that would distinguish between these possibilities. The first involves modeling the properties of the avian vocal tract, both physically and mathematically, to determine the passive acoustic properties of the vocal tract and whether these properties can account for the attenuation of overtones occurring in normal song.

The second approach is experimental and involves testing song production in an atmosphere that is *heavier* rather than lighter than normal. In this manipulation, the passive acoustic properties (filtering effects) of the vocal tract are expected to shift to lower frequencies. The fundamental frequency should be attenuated, as in heliox, because it is no longer squarely centered under the passband of a vocal tract resonance. Unlike the case of singing in heliox, however, one might predict, at first, that this attenuation would *not* be accompanied by the appearance of higher-frequency overtones, because the filter function has been shifted to lower frequencies (but see below). If, on the other hand, the vocal tract is normally coupled in some way to the source, then a shift in vocal tract resonances to either higher or lower frequencies should both produce a decoupling effect. Overtones would appear because the source is no longer constrained to produce pure tones, irrespective of the fact that the filter function has been shifted to lower frequencies.

Tetrafluoromethane (TFM) is a gas of sufficiently greater density than nitrogen to produce a strong effect on vocal tract filtering and is also biologically inert. The speed of sound is decreased by about 35 % in an 80:20 mixture of TFM : O^2. Thus, a lowering of resonance frequencies by 35 % would be predicted, as compared to an increase of 70 % in the case of heliox (a total difference of about an octave).

Preliminary work with birds recorded in TFM has shown an effect that is qualitatively similar to that observed in helium, with the appearance of higher-frequency harmonics above the fundamental (Nowicki and Lohr, unpublished data). This result suggests that the passive filter model is at least insufficient, if not incorrect, but closer examination reveals additional complications. Specifically, the lowering of the first resonance of the vocal tract in TFM : O^2 would be expected to be accompanied by a proportional decrease in the higher-frequency resonances! Consider a hypothetical case in which the dominant resonance of a bird's vocal tract is 4 kHz. In heliox, this resonance is predicted to shift upward to about 6.8 kHz. In TFM : O^2, it should shift downward to about 2.6 kHz. But if the vocal tract is modeled as a tube open at one end, an additional resonance is expected to occur at the three times the frequency of the lowest resonance (Benade 1976). This second resonance, normally at 12 kHz and therefore beyond the upper frequency limit of most birdsongs, will also shift downward, to about 7.8 kHz. In other words, in both heliox *and* TFM : O^2 we expect a resonance in the upper-frequency range. This unfortunate complication does not rule out the usefulness of the TFM : O^2 experiment. However, the analysis must include careful quantification of the relative amplitudes of frequency components and more accurate models of vocal tract acoustic properties.

Irrespective of which alternative proves to be the more accurate model of the vocal tract acoustic effects tract on song production, there is also much to learn about the peripheral effectors that control vocal tract acoustic properties and how those effectors are coordinated with the activity of the syrinx. Westneat et al.'s (1993) kinematic analysis of birdsong, which emphasized the role of the beak, merely scratched the surface of the functional morphology that must be considered. In addition, the use of x-ray cinematography to image the motions of internal structures, especially the opening and closing of the glottis, will provide a much broader view of the various effectors that contribute to a bird's ability to dynamically vary the acoustic properties of its vocal tract while singing. Finally, we deem it likely that it will be a futile exercise to consider "the function" of "the vocal tract" in birds. As with syringeal function, there are likely to be multiple modes of operation for the vocal tract and the way in which it interacts with the source signal. One has to consider only the call of the black-capped chickadee, which includes notes that change from extremely tonal frequency sweeps to wideband, harmonically rich sounds within a matter of milliseconds (e.g., Figure 3A; Ficken et al. 1978; Nowicki and Nelson 1990) to gain an appreciation for the possibility of qualitatively distinct modes of operation of a bird's vocal tract filter.

4
Summary

The last several years have seen significant progress in our understanding of some of the mechanisms used by birds to produce sound. Nonetheless, our knowledge in this area is rudimentary and our techniques primitive when compared to the accomplishments and approaches routinely employed by our colleagues who study human speech. We might suggest that birdsong production is more difficult to investigate because the signal is acoustically more variable and involves less accessible anatomical structures, but this argument would be specious. There is, however, one essential difference between birdsong and speech that bears mention. The important outcome of both speech and song is communication. Questions about the physiology underlying production are restricted, if not informed to some extent, by an understanding of which aspects of the signal being produced are important for communicating information. In studies of speech, a human subject can be asked directly to report their perception or interpretation of a speech feature. We are not so fortunate when analyzing birdsong or other animal communication signals. Instead, clever tests must be devised both for the field and the laboratory that can provide glimpses into the relationship between structure and function in such signals. Recent progress in the use of articulatory synthesis in speech demonstrates the utility of understanding perception and meaning in analyzing the production mechanisms involved. Perhaps the next great advances in our understanding of birdsong physiology await the development, not of more sensitive or precise physiological techniques, but of methods for better understanding the relationship between structure and function in the songs themselves.

References

Allen SE, Suthers RA (1994) Lateralization and motor stereotypy of song production in the brown-head ed cowbird. J Neurobiol 25: 1154-1166

Ames P (1971) The morphology of the syrinx in passerine birds. Peabody Mus Nat Hist Yale Univ Bull 37: 1-194

Baptista LF and Trail PW (1992) The role of song in the evolution of passerine diversity. Syst Biol 41: 242-247

Beddard FE (1898) The structure and classification of birds. Longman, Green, New York

Beebe C (1925) The variegated tinamou, *Crypturus variegatus variegatus* (Gmelin). Zoologica 6: 195-227

Benade AH (1976) Fundamentals of musical acoustics. Oxford University Press, London

Berger M, Hart JS (1968) Ein Beitrag zun Zusammenhang zwischen Stimme und Atmung bei Vögeln. J Ornithol 109: 421- 424

Borror DJ, Reese CR (1956) Vocal gymnastics in wood thrush songs. Ohio J Sci 56: 177-182

Brackenbury JH (1978a) A comparison of the origin and temporal arrangement of pulsed sounds in the songs of the grasshopper and sedge warblers, *Locustella naevia* and *Acrocephalus schoenobaenus*. J Zool (Lond) 184: 187 206

Brackenbury JH (1978b) Respiratory mechanics of sound production in chickens and geese. J Exp Biol 72: 229-250

Brackenbury JH (1979) Aeroacoustics of the vocal organ of birds. J Theor Biol 81: 341-349

Brackenbury JH (1982) The structural basis of voice production and its relation to sound characteristics. In: Kroodsma DE, Miller EH, Ouellet H (eds) Acoustic communication in birds, vol 1. Academic Press, New York, p 53

Brackenbury JH (1989) Functions of the syrinx and the control of sound production. In: King AS, McLelland J (eds) Form and function in birds, vol 4. Academic Press, London, p 193

Calder WA III (1970) Respiraton during song in the canary (*Serinus canarius*). Comp Biochem Physiol 32: 251-258

Casey RM, Gaunt AS (1985) Theoretical models of the avian syrinx. J Theor Biol 116: 45-64

Chaunaud RC (1970) Aerodynamic whistles. Sci Am 222: 40-46

Cherry C (1966) On human communication, 2nd ed. MIT Press, Cambridge

Clark RB (1949) Some statistical information about bird song. Br Birds 42: 337-346

Clark CW, Marler P, Beeman K (1987) Quantitative analysis of animal vocal phonology: an application to swamp sparrow song. Ethology 76: 101-115

Dabelsteen T, Pedersen SB (1985) Correspondence between messages in the full song of the blackbird *Turdus merula* and meanings to territorial males, as inferred from responses to computerized modifications of natural song. Z Tierpsychol 69: 149-165

Fant G (1960) Acoustic theory of speech production. Mouton, The Hague

Ficken MS, Ficken RW, Witkin SR (1978) Vocal repertoire of the black-capped chickadee. Auk 45: 34-48

Fletcher NH (1988) Bird song - a quantitative acoustic model. J Theor Biol 135: 455- 482

Fletcher NH (1989) Acoustics of bird song - some unresolved problems. Comments Theor Biol 1: 237-251

Fletcher NH (1992) Acoustic systems in biology. Oxford University Press, NY

Gaunt AS (1983) An hypothesis concerning the relationship of syringeal structure to vocal abilities. Auk 100: 853- 862

Gaunt AS (1987) Phonation. In: Seller TJ (ed) Bird respiration, vol 1. CRC Press, Boca Raton, FL, p 71

Gaunt AS, Gaunt SLL (1977) Mechanics of the syrinx in *Gallus gallus*; II. Electromyographic studies of ad libitum vocalizations. J Morphol 152: 1-20

Gaunt AS, Gaunt SLL (1985a) Electromyographic studies of the syrinx in parrots (Aves, Psittacidae). Zoomorphology (Berl)105: 1-11

Gaunt AS, Gaunt SLL (1985b) Syringeal structure and avian phonation. In: Johnston RF (ed) Current ornithology, vol II. Plenum Press, New York, p 213

Gaunt AS, Wells MK (1973) Models of syringeal mechanisms. Am Zool 13:1227-1247

Gaunt AS, Stein RC, Gaunt SLL (1973) Pressure and air flow during distress calls of the Starling, *Sturnus vulgaris* (Aves, Passeriformes). J Exp Zool 183: 241-261

Gaunt AS, Gaunt SLL, Hector DH (1976) Mechanics of the syrinx in *Gallus gallus*, I. A comparison of pressure events in chickens with those in oscines. Condor 78: 208-223

Gaunt AS, Gaunt SLL, Casey RM (1982) Syringeal mechanics reassessed: evidence from *Streptopelia*. Auk 99: 474-494

Gaunt AS, Gaunt SLL, Prange HD, Wasser JS (1987) The effects of tracheal coiling on the vocalizations of cranes (Aves; Gruidae). J Comp Physiol 161A: 43-58

Goldspink G (1983) Alterations in myofibril size and structure during growth, exercise, and changes in environmental temperature. In: Peachy LD (ed) Handbook of physiology. Sec 10, Skeletal muscle. American Physiological Society, Baltimore

Goller F, Suthers RA (1995) Implications for lateralization of bird song from unilateral gating of bilateral motor patterns. Nature 373: 63-66

Greenewalt CH (1968) Bird song: acoustics and physiology. Smithsonian Institution Press, Washington, DC

Hartley RS, Suthers RA (1989) Airflow and pressure during canary song: direct evidence for mini-breaths. J Comp Physiol A 165: 15-26

Hartley RS, Suthers RA (1990) Lateralization of syringeal function during song production in the canary. J Neurobiol 21: 1236-1248

Hunter ML Jr (1980) Vocalization during inhalation in a nightjar. Condor 82: 101-103

King AS (1989) Functional anatomy of the syrinx. In: King AS, McLelland J (eds) Form and function in birds, vol 4. Academic Press, London, p 105

Koniski: M (1989) Birdsong for nueurobiologists Neuron 3:541-549

Lieberman P (1977) Speech physiology and acoustic phonetics. Macmillan, New York

Marler P (1969) Tonal quality of bird songs. In: Hinde RA (ed) Bird vocalizations. Cambridge University Press, Cambridge, p 5

Marler P (1970) Birdsong and speech: could there be parallels? Am Sci 58: 669-673

Martin WF, Gans C (1972) Muscular control of the vocal tract during release signaling in the toad *Bufo valliceps*. J Morphol 137: 1-28

McCasland JS (1987) Neuronal control of bird song production. J Neurosci 7: 23-39

Nelson DA (1989) The importance of invariant and distinctive features in species-song recognition in birds. Condor 91: 120-130

Nelson DA, Marler P (1990) The perception of birdsong and an ecological concept of signal space. In: Stebbins WC, Berkley MA (eds) Comparative perception, vol 2. Complex signals. John Wiley, New York, p 443

Nottebohm F (1971) Neural lateralization of vocal control in a passerine bird. I. Song. J Exp Zool 177: 229-261

Nottebohm F (1976) Phonation in the orange-crowned amazon parrot, *Amazona amazonica*. J Comp Physiol A 108: 157-170

Nottebohm F (1977) Asymmetries in neural control of vocalization in the canary. In: Harnad F, Doty RW, Goldstein L, Jaynes J, Krauthamer G (eds) Lateralization in the nervous system. Academic Press, New York, p 23

Nottebohm F (1980) Brain pathways for vocal learning in birds: A review of the first 10 years. Prog Psychobiol Physiol Psychol 9: 85-124

Nottebohm F (1991) Reassessing the mechanisms and origins of vocal learning in birds. Trends Neurosci 14: 206-211

Nottebohm F, Nottebohm ME (1976) Left hypoglossal dominance in the control of canary and white-crowned sparrow song. J Comp Physiol A 108: 171-192

Nottebohm F, Kelley DB, Paton JA (1982) Connections of vocal control nuclei in the canary telencephalon. J Comp Neurol 207: 344-357

Nowicki S (1987) Vocal tract resonances in oscine bird sound production: evidence from bird songs in a helium atmosphere. Nature 325: 53-55

Nowicki S (1989) Peripheral lateralization of birdsong reanalyzed: comparison of multiple techniques for unilateral disablement of syringeal function in sparrows. In: Erber J, Menzel R, Pflger H-J, Todt D (eds) Neural mechanisms of behavior, Proc 2nd Int Congr of Neuroethology. Georg Thieme, Stuttgart, p 121

Nowicki S, Capranica RR (1986a) Bilateral syringeal coupling during phonation in a song bird. J Neurosci 6: 3595-3610

Nowicki S, Capranica RR (1986b) Bilateral syringeal interaction in vocal production of an oscine bird sound. Science 231: 1297 1299

Nowicki S, Marler P (1988) How do birds sing? Music Percep 5: 391-426

Nowicki S, Nelson DA (1990) Defining natural categories in acoustic signals: comparison of three methods applied to 'chick-a-dee' call notes. Ethology 86: 89-101

Nowicki S, Westneat M, Hoese W (1992) Birdsong: motor function and the evolution of communication. Semin Neurosci 4: 385-390

Paulsen K (1967) Das Prinzip der Stimmbildung in der Wirbeltierreihe und beim Menschen. Akademische Verlagsgesellschaft, Frankfurt am Main

Phillips RE, Youngren OM (1981) Effects of denervation of the tracheo syringeal muscles on frequency control in vocalizations of chicks. Auk 98: 299-306

Potter RK, Kopp GA, Green HC (1947) Visible speech. Van Nostrand, Princeton

Rayleigh JWS (1896) The theory of sound (1945 reprint of 2nd edn). Dover, New York

Rothenberg M (1981) Acoustic interaction between the glottal source and the vocal tract. In: Stevens KN, Hirano M (eds) Vocal fold physiology. University Tokyo Press, Tokyo, p 305

Rothenberg M (1987a) Cosi Fan Tuti and what it means, or, nonlinear source-tract interaction in the soprano voice and some implications for the definition of vocal efficiency. In: Baer T, Sasaki C, Harris K (eds) Laryngeal function in phonation and respiration. College-Hill, Boston, p 255

Rothenberg M (1987b) The control of air flow during loud soprano singing. J Voice 1: 262-268

Ryan MJ, Brenowitz EA (1985) The role of body size, phylogeny, and ambient noise in the evolution of bird song. Am Nat 126: 87-100

Schild D (1986) Syringeale Kippschwingungen und Klangerzeugen bein Feldschwirl (Locustella naevia). J Ornithol 127: 331-336

Shannon CE, Weaver W (1949) The mathematical theory of communication. University Illinois Press, Urbana

Stein RC (1968) Modulation in bird sounds. Auk 85: 229-243

Suthers RA (1990) Contributions to bird song from the left and right sides of the intact syrinx. Nature 347: 473-477

Suthers RA, Hector DH (1982) Mechanism for the production of echolocating clicks by the grey swiftlet, Collocalia spodiopygia. J Comp Physiol A 148: 457-470

Suthers RA, Hector DH (1985) The physiology of vocalization by the echolocating oilbird, Steatornis caripensis. J Comp Physiol A 156: 243-266

Suthers RA, Goller F, Hartley RS (1994) Motor dynamics of song production by mimic thrushes. J Neurobiol 25: 917-936

Vicario, DS (1991) Contributions of syringeal muscles to respiration and vocalization in the zebra finch. J Neurobiol 22: 63-73

Vogel S (1994) Life in moving fluids, 2nd ed, Princeton University Press, Princeton

Wainwright S (1988) Axis and circumference. Harvard University Press, Cambridge

Westneat MW, Long JH Jr., Hoese W, Nowicki S (1993) Kinematics of birdsong: Functional correlation of cranial movements and acoustic features in sparrows. J Exp Biol 182: 147-171

Youngren OM, Peek FW, Phillips RE (1974) Repetitive vocalizations evoked by local electrical stimulation of avain brains. III. Evoked activity in the tracheal muscles of the chicken (Gallus gallus). Brain Behav Evol 9: 393-421

SECTION III

ASSESSING BIOLOGICALLY IMPORTANT RESPONSES

CHAPTER 10

Sound Playback Studies

S. L. HOPP AND E. S. MORTON

1
Introduction

In the early 1890s, before Edison publicly released his gramophone, R.L. Garner, a researcher at the newly established National Zoological Park in Washington, D.C., had begun recording animal vocalizations. As part of his effort to decipher "the speech of monkeys", Garner conducted some of the first sound playback experiments. He reasoned that by replaying the vocalizations to animals and watching their reactions, he could learn the functions of the sounds. In one experiment he replayed the sound of a female rhesus monkey to her mate, who had been separated from visual contact. Garner recounts, "He gave evident signs of recognizing the sounds, and at once began a search for the mysterious monkey doing the talking. His perplexity at this strange affair cannot well be described. The familiar voice of his mate would induce him to approach, but that squeaking, chattering horn was a feature which he could not comprehend. He traced the sounds, however, to the source from which they came, and failing to find his mate, thrust his arm into the horn quite up to his shoulder, then withdrew it, and peeped into it again and again" (Garner 1892, p. 6). Although his methods were simple by today's standards, and his interpretation of the response not exactly conclusive, Garner nevertheless anticipated what has become one of the most important techniques for studying animal communication, the sound playback experiment.

Despite their potential, sound playbacks were not used on a large scale until the middle part of the present century (see historical account by Falls 1992). It was not until the 1950s and 60s, that high-fidelity portable tape recording and sound spectrography for signal classification and analysis became widely available. These tools, coupled with a strong interest in the use of visual and vocal models for studying animal communication, set the stage for growth in both the quantity and quality of playback studies. The applications of playback studies have expanded considerably since then, and a recently released volume devoted to the subject gives accounts of the extremely diverse uses of sound playback studies (McGregor 1992a). In addition, computers have broadened the potential of playback studies by making it possible to manipulate and synthesize acoustic signals, construct precise stimulus sequences on tape, and directly control stimulus presentation in field trials. Continuing technological improvements and innovations promise to contribute to even further growth in the sophistication of playback studies.

In spite of recent rapid changes in the instruments used for playbacks, the logic and mechanics of the research design underlying playback studies remain largely unchanged. It is this aspect of playback studies, the design and execution, on which this chapter focuses. For our purposes playback studies (*playbacks*) are broadly defined as the use of broadcast sound with accompanying bioassay to address questions related to animal behavior and acoustic communication. Typical playbacks are conducted in the field with naturally occurring behaviors as dependent measures. However, the concepts discussed here are not specifically limited to these types of studies. Our intent is to provide an overview of common approaches used in the manipulation, measurement, and control of variables used in playbacks, and how these approaches can be used in testing hypotheses. The chapter by Cynx and Clark (this Volume) primarily addresses experiments in which the subjects have been trained. However, the theoretical distinctions between that discussion and this one are not absolute. Field playbacks can demonstrably induce an element of learned behavior in the subjects (e.g., Petrinovich and Patterson 1980; Cheney and Seyfarth 1988; Searcy et al. 1995). Alternatively, playbacks can be used to address questions about the perception of signals (e.g., Horn 1992; Evans 1993). The discussions offered in these two chapters are meant to be complementary.

Our individual backgrounds predispose us to an emphasis on avian studies, though this emphasis should not imply the exclusion of other taxonomic groups in the application of the ideas we discuss. Broad differences in signal structure, social systems, natural history, and environmental acoustics prevent an exhaustive coverage of the details involved in implementing playbacks for individual groups. Indeed, much of the success of a playback study is determined by a thorough knowledge of these aspects of the target species.

We intend to focus on aspects of design and methodology that can be transferred conceptually across taxa, and to provide guidance in translating hypotheses into practice. Since use of sound playbacks requires a thorough understanding of sound transmission, we begin with an overview of sound propagation and acoustics (see also Gerhardt, this Volume). An understanding of the transmission of complex sounds in heterogeneous environments involves sound playbacks coupled with careful acoustic measurements. In the next section we consider several topics related to the design of playback studies. To begin, we outline the advantages and disadvantages of different stimulus presentation techniques. These include procedures for the timing and structure of observation and stimulus presentation periods. We then offer some ideas for increasing the realism of playbacks. In the section that follows we discuss aspects of the bioassay and considerations for the selection of appropriate response measures. We then discuss the analysis of data from playbacks. In the last section under design considerations we outline some problems that recur in playbacks, and suggest solutions. In the final major section of the chapter we consider the range of uses for studies using sound playback, and outline some of the common designs used to address different categories of questions or hypotheses. This section details various lines of research where playbacks have played a significant role. For example, sound playbacks are used to study territoriality or conduct animal censuses. Also, different categories of questions about the functions of communicative signals are outlined, and we provide examples of these areas.

2
Sound Transmission and Habitat Acoustics

The distance across which animals can communicate by sound is enormously variable, and depends not only on the source amplitude, but also on the penetrating capability of the sound in the environment, the level of intervening ambient sound or noise, and the auditory sensitivity of the individual receiving the sound. Playbacks of natural and synthetic sounds have been used to study all three aspects in natural habitats, and careful consideration of these factors should accompany studies using broadcast sound. Differences in signal frequency and the nature of the transmission medium also contribute to the broadcast properties of signals (see related discussions by Gerhardt, Pye and Langbauer, and Tyack, this Volume). Here we consider the physical factors influencing sound transmission in air, and discuss some of the factors that bear on playbacks.

The measurement of sound attenuation in natural environments involves a simple playback procedure, but the results are derived from variables generally too complex to be adequately modelled (cf. Roberts et al. 1981; Martin 1981). In the absence of interference, sound spreads spherically, with energy decline proportional to the increasing surface of the sphere formed by the advancing sound front; the attenuation is 6 dB per doubling of distance. A calculation of this *spherical spreading* provides a reference level for evaluating other effects that occur in a natural environment. *Excess attenuation* (EA) is thus a comparison of the attenuation observed in a natural habitat minus that due to spreading effects.

Sound attenuation has two components. One is directly proportional to distance, and one is of unpredictable magnitude. The total EA is their sum, which can be calibrated by using a sound source of known acoustical characteristics. It is then possible to calculate sound level at any point in space under the assumption of an infinite, homogeneous, frictionless medium. Measured sound levels can be compared with these predictions in order to determine EA (e.g., Waser and Brown 1984).

Another approach, more widely used, is to approximate a calibration by measuring the sound level from the speaker at a relatively close distance, and comparing this level with measurements taken farther away. Consequently, if the speaker's sound level varies from test to test, or if some frequencies are produced with more amplitude than others, these can be taken into account in the data reduction. Specifically, the subtraction of sound levels in decibels is equivalent to taking energy ratios, so the frequency response characteristics of the speaker do not really enter into the calculations (e.g., Morton 1975, Marten and Marler 1977).

The importance of air homogeneity to sound propagation is, in part, due to the speed of sound relative to air temperature. The velocity of sound in air is approximately $(331.4 + 0.607C)$ m/s, where C is the temperature in degrees Celsius. Sound velocity also increases slightly with increasing humidity, e.g. at 20°C the sound velocity at 100% relative humidity is about 0.3% greater than at 30% relative humidity (Michelsen 1978). One important consequence of sound velocities varying as a function of temperature is that a sound wave front conducting through a temperature gradient will be refracted from the warmer air toward cooler air. This refraction causes a regular but fluctuating feature of open habitats called a *shadow zone*. A wave front advancing parallel with the earth is defracted upward due to air temperature and wind gradients (in the upwind direction). This defraction leaves an area of attenuated sound under the wave front, which is why it is so difficult to hear voices across a heated field. By contrast, if the surface cools the air immediately above it, such as over a cold lake, it is possible to hear voices at much greater distances. The best way for an animal in an open habitat to avoid such *sound shadows* is to call from an elevated spot or fly above the ground. Since the benefits of increased broadcast area increase exponentially with height, as little as 3 to 6 m above the ground will erase most of the shadow zone effects (Pridmore-Brown and Ingard 1955).

Forests, especially tropical ones, have relatively homogeneous air below the canopy. Thus, at lower levels, they differ from open habitats in lacking shadow zones for calling animals, showing the shadow-zone effect only at tree-crown levels. Above the tree crowns, however, temperature gradients may again become steep. If the air within the canopy is cooler than the air above it, the canopy and ground may act as a wave guide on sounds of some frequencies emanating from within the forest (Waser and Waser 1977). These sounds will not diminish in amplitude at the rate predicted by the inverse square law (6 dB per doubling of distance). Instead, by not spreading geometrically, they will experience *negative EA* and attenuate less than expected.

The presence and amount of reverberation constitutes another major difference between open habitats, such as grasslands, and forests (Morton 1975; Richards and Wiley 1980). Tree trunks and the forest canopy may reflect and scatter some frequencies, such that the sound reaching the receiver has components that arrive at different times. This scattering from multiple surfaces causes a decay of sharp signal onsets and offsets. The standard measure of *reverberation time, RT60*, is the amount of time it would take a signal to decay by 60 dB after its termination. Waser and Brown (1986) found essentially no reverberation in African savanna, whereas RT60s in rain and riverine forest were much longer (ca. 0.44 s), and increased rapidly and linearly with distance.

An important addition to frequency-dependent attenuation is a consideration of what happens to the waveform and frequency content of a signal as it travels through the environment. Reverberations, amplitude fluctuations, and differential frequency attenuation cause changes in the signal that are related to the distance the signal has traveled. Since the pioneering studies of Richards and Wiley (1980), several studies have sought to incorporate signal degradation into evolutionary considerations of signal design and long-distance communication. Further growth in our understanding of signal transmission in complex habitats will contribute to improvements in the use and design of playbacks.

3
Design Considerations

3.1
Trial Structuring

Much of the structuring of acoustic playbacks finds roots in the studies conducted with models of animals in the visual realm, and much of the logic of trial structures can be found in studies conducted with other modalities, for instance in olfaction (which might be called "spraybacks"; see Stevens 1975) or vision (e.g., Evans and Marler 1991; Rowland et al. 1995). While the goal is quite similar, to assess differential responsivity to communication stimuli, a major difference in the approaches lies in the fact that visual and olfactory models persist in time and can be presented simultaneously, while acoustic stimuli are finite in duration.

The simplest trial structure is probably that in which a single stimulus is presented to the subject(s) and an elicited response is measured. For example, Chapman et al. (1990) presented an alarm-call segment to sleeping groups of spider monkeys (*Ateles geoffroyi*) and measured their resulting responses. Such a design proves valuable in situations where observable responses are not occurring, or are at low levels, at the onset of the trial. The format is appropriate for assessing the effectiveness of particular stimuli in eliciting responses of interest.

In situations where one tests the relative effectiveness of two or more stimuli, the trials can be constructed in the same manner but with two stimuli presented to the subject. In this situation, the stimulus pair is presented alternately from two speakers placed equidistant from the subject (simultaneous presentation, which is frequently used with visual stimuli, would result in acoustic interference). The relative orientation or approach of the subject is then assessed. Nowicki et al. (1989) employed this design in testing the relative responsiveness of swamp sparrows (*Melospiza georgiana*) to normal songs, computer altered songs, and songs produced in helium. They measured the average approach distance to songs from two of the three stimulus categories, when presented in pairs. They then reversed the location of each stimulus pair by switching speaker wires, to control for potential position effects. In another two-speaker design, Pallett and Passmore (1988) presented paired calls to female reed frogs (*Hyperolius tuberilinguis*) positioned in an arena, and scored a positive response when the subject approached within 10 cm of a speaker. The primary advantage of this type of design is that simultaneous presentation controls for individual variability in responsiveness; relative effectiveness of a stimulus pair is directly assessed.

In situations where the responses of the subject may preexist at some nonzero level, the effectiveness of the signal can be assessed by recording the extent to which the playback induces a change in some target behavior(s). This necessitates a trial format where time intervals alternately do and do not include a playback stimulus. In the simplest case, measures of responses occurring during a silent or baseline interval are compared to measures of the same responses during a subsequent interval in which stimuli are presented. This two-part (or *AB*) design can be expanded to include silent intervals both before and after stimulus presentation periods (*ABA* design), multiple periods of silent and playback intervals (*ABAB* design), or periods wherein different stimuli are presented, separated by silent intervals (e.g. *ABAC* design). This latter design is not often used in playback studies because of *carryover effects*, where the subject's behavior is altered in the first playback period, which persists through later periods.

The ABA design has emerged as perhaps the most frequently used format for acoustic playbacks and many examples can be cited. One example is found in Falls et al. (1988). They conducted trials with western meadowlarks (*Sturnella neglecta*) where an initial 5-min silent period was followed by 2 min of playback, then followed by a 3-min *post-playback* or *residual* period. The duration of the different intervals (baseline, playback, and postplayback) is somewhat arbitrary — choice of appropriate time intervals typically depends on a thorough knowledge of the subject(s) being studied, and their natural time course for responding.

Another type of design uses *recovery-from-habituation*[1] to assess whether subjects detect a change in stimulus condition. Typically, a stimulus that elicits a given response is presented repeatedly to a subject until the response declines. Following the decline, the subject is presented with a new stimulus, or a stimulus from a new category. An increase or recovery of the habituated response indicates that the subject detects the stimulus change. This design is useful in detecting a subject's ability to distinguish among stimuli that would not necessarily result in differential response using either an ABA or a *between-subjects* design. For example, Searcy et al. (1994) presented songs to red-winged blackbirds (*Agelaius phoeniceus*), who responded by showing wing-spread displays. Following habituation of the response, presentation of a different song type resulted in recovery of display intensity, indicating that the subjects discriminated between song types. Other examples include work by Cheney and Seyfarth (1988) and Searcy et al. (1995). This design closely parallels the logic of *transfer-of-training* paradigms in laboratory studies of learning, and discussions of these designs might suggest further possibilities for sound playbacks (e.g., see Cormier and Hagman 1987).

3.2
Realism in Design

Increasingly, field studies using sound playback are being designed to carefully mimic an intruding animal. In the past, studies have typically relied on timed repetitions of a single stimulus, presented from tape loops or pre-recorded onto tape. The result is often a mechanical-sounding stimulus that bears only marginal resemblance to vocal behavior of an actual animal, and sometimes results in poor response by the subject. Efforts to improve the realism of playbacks have included placing the speaker in an anechoic

---

1   Some authors have referred to this type of trial as *dishabituation*. The original use of the term *dishabituation* was to refer to recovery of response to the <u>original</u> stimulus, rather than a new one (Thompson and Spencer 1966). Using their terminology, the trials described here would be a *novelty* response. Thus, here we use *recovery-from-habituation*. However, the use of *dishabituation* in the context of playback studies has become common and will likely persist.

container to reduce the influence on neighbouring individuals (e.g., Stoddard et al. 1992). Other possibilities for improving realism include using naturally recorded sequences of song to provide stimulus variability in playback sequences, or using multiple speakers sequentially to mimic the movements of an animal.

One increasingly used procedure involves presentations of playback stimuli that are contingent on what the subject is doing, rather than with a predetermined timing. Although not a new technique (e.g., Petrinovich and Patterson 1980), the difficulty of accurate control of stimulus presentation in the field has prevented widespread use of response-dependent playbacks. Recent developments in technology have allowed for more such studies (Dabelsteen and Pedersen 1991; Bradbury and Vehrencamp 1993; Otter et al. 1993; McGregor and Ranft 1994), and use of this technique has proven fruitful (e.g., Bremond and Aubin 1992; Dabelsteen 1992; Smith 1996).

3.3
Dependent Measures

The choice of what to measure as an outcome in a sound-playback experiment may not be straightforward. It typically involves decisions made at each of several levels. A thorough discussion of dependent measures is well beyond the scope of this chapter, instead we simply offer several considerations to help guide choices for an appropriate bioassay. Many of the considerations reflect choices inherent in all studies of animal behavior, and several sources of discussion are available (e.g., Lehner 1979; Bakeman and Gottman 1986; Martin and Bateson 1986). These considerations include the reliability and validity of the measures used, the ease of observation, the method of coding, and whether or not to use single or multiple response measures. In many playback studies, video or audio recordings of behavior are made and used for later coding. Ultimately, the choice of a particular response or set of responses will be determined by a combination of the question being addressed and a working knowledge of the subjects being studied.

Part of the rationale for using a playback design with a particular species is the assumption or recognition that some overt response will occur in response to the sound stimulus. Many — perhaps most — playbacks use some variation of orientation or movement (*phonotaxis*) relative to a sound speaker. Examples include the presence of postures interpreted as "alert" behavior (Langbauer et al.1991; Weary and Kramer 1995), whether or not the subject approaches the speaker within a certain distance (Gerhardt 1991; Greenberg et al. 1993), coding categories of distance from the speaker (Nelson 1989; Searcy et al. 1995), and latency to approach within a given distance to the speaker (Nelson 1989; Greenberg et al. 1993). Another frequently used category of behavior is the vocalization of subjects in response to the sound playback, or *antiphonal* response. Again, many variations of measures have been used, including the presence of antiphonal vocalizing (Langbauer et al. 1991), and the rate or latency to vocalize (Nelson 1989; Greenberg et al. 1993). Others have recorded aspects of the timing of vocal signals produced, such as song duration or length of pauses between songs (Klump and Gerhardt 1992; Adhikerana and Slater 1993). When subjects have many vocalizations in their repertoire, researchers can assess *matching*, or whether the choice of a response signal by the subject is the same as the signal being broadcast (Krebs et al.1981; Falls et al. 1988; Horn 1992; stoddard et al. 1992).

In addition to phonotaxic and antiphonal responses, some other behaviors can be coded as being specific to the sound stimulus. Examples include behaviors taken to be aggressive, such as wing-spread displays by red-winged blackbirds (*Agelaius phoeniceus*; Searcy et al. 1994), behaviors considered to be sexual responses by females (notably copulation solicitation displays; Vallet and Kreutzer 1995; see reviews by Searcy 1992; Searcy and Yasukawa 1996), and escape or avoidance behaviors in response to alarm vocalizations (Weary and Kramer 1995). In studies hypothesizing specific referential categories of sounds, behaviors appropriate to those references are recorded (e.g., Cheney and Seyfarth 1988).

Finally, physiological measures have been used infrequently, but represent an underexploited means of measuring subtle changes in behavior related to playbacks. These measures include things such as changes in heart rate in response to acoustic stimuli (Davis 1986, Diehl 1992). Similarly, hormonal changes in response to acoustic stimuli provide a tool for studies of neuroendocrine involvement in communication (see Ball and Dufty, this Volume).

3.4
Analysis

As with all research, the choice of a statistical procedure is based on the design of the study, specifically, the comparisons being made and the measurements used. A complete discussion of these considerations is obviously beyond the scope of this chapter and the reader is referred to any good statistics or research methods text. Here we discuss data analysis of playbacks and issues that deserve specific attention.

The choice of an appropriate method of data analysis must be dictated by the hypothesis being tested, and in many cases is also influenced by the experiment's trial structure. The ABA structure which is often used for playbacks allows one to ask first whether the treatment has an effect (comparison of treatment to baseline conditions), whether this effect has any carryover (comparison of residual to either treatment or baseline), and finally, if more than one treatment is included, whether the two (or more) groups differ. This last comparison is typically the target of the study, and often the pre-trial and posttrial baseline periods are removed or ignored, and the response levels during playback periods are compared. This approach to analysis is problematic. While the ABA structure has emerged as one of the most common designs in playbacks, analysis has been an unfortunate reinvention of the wheel, with many different methods employed. Well-established techniques have long existed for analysis of time-series designs of this nature. These methods not only compare across the two groups, but also address issues such as whether the initial baseline periods are equivalent (e.g., Chatfield 1975; Rushe and Gottman 1993). The use of these time-series analyses would be beneficial in this area.

A second issue that influences the approach in analysis is the decision of which and how many dependent measures are used. McGregor (1992b) discusses this issue at length and we refer the reader to his treatment. Simply put, the trade-off between one or many dependent measures finds the researcher choosing between oversimplifying the complexity of response on the one hand, and engendering unmanageable confusion and intermeasure conflict in analysis and data presentation, on the other. McGregor (1992b) concludes that the solution to this dilemma is to assess multiple measures, but to simplify the analysis and data summary by use of principal components analysis (PCA) to first transform the data, then conduct inferential tests on the result.

We agree with McGregor's conclusion that multiple analysis is a flawed approach, but we disagree with the value of PCA as a substitute, for three primary reasons. First, PCA is of questionable value when analyzing multivariate playback data. The procedure is most effective when applied to multiple measures in the same scale, e.g., multiple measures of length, or a series of strongly correlated variables (Jackson 1991). In the case presented by McGregor (1992b), there appear to be two clusters of measures, one related to approach or other forms of movement and the other related to vocal behavior. This clustering effect suggests that PCA is not appropriate. A second, related, point is that since the original measures were standardized, and the analysis then conducted on the result, it is unclear what exactly is being compared in the final analysis. Because the original scales of measurement are removed one is unable to tell whether a significant outcome occurs because one or a few of the measures are strongly different, while others are only trivially different, or whether the significant outcome reflects a combined effect of multiple variables. This loss of information in the analysis prevents a clear interpretation of what the results reflect in the subjects' actual behaviors. Finally, information about individual behaviors is lost, providing only a weak foundation on which to base future work.

To overcome these drawbacks, a researcher should ideally consider and present results from both individual variables and the multivariate results. This approach actually adds more information than has been typical in playback studies, contrary to McGregor's assertion. McGregor (1992b) proposes that the observed clarity of PCA-based results (data from McGregor and Krebs 1984) implies that such results will occur in other applications. Such clarity may or may not result.

We suggest that studies using multiple measures should employ a multiple analysis of variance (MANOVA) as the most appropriate multivariate technique. Many models appropriate to different designs are discussed in the literature (e.g., Timm 1993) which adjust significance levels for multiple comparisons. Such procedures provide for a conclusion of a significant overall effect as well as assessing the relative importance of the various individual measures.

A third issue in the analysis of playbacks concerns the application of *alternative probability* models. Gerhardt (1992) has argued that *Bayesian* statistics would be quite useful in analysis of data from playbacks. Several treatments of this statistical strategy are available, including general introductions (Winkler 1993), applications in the context of playbacks (Gerhardt 1992), and models for conducting Bayesian analysis of variance (Lewis 1993). These techniques have not yet been extensively used, though this has less to do with their potential utility than with the fact that the methods are unfamiliar to most researchers.

3.5
Common Problems with Playback Studies

Several problems are commonly encountered by those who design and employ sound playbacks. Many of these are problems not unique to playbacks per se, but are associated generally with field studies in animal behavior. Several sources outline common problems, such as observer bias, reliability of behavioral coding, and analysis of data (e.g., James and McCulloch 1985; Kroodsma 1986; Martin and Bateson 1986; Gerhardt 1992). Among the problems faced in studies using sound playback, the issue of *pseudoreplication* has received considerable attention. Pseudoreplication, to summarize a complex idea, involves inadequate representation or sampling of the levels of an independent variable in the design of the study (see Hurlbert 1984). Because it has been presented (Kroodsma 1989a; 1990), argued (Catchpole 1989; Kroodsma 1989b; Searcy 1989), and summarized (McGregor et al. 1992) elsewhere, we offer only a brief discussion here (see also discussion by Cynx and Clark, this Volume).

Hurlbert (1984) initially expressed concern about the design of field experiments in general, and discussed several issues related to sampling, control and manipulation of variables, and statistical independence. Kroodsma (1989a, 1990) applied some of these ideas to playbacks, with the specific recommendation that, in practice, researchers should use multiple examples of song stimuli rather than including only a single representative from each stimulus category, as had been a common practice. The discussion resulting from this article (Catchpole 1989; Kroodsma 1989b; Searcy 1989), and the so-called consensus (McGregor 1992) that resulted, identified two main design issues important to playbacks, those of *internal* and *external validity*. Both of these are commonly jeopardized in poorly designed studies, and both can, in part, be addressed by use of multiple sound stimuli. Here we briefly outline both concepts.

Internal validity refers to the isolation of the independent variable from the influence of confounding variables (Campbell and Stanley 1963). The presence of internal validity comes from the design of the study and is separate and independent from the outcome of the study. In studies where the hypothesis predicts differential responding to different sound categories, it is necessary to separate the conceptual aspects of the hypothesized category from other potential confounding variables. To address this, Kroodsma (1989a, 1990) proposed that researchers use multiple sound stimuli from the hypothetical categories, in effect sampling from the variation within the category, but also randomizing other variables that might be confounding. Kroodsma (1989a, 1990) presents several different options for use of multiple stimuli, but optimally each subject would be presented with a different sound stimulus appropriate to the proposed category.

Among different organisms, stimulus characteristics exist along continua that range from simple or unidimensional (e.g., the intensity or duration of a stimulus), to complex (e.g., song dialects in birds). When using a simple dimension as an independent variable, such as intensity, the levels of the variable can be under direct control of the experimenter, and thereby be less likely to be affected by external variables. For example, Nowicki et al. (1989) conducted playbacks to song sparrows (*Melospiza melodia*) of normal song sparrow songs and the same songs produced in a helium-based atmosphere. Since the two sound categories (normal and helium-altered) were represented by songs produced by the same individual birds, the problem of pseudoreplication is minimized. Similarly, where synthetic stimuli are used, the experimenter has control over the differences between stimulus categories, minimizing the need for multiple stimuli (e.g., Gerhardt et al. 1989; see discussion in McGregor et al. 1992).

In contrast, in cases where the experimenter does not know which particular stimulus characteristics are used to code that category, it becomes difficult to experimentally isolate the categories of stimuli. In these cases, the potential for confounding is much greater, and the potential for pseudoreplication effects is greatest. To isolate complex sound categories, it is necessary to use multiple examples from each proposed conceptual stimulus category (Kroodsma 1990).

External validity, also called *generalization* or *ecological validity*, refers to the ability of a researcher to apply the results of a study to circumstances outside the experimental context. Like internal validity, generalization is, to a certain degree, a problem with the application of the results of all controlled studies. However, relevant to the issue of multiple sound stimuli, using a broad sample of playback stimuli allows for maximum application of results beyond a given study. If the number of sound stimuli are limited, the applicability of the results may also be limited. This applicability is true even if the hypothesis being tested has nothing to do with sound-stimulus categories. Thus, both internal and external validities are increased by use of multiple stimuli.

Previous coverage of pseudoreplication has made it familiar to most researchers using playback designs. In a sample of recent articles from several major journals, we found that of 20 playback studies reported between 1993 and 1995, 16 investigators used multiple stimuli in their trials, and 14 of these acknowledged that they did so in order to prevent pseudoreplication effects. Of the four studies that did not use multiple stimuli to sample from vocal categories, all used synthetic stimuli, i.e., stimuli that were either synthesized or reconstructed using a computer, which presumably minimizes pseudoreplication effects (see discussions above and in McGregor et al. 1992). The ease with which multiple stimulus tapes can be produced with digitally based sound-processing programs (Stoddard 1990), and the apparent pressure from journal reviewers, will likely insure that use of multiple stimuli will remain a standard part of the design of playback studies.

A final problem area in playback designs is the procedures used in sampling individuals for inclusion in studies. Often, as a technique to save time or resources, a single subject is used in more than one trial for a study. This problem is not specific to playback studies and has been discussed in the context of ethological studies in general (Machlis et al. 1985). Use of a single subject on repeated trials violates statistical assumptions of independence. Suggestions given by Machlis et al. (1985) are relevant here and we refer the reader to that article.

A similar issue arises when groups of animals are treated as a single subject or trial. Studies wherein sounds are broadcast to groups of animals become difficult to interpret. The probability of response by any given individual may be tied to the response of other members of the group. Once a given individual responds, subsequent response can be either to the presented sounds or to the responding individuals. Presence of other individuals can induce audience effects, facilitation, or inhibition of response, even if no response is observed. Further, the tendency to respond may be quite variable, but the coding of response is often done in an all-or-none fashion. Thus varying degrees of response are coded as being similar. It is important that researchers consider the effects of multiple animals on the outcome of playback studies, and either attempt to control for or assess these effects.

4
Uses of Playback

Playbacks take advantage of naturally occurring responses of animals to communication signals. These responses derive from the many natural functions of acoustic signaling, including territoriality, response to alarm calling, or responses to members of social groups such as mates, rivals, or offspring. Given this tendency to respond, the aim of a playback study is to mimic the natural signal presentation of a caller in such a way that a recipient's response reveals something about how the animal perceives and categorizes the signal. The crucial element in interpreting the outcome of such work is the occurrence of differential, stimulus-dependent response. A simple example is the presence or absence of a response, from which the experimenter can infer which features of the sounds are actually essential in eliciting the response. A more complex experiment might measure variations in the magnitude or intensity of responses to the presentation of different sounds or categories of sounds, which can then be used to make inferences about how the subject perceives such variation or discriminates among sound stimuli.

Interpreting the results of playback experiments requires recognition of four important limitations of this general approach. First, not all signals elicit a predictable response from receivers. The questions that can be asked by using playbacks are therefore limited to situations in which the animal actually responds reliably. Second, an animal's response in the field represents a combination of its perception of the stimulus on the one hand, and its tendency to respond to that stimulus on the other. An experimenter cannot control the factors influencing when and why an animal responds, nor how it categorizes sounds. Further, the experimenter must take care to distinguish between behavior induced by the sound that is played back and the underlying motivation for responding — the animal provides the response criteria. We discuss this issue more thoroughly later in the chapter. Third, when a study fails to show differential responding to various stimuli, the researcher cannot know whether the subjects were unable to perceive the acoustic differences in the sounds, or perceived the stimuli as being acoustically distinguishable but biologically equivalent (see discussions by Irwin 1958; Beecher and Stoddard 1990; Stoddard 1996). In other words, in studies where no differences in response are measured to different stimuli, it is difficult to unambiguously account for this result. Finally, playbacks induce responses only in receivers of stimuli. These responses to sound playback do not necessarily provide information about the function of the sound or the motivation of the animal that originally produced it. For example, a rapid-approach response is often interpreted as being aggressive. This does not imply that the signal can be interpreted as carrying an aggressive message nor that the signaler had an aggressive intent.

In spite of these potential difficulties, cleverly designed playback studies have been used to answer a wide variety of questions about animal acoustic communication. Placing these questions in categories provides a means for organizing our discussion of experimental design, although any attempt at a simple classification would fail to characterize the complexity and number of hypotheses tested with playback protocols. In the following, we present nine categories of study areas, generally arranged from simple to complex in use, design, or analysis. We provide examples to illustrate the techniques that are characteristic of each category and offer citations to sources providing more thorough discussions. See also Falls (1992) and other chapters in McGregor (1992a) for discussions of the range of uses for sound playbacks.

4.1
Censusing and Demographic Studies

The simplest use of a playback is to induce an animal to make its presence known by responding to the sound as if an intruder had invaded its territory. In this way, playbacks can be used to count animals (e.g., Gilbert et al. 1994; Hambler et al. 1994; Hawkins 1994). Documenting the presence or absence of individuals in a range of habitats or over larger geographic areas contributes to understanding the habitat preferences and distribution of a given species. Adding geographic area as a variable, and using playbacks systematically (e.g., presenting a stimulus every 50 m along transects), allows researchers to census animals that use sound for territorial defense, and determine their population densities. Playbacks of appropriate sounds can bring animals into view, or induce them to vocalize so that they may be counted and mapped. Because playbacks are so effective, they can also be used to help document the absence of a particular species from an area (Villard et al. 1995).

However, this approach also has some limitations. Playbacks used for censusing are limited to one or a few species per playback, and often only one sex will respond. Further, animals may show seasonal and circadian variation in response, and non-territorial (floating) individuals may tend not to respond and therefore be missed.

Examples of the use of playbacks to census birds on non-breeding territories are provided by Lynch (1989) and Sliwa and Sherry (1992). Lynch and colleagues played species-specific *chip* calls of hooded warblers (*Wilsonia citrina*), followed by Kentucky warblers (*Oporornis formosus*), over a large area of Mexico's Yucatan Peninsula. Kentucky warblers were found only in wet forests, whereas hooded warblers were found in all forest types and in non forest scrub. Lynch also performed vegetation analyses at each census point in order to add habitat variables to the census results. Female and male hooded warblers were found to occupy different habitats, though both responded equally to playbacks of chips recorded from either sex (Lynch et al. 1985). This habitat separation was found to be due to female preference for oblique vegetative stems and male preference for vertical stems (Morton 1990). After a hurricane broke most of the branches of forest trees, females were found in territories where slash produced a predominance of oblique stems, which were areas in which they had not been recorded in censuses conducted over the previous decade (Morton et al. 1993).

Playbacks are also useful in marking studies, in which individuals of a species are permanently tagged to monitor their movement or survival. Setting playback speakers near mist nets is one of the most time-efficient means of capturing a breeding population of birds for marking. Marking individuals in a population provides a way to study important aspects of demography such as reproductive success, genetic mating systems, longevity, and sexual and age differences in survival and dispersal (e.g., Rowley and Russell 1993). Studies reported by Mabey and Morton (1990) exemplify the use of playback for both censusing and demography. These investigators used playbacks to count Kentucky warblers on non-breeding territories in Panama, but also placed playback speakers near mist nets to capture and mark individuals for identification and survival estimation. In species with sex-specific vocalizations, males and females may be captured independently. In Carolina wrens (*Thryothorus ludovicianus*), only males sing and respond to song playbacks, whereas only females respond to female *chatter* calls. Playing tapes with only one of these calls will therefore only attract individuals of the sex represented by the sound; playbacks containing both calls capture mated pairs (Morton unpubl. data). Marking breeding individuals using playback near mist nets, and recapturing or resighting them at intervals, adds a demographic dimension to population censuses.

4.2
Territory Studies

Playbacks can also be used for mapping territorial boundaries and studying territorial behavior. In territory mapping, playbacks are used to induce movement toward the sounds. Speaker locations are moved away from the target animal until it no longer moves toward the speaker location, or until a second individual confronts the target animal. This response allows the researcher to estimate the territorial boundary and the playback is repeated at other locations until enough points are known to define the territory (Falls 1981). A potential drawback to this technique is that playback may sometimes stimulate animals to go beyond their normal territorial boundaries, resulting in an exaggerated or distorted estimate of territory area. Territories mapped using playbacks should be compared with boundaries observed without playback, once animals are marked for individual identification. Again, this technique is limited to species or individuals that respond to this treatment, which generally means males only.

Studies of territorial behavior form a large and conceptually important area of behavioral ecology. Many studies using sound playbacks test specific hypotheses concerning how an animal assesses another's ability to defend a resource. Adhikerana and Slater (1993) designed an experiment to determine whether singing rate was related to assessment of "vigor" among rival male coal tits (*Parus ater*). They found that the birds treated a playback of a singing bout with an increased production rate as a more serious threat than one with a decreased singing rate. McGregor and Horn (1992) found significant correlations between the number of repetitions of phrases within a song produced by a male in response to a playback and the strength of his response to playbacks on the edge or center of his territory in great tits (*Parus major*). This result is interesting, because playback of only a single song phrase is sufficient for species-recognition in great tits (Cheon and Park 1995). McComb et al. (1994) played back recordings of both individual and groups of three female lions (*Panthera leo*) roaring. Defending adult females were less likely to approach playbacks of three intruders than a single one, and apparently adjusted their decision to approach according to the size and composition of their own group. Coalitions of male lions cooperate by approaching broadcast roars of stranger males based on mutualism (Grinnell et al. 1995). Similar studies with non-human primates have shown that animals use long-distance signals to adjust inter-group spacing (e.g., Whitehead 1987).

Studies of intersexual, and interspecific territoriality are greatly benefited by playbacks. Duetting in mated pairs of birds, initially thought to represent cooperative territorial defense, is now known to be based upon individual, sex-specific goals (Morton 1996). Farabaugh (1982) and Levin (unpubl) have shown that complex antiphonal duets in Thryothorus wrens occur immediately and spontaneously during the first encounters of potential mates, and do not require lengthy periods of coordinated learning in which the male and female create songs unique to the pair, as was once thought. Playbacks have also been used in the study of interspecific territoriality, documenting extensive interspecific responses among congeneric species, reflecting competition that may contribute to habitat selection in neotropical birds (Robinson and Terborgh 1995).

4.3
Inter- and Intraspecific Acoustic Interference

Playbacks of vocal signals can be used to study the interference of ambient sound on the efficiency of vocal sound transmission. Not all ambient sound energy reduces the conspicuousness of a given signal; the greatest masking effect is caused by a band of frequencies on either side of and including the signal frequency. This *critical band* (Fletcher and Munson 1937; Fletcher 1940; Bilger and Hirsh 1956) is not an absolute entity, but reflects the characteristic auditory processing in the receiver (see Dooling 1982). Conspicuousness of a signal is thus a complex phenomenon that depends upon both the ambient noise levels and the hearing physiology of the listener, each of which can affect playback design. Overall, however, it is an individual's conspecifics that are most apt to cause significant ambient masking sound. These sounds typically exhibit closely matched acoustic characteristics and are therefore especially likely to produce interference.

Intra- and interspecific communication interference has been studied most thoroughly in frogs, who often call in close proximity to one another, and whose hearing is typically very finely tuned to the frequency components of conspecific calls (e.g., Capranica et al. 1973). In a series of experiments on several species of neotropical *Hyla*, Schwartz and Wells (1983a,b, 1984a,b, 1985) documented both the occurrence and perception of intra- and interspecific calling in individual frogs. Their study of *H. ebraccata* and *H. microcephala* serves as an example. Both species are found in the same microhabitats and their sounds overlap in spectral frequency. They measured an SPL of 107 dB at 50 cm from *H. microcephala*, about 6 dB greater than the other species' call. Background noise generated by *H. microcephala* choruses were found to cause a shift in the timing and type of calls given by nearby *H. ebraccata*, reduced their calling rates and the proportion of multi-note calls and aggressive calls (Schwartz and Wells 1983a).

Using playbacks of artificial sound, the study showed that only signals that included energy at 3 kHz sound, the most important masking frequency, was effective in changing a male *H. ebraccata's* calling. This finding suggests a simple proximate mechanism may be causing the calling changes observed in *H. ebraccata* under natural conditions. If a male of this species cannot hear conspecific males because of the masking effect of *H. microcephala* calls, it calls as though it were alone. There is a good reason for *H. ebraccata* males to avoid overlap with the louder species: female *H. ebraccata* are apparently not attracted to conspecific advertisement calls when they overlap with a chorus of *H. microcephala* (Schwartz and Wells 1983b). In contrast, a third sympatric species, *H. phlebodes,* increases its calling rate and adds click notes to its advertising calls in response to playbacks of conspecific and heterospecific calls (Schwartz and Wells 1984a). Clearly, these species of frogs make changes in their calls as adaptations to regularly occurring acoustic interference they encounter in their complex tropical communities. Note, however, that selection underlying call adaptations is generated intraspecifically through the agency of female mate choice.

Studies of anuran communities suggest that there is strong competition among individuals for what might be called *acoustic space*. Hodl (1977) studied 15 species of synchronously breeding frogs that are restricted to floating islands of vegetation in the Amazon river. Most of the species produced mating calls with unique dominant frequencies. Those whose calls showed overlapping spectral content and gave vocalizations from similar sites used sounds with different temporal features. In similar studies, Duellman and Pyles (1983) found patterns of total acoustic variation to be similar in three frog communities, but calls of closely related allopatric species were more similar than those of closely related sympatric species. Closely related sympatric species with similar calls either breed at different times or do not coexist in the same microhabitats.

The Puerto Rican coqui (*Eleutherodactylus coqui*) is an arboreal frog that provides an example of another aspect of acoustic interference. The species is named after the males' characteristic *coqui* call. The first note, *co*, is a constant-frequency tone, about 100 ms long, that is used in male-male territorial interactions (Narins and Capranica 1976). The second note, *qui*, sweeps upward in frequency and functions to attract females. The frequency of the *co* segment varies inversely with the size of the male. If a *co* sound is played to a male, he responds by dropping the *qui* segment of his own call. An experimenter can therefore learn whether a frog has perceived a *co* stimulus presented to it by the subsequent presence or absence of the *qui* sound. This response was used by Narins and Capranica (1976) to determine that the intensity of a *co* playback at a calling male's own *co* frequency is about 30 dB lower than the level needed when it is tested with calls 200 Hz lower or higher than its own call. Neighboring males usually do have *cos* of the same frequency. Neighbors typically are spaced about 2 m apart, and experience their neighbors' *co* calls at about 83 dB SPL, roughly 30 dB above the threshold for the *co* note response. It follows that an individual male will not hear a neighbor whose *co* note differs in frequency by about 200 Hz from his own (Narins and Smith 1986).

Many species of birds similarly show behaviors that apparently minimize acoustic interference with both inter- and intraspecific competitors, for instance by altering the timing of their singing . For example, Ficken et al. (1974) studied singing of two sympatric North American bird species, the least flycatcher (*Empidonax minimus*) and the red-eyed vireo (*Vireo olivaceus*). Recordings of adjacent males of these two species were analyzed for temporal overlap in songs to determine if either was influencing the timing of song delivery by the other (see Ficken et al. 1974; Wasserman 1977, for statistical methods). The results showed that the flycatchers avoided beginning a song while a vireo was singing, but vireos began songs regardless of flycatcher singing. One possible interpretation of this pattern is that the flycatcher was sensitive to the vireo singing because its shorter song could be masked by the longer vireo song, but not vice versa. In another study of interspecific effects, ovenbirds (*Seiurus aurocapillus*) shortened their songs and sang less frequently and at more irregular intervals when songs from other species were played through a loudspeaker (Popp and Ficken 1987). These examples show that the timing of singing in birds may be altered in order to avoid acoustic interference, just as in frogs. Masking effects are thus reduced by separation in both time and frequency.

Acoustic masking is difficult to achieve experimentally in free-living animals. Bremond (1978) attempted to mask songs of the northern wren (*Troglodytes troglodytes*) by combining sympatric heterospecific songs or altered wren songs with playback of the wren's normal territorial song. When the songs were mixed with those from two other species, the wrens delayed their approach to the loudspeaker. However, he was unable to mask the wrens' song by mixing it with tapes of altered wren song. Thus the effect of a rich acoustic environment, one in which several other species are also singing, seems to hinder the wrens' ability to respond quickly. Bremond attributed this outcome to attentional factors rather than to simple acoustic masking, proposing that the animal hears the signal but takes longer to decide whether or not to attack.

4.4
Distance Estimation and Ranging Theory

Recognition that birds are sensitive to the degradation of sound caused by transmission through heterogeneous environments (e.g., Richards 1981) led to the formulation of *ranging theory* (Morton 1982, 1986, 1996). Ranging theory considers the different demands on both the singer and the recipient, and how these demands relate to the transmission of sound in the environment and the evolution of signal structure. Since the essential components of ranging theory consider the transmission of signals through the environment, and the reception of and response to these signals, the use of playbacks to test the ideas is fundamental.

Briefly, the theory considers that territorial male birds benefit from the disruption they cause to neighboring rivals. The strategies from the point of view of the two roles, sender and recipient, then, are quite different. For the listener, an ability to accurately gauge the distance of a potential intruder is essential. This ability can be done by learning both the song types neighboring birds have, in conjunction with becoming familiar with the degradation effects on sound transmission by the local environment. For the singer, choosing songs that resist degradation in the local environment, and/or use of songs unfamiliar to the neighbors, will result in greater disruption to the neighbor because of his difficulty in judging the actual distance of the singer. The different and opposing strategies of the signaler and listener result in an escalating contest between them (Godard 1993).

The results from a number of different studies are consistent with predictions from ranging theory. Several species have been shown to differentially respond to degraded and undegraded (far and near) songs, including Carolina wrens (*Thryothorus ludovicianus*; Richards 1981) and western meadowlarks (*Sturnella neglecta*; McGregor and Falls 1984). In a reciprocal playback study, Gish and Morton (1981) showed that Carolina wren songs recorded in two different local habitats, deciduous forest in Maryland and palmetto hummocks in Florida, showed less degradation when played back in the original, rather than the other, environment. This finding is consistent with the idea that song learning could provide a means for birds to quickly adapt songs to local habitat acoustics (Hansen 1979). A related finding is that young Carolina wrens preferentially learn undegraded songs in laboratory-based learning studies (Morton et al. 1986).

Several studies have looked at the ability of a bird to accurately judge the distance of a song as a function of the presence or absence of that song in the bird's repertoire (e.g., McGregor et al. 1983; Sorjonen 1983; McGregor and Falls 1984; McGregor and Krebs 1984; Shy and Morton 1986a, 1986b; Morton and Young 1986). Most of these studies have shown that familiarity by comparison to one's own repertoire is important to a listener's ability to accurately range signals (although see McGregor and Krebs 1984; McGregor 1992b). A clearer understanding of this ability may result from a careful theoretical distinction between familiarity based on a bird having a particular song in its own repertoire, and familiarity as a result of repeated exposure to the songs of neighbors (see McGregor and Avery 1986). Another area of predictions from ranging theory addresses the relationship between resource availability and song output (e.g., Gottlander 1987; Reid 1987; Strain and Mumme 1988; Alatalo et al. 1990; Arvidsson and Neergaard 1991).

Ranging theory provides a broad theoretical framework for understanding the evolution of song repertoires and intraspecific communication in songbirds. Because this theory involves integrating sound transmission with both proximate and ultimate understanding of long-distance communication, investigators necessarily rely heavily on sound playbacks for testing its predictions (see Morton 1986, 1996).

4.5
Cognition and Referential Studies

Strong interests in referential communication in animals and the closely related topic of animal cognition have led to a number of empirical and theoretical papers on these topics. Playbacks have been used extensively for testing hypotheses in these areas. The use of playbacks for studies of signal meaning in animals has been reviewed by Macedonia and Evans (1993) in the context of referential signaling. Once a relationship between acoustic structure and an eliciting stimulus is established, then responses to playback can be used to gauge stimulus meaning. The confounding issues surrounding these cognitive approaches using playbacks are either conceptual or logical. A strong conceptual concern is how much does the question of "information conveyed" as opposed to "information extracted" influence the design and interpretation of experiments. A strong logical concern is whether the signaling driven by motivation is due to the urgency of the situation, or the meaning of a signal apart from motivation. These two positions can be viewed as endpoints of a continuum ranging from motivational to referential signaling. Theoretical and methodological issues concerning the study of these questions are discussed by Marler et al. (1992) and Macedonia and Evans (1993).

Questions about message and meaning that concern what information is conveyed and what is decoded can be addressed when playbacks are combined with field observations of natural signal use. Smith and Smith (1992), for instance, used playbacks to test whether eastern kingbird (*Tyrannus tyrannus*) territory defenders would approach an intruder being simulated using *chatter-zeer* or the "less provocative" *zeer* vocalizations. The birds chose the former, as field observations had predicted.

Similarly, Cheney et al. (1995) used playbacks to study "cause-effect relations in the context of social interactions" in baboons (*Papio cynocephalus*). The researchers played back call sequences, the order of which was either appropriate (i.e., a dominant females *threat call,* followed by a subordinate animal's *fear bark*) or anomalous (i.e., a subordinate female's *threat call* followed by a dominant individual's *fear bark*). Subjects responded more strongly to the anomalous sequences, which Cheney et al. interpreted as evidence of sensitivity in these animals to the apparently disordered cause-and-effect relationship being shown.

Playbacks have also been widely used to study what are termed *referential signals* (Marler et al. 1992). Referential signals are typically defined by criteria that combine an examination of the specificity of both the stimuli that elicit the vocalizations, and the resulting behavior in a receiver of such signals. In the former, referential signals are those confined to specific contexts, or with such high situational specificity, that they are evoked only in response to one or a few items in the environment. In the latter, responses given to signals are judged as specifically appropriate to the referent (external context) of the signal. For example, vervet monkey (*Cercopithecus aethiops*) calls given to different classes of predators are thought to be referential (Seyfarth et al. 1980). In these, predators (referents) such as an eagle, snake, or leopard, elicit signals with specific acoustic structures from the monkeys (Struhsaker 1967). Playbacks can be presented in the absence of the predators to see if appropriate escape behaviors are induced by the signal alone, in other words as a signal that functions in the absence of contextual cues.

Seyfarth and Cheney (1990) used a recovery-from-habituation design, described above, to determine whether vervet monkeys transfer habituation to signals evoked by similar or different predators by another species, the superb starling (*Spreo superbus*). Vervets transferred habituation between a starling raptor alarm call and a vervet eagle alarm, but not between vervet leopard alarms and starling raptor alarms. This was taken to indicate that the vervets' responses to the starling alarm calls were based on a presumed relationship between the acoustic categories and the likely eliciting stimuli.

4.6
Studies of Vocal Development

Vocal development studies, usually done in the laboratory, use playbacks that are exact models of natural vocalizations. In bird-song research, experiments in which hatchlings are raised in isolation show that exposure to natural song is necessary for normal song development to occur in oscine passerines (reviewed in Baptista and Gaunt 1994). Early studies produced new questions and a rich era of research into how and when species-typical songs are acquired. Extensive work by Peter Marler and his many colleagues on how song and swamp sparrows learn appropriate species-typical songs provide many examples (e.g., Marler and Pichert 1984). These studies showed that some species learn songs from playback tapes in the absence of social interaction with live tutors, whereas others fail to learn from taped stimuli. Still others can learn songs from tapes, but preferentially learn songs from live tutors rather than from tapes when both are available, even if the tutor belongs to the wrong species (Baptista and Petrinovich 1986). Much current research focuses on the timing of song learning in relation to the sharing of song-types among breeding populations (Nelson and Marler 1994). Studies of song acquisition in nonoscine passerines (e.g. flycatchers) have shown them to have innate singing abilities (Kroodsma 1984, 1985, 1989c), as was expected since their songs vary little geographically. Kroodsma and Konishi (1991) showed that deafened eastern phoebes (*Sayornis phoebe*) produced normal songs. Morton et al. (1986) studied the contribution of song degradation due to sound transmission to songs learned by Carolina wrens, using playbacks of both degraded and undegraded songs to tutor naive males in the laboratory. They found that the subjects showed a strong preference for learning undegraded songs. The ratio of undegraded to degraded song types learned was remarkably similar to the proportion of songs typically shared by neighboring males under natural circumstances.

4.7
Differential Responses to Signal Categories

By designing studies to reveal whether subjects respond differentially to different stimuli, researchers have been able to study various functional categories of sounds. This is likely the largest category of playbacks, in which the experimenter assesses differential responses to stimuli sampled from each of two (or more) acoustic categories. These studies are also likely to be the most susceptible to problems with pseudoreplication (see discussion above and in McGregor et al. 1992). The more common of these studies are presented here.

4.7.1
Neighbor/Stranger Recognition

The *neighbor/stranger hypothesis* is the most widely accepted song function based upon the responses of resident birds to song playbacks. Unfortunately, in the history of neighbor-stranger experiments, the song structures used were rarely specified exactly, degradation in the test songs was not standardized, and experimenters did not report whether the song type used was also in the repertoire of the birds under study (Shy and Morton 1986a). Overall, researchers have consistently found that the response of resident birds is greater to the playback of songs from a stranger (a noncontiguous neighbor) than to playback of songs from a firmly established territorial neighbor. By discriminating stranger from neighbor songs, males with well-established territories are thought to reduce ongoing strife, thereby reducing pointless energy expenditure (Weeden and Falls 1959). This proposal is also called the "dear enemy" hypothesis (Fisher 1954; Wilson 1975). Designating signals as belonging to either a neighbor or a stranger, however, does not allow a rigorous discussion of the effects of signal structure on responses, even though such discriminations may exist along with ranging (Falls 1992). Godard (1991), using this design protocol, suggested that neighbor recognition resulted in lower levels of aggression in response to songs of former neighbors than newcomers in hooded warblers (*Wilsonia citrina*).

The biological significance of sharing one or more song-type with a neighbor is to permit a singing individual to threaten others in a very specific sense. The singer is not simply proclaiming its presence, which could be accomplished through production of any species specific signal, rather, it is defending its territorial space more effectively by indicating its distance from listeners. When a territorial male hears a competitor, it can deliver a song that matches its rival's song, thereby decreasing its apparent distance from the competitor without bearing the energy cost of locomotion. This function of song matching was explained by Krebs et al. (1981), using the ranging hypothesis.

4.7.2
Geographically Based Dialects

Systematic regional or geographic differences in song structures has been well documented in oscines. Differential response to these regional differences, or *dialects*, has also been studied using playback experiments. As is the case with other sound categories discussed here, differential responding to two (or more) different sound variants presumably suggests that dialect information serves an important communicative function. In this particular case, care must be taken to use enough stimuli to adequately sample from each dialect being tested (Kroodsma 1989a). A summary of empirical data concerning song dialects in birds, including the responses of birds to dialects, is given by Catchpole and Slater (1995).

4.7.3
Species Recognition

Species recognition based on acoustic cues is said to occur when animals respond to sounds from their own species but not to those of other species. One can use such differences in response to ask what features of sounds serve to encode species identity. By systematically varying different acoustic features of vocalizations, one can assess the relative effectiveness of each to species recognition. The classic study by Emlen (1972) serves as a prototype, and reviews by Becker (1982) and Payne (1986) provide outlines of current studies with birds. Snowdon et al. (1986) provide an example of work with nonhuman primates.

4.7.4
Mate Choice, Attraction, and Stimulation

Playbacks can be used to assess which aspects of signals are important for attraction of a potential mate, to what extent sound signals are used in the selection of sexual partners, and the effectiveness of songs in stimulation of mates. One function of song production in male birds is understood to be attracting females. Empirical tests of the relative effectiveness of signals in mate attraction are lacking. The problem is one of designing a study sensitive enough to assess differential responding by females, as much of the choice process is either too subtle to detect or too complex to adequately assess in a playback study. Searcy (1992) provides an excellent summary of the techniques used to assess female responses to male song, and outlines the relative merits of several different types of measures used. The latter include measuring physiological changes, such as in heart rate, elicitation of copulation postures, and phonotaxis responses.

An increasing amount of relevant work is occurring in the area of behavioral neuroendocrinology, where it is of interest to measure endocrine or hormonal changes that can be observed in response to playbacks of communication signals (see Ball and Dufty, this Volume). While much of this work has been done in laboratory settings, the relevance of playback study designs to those studies is meaningful. Further, the success of this approach in laboratory-based studies suggests a great potential for greater applications in field settings.

4.7.5
Recognition of Kin, Social Group, Individual, and Offspring

In many circumstances, animals respond differently to different individuals. These response categories can include behavior toward a mate, individuals in a social group, or behavior of parents toward offspring. In all cases, demonstrating recognition is similar to the requirements of studies outlined above, i.e. to design a situation in which the subject responds differently to two (or more) different stimuli, one familiar, one not. The main limitations are in the natural response repertoire of the animal, and the difficulty in separating commonly confounded variables (see discussion by Hopp et al. 1985). For example, an animal may show quite different behavior to its own offspring versus those of another animal, but whether it distinguishes among its own young is more difficult to determine. Further, many studies fail to include control procedures that allow the researcher to separate the occurrence of true individual recognition from other, less interesting, effects, like familiarity or habituation. There are many examples of studies demonstrating differential recognition of offspring (Symmes and Biben 1986), mates (Wooller 1978), and kin groups (Rendall et al. 1996). However, recognition of individuals, like mates or offspring, must be considered to be separate from the kind of differential response seen in territorial songbirds to neighbors and noncontiguous territory owners, as discussed above. Discussions of the ability of birds to distinguish among different categories of individuals, and design considerations for empirical work in this area are provided by Lambrechts and Dhondt (1995). Marler and Mitani (1988) describe several analogous lines of research with nonhuman primates, and Tyack (this Volume) discusses work with marine mammals.

In a strong test for individual recognition in rhesus monkeys (*Macaca mulatta*), Rendall et al. (1996) used a recovery-from-habituation design (see description above) to test free-ranging animals. Briefly, animals were exposed to repeated presentations of calls from particular individuals until the subjects stopped looking in the direction of the playback speaker when hearing the sounds. This response recovered when new stimuli, recorded from another individual, were presented. Further, and importantly, the calls they used for the habituation trials were of different vocalizations recorded from the same individual, rather than the same recording presented repeatedly. Thus, the researchers could demonstrate that the recovery of the habituated response was not simply due to a change in the particular stimulus that was presented, but rather to a change in the identity of the individual that had originally produced the sounds.

4.8
Subject Variables

In contrast to the types of studies discussed above, where the underlying hypotheses involve stimulus-based variations, researchers often also seek to learn about differences in response to signals as a function of variation in the characteristics or qualities of the subjects themselves. By definition these studies often involve an ex post facto design in which subjects cannot be randomly assigned to groups. The researcher's main effort then is in isolating the subjects being tested from the potential influence of confounding variables. This consideration is typically addressed by matching or randomizing such variables. Examples of these types of studies include placing the playback speaker in different areas of the subject's territory (Price 1994), conducting playbacks in a variety of habitats (Gish and Morton 1981), or presenting stimuli at different times during the day (Shy and Morton 1986b).

4.9
Perceptual Studies

To a certain degree, playback designs can be used to address questions about the sensory capabilities of subjects, although the conclusions drawn from such studies must be made with great care. Since most field studies rely on a subject's naturally occurring behaviors, both the "rules" of differential response as well as the subject's motivation for response are not under the experimenter's control. Thus, in situations where a subject shows reliable differential responding to two different stimuli, one can conclude that the subject distinguishes or discriminates between them. However, the lack of differential responding does not mean that the subject is unable to distinguish between the stimuli, only that this particular circumstance does not lead to differential responding. The distinction between what an animal is capable of doing and what an animal actually does is an important one and has been discussed in detail elsewhere (Irwin 1958;

Beecher and Stoddard 1990; Nelson and Marler 1990; Owren et al. 1992, see also Cynx and Clark, this Volume).

In spite of these potential difficulties, field playbacks have been effectively used to assess perceptual abilities such as localization of signals in space (Brown and May 1990) and organization of signal categories (e.g., Falls et al. 1988; Nelson and Marler 1990). The growing use of relatively sensitive paradigms such as recovery-from-habituation (e.g., Cheney and Seyfarth 1988; see discussion above) and the development of computer techniques for precise control of sound synthesis (e.g., Beeman, and Owren and Bernacki, this Volume) will be likely to lead to a greater use of field playback techniques for asking questions about animal perception.

Summary

The sound playback technique has become one of the standard tools used by bioacousticians to study many aspects of both sound transmission and acoustic communication in animals. The robust responses produced by a wide variety of organisms to acoustic signals makes sound playback an effective and practical technique for testing a variety of hypotheses about behavior and the communication process. This chapter introduced the basics of playback studies. We first provided an overview of some of the characteristics of acoustics that influence the transmission of sound in field studies. Following this we discussed several topics related to the design of playbacks. The attention given to the design of studies and some of the problems commonly encountered in playback designs (Kroodsma 1986, 1989a, 1990) has led to refinements in the application of this technique. Recent advances in telemetric and interactive procedures suggest that more sophisticated uses of sound playback are likely to emerge in the future. Finally, we presented discussions of the application of playbacks to specific issues. Here we have provided guidelines for directing inquiry in these areas and provide examples of studies to exemplify different lines of inquiry. It is hoped that this chapter provides guidelines in helping to direct design of playbacks.

References

Adhikerana AS, Slater PJB (1993) Singing interactions in coal tits, *Parus ater*: an experimental approach. Anim Behav 46: 1205–1211
Alatalo RV, Glynn C, Lundberg A (1990) Singing rate and female attraction in the pied flycatcher: an experiment. Anim Behav 39: 601–603
Arvidsson BL, Neergaard R (1991) Mate choice in the willow warbler–a field experiment. Behaviour Ecol Sociobiol 29: 2251–229
Bakeman R, Gottman JM (1986) Observing interaction. Cambridge University Press, New York
Baptista LF, Gaunt SLL (1994) Advances in studies of avian sound communication. Condor 96: 817–830
Baptista LF, Petrinovich L (1986) Song development in the white-crowned sparrow: social factors and sex differences. Anim Behav 34: 1359–1371
Becker PH (1982) The coding of species-specific characteristics in bird sounds. In: Kroodsma DE, Miller EH, Ouellet H (eds) Acoustic communication in birds. Academic Press, New York, p 213
Beecher MD, Stoddard PK (1990) The role of bird song and calls in individual recognition: contrasting field and laboratory perspectives. In: Stebbins WC, Berkley MA (eds) Comparative perception, vol II. Wiley, New York, p 375
Bilger RC, Hirsh IJ (1956) Masking of tones by bands of noise. J Acoust Soc Am 28: 623–630

Bradbury J, Vehrencamp S (1993) SingIt! A program for interactive playback on the Macintosh. Bioacoustics 5: 308–310

Bremond JC (1978) Acoustic competition between the song of the wren (*Troglodytes troglodytes*) and the songs of other species. Behaviour 65: 89–098

Bremond JC, Aubin T (1992) Temporal matching of song: an interactive playback experiment with territorial wrens (*Troglodytes troglodytes*). C R Acad Sci Ser III Sci Vie 314: 37–42

Brown CH, May BJ (1990) Sound localization and binaural processes. In: Berkley MA, Stebbins WC (eds) Comparative perception, vol I. Basic mechanisms. Wiley, New York, p 247

Campbell DT, Stanley JC (1963) Experimental and quasi-experimental designs for research. Rand McNally, Chicago

Capranica RR, Frishkopf LS, Nevo E (1973) Encoding of geographic dialects in the auditory system of the cricket frog. Science 182: 1272–1275

Catchpole CK (1989) Pseudoreplication and external validity: playback experiments in avian bioacoustics. Trends Ecol Evol 4: 286–287

Chapman CA, Chapman LJ, Lefebvre L (1990) Spider monkey alarm calls: honest advertisement or warning kin? Anim Behav 39: 197–198

Chatfield C (1975) The analysis of time-series: theory and practice. Chapman and Hall, London

Cheney DL, Seyfarth RM (1988) Assessment of meaning and the detection of unreliable signals by vervet monkeys. Anim Behav 36: 477–486

Cheney DL, Seyfarth RM, Silk JB (1995) The responses of female baboons (*Papio cyanocephalus ursinus*) to anomalous social interactions: evidence for causal reasoning? J Comp Psychol 109: 134–141

Cheon SM, Park SR (1995) Signal value of partial song (composed of 1 phrase unit) in great tits, *Parus major*: evidence from playback experiments. Korean J Zool 38: 230–237

Cormier SM, Hagman JD (eds) (1987) Transfer of learning: contemporary research and applications. Academic Press, New York

Dabelsteen T (1992) Interactive playback: a finely tuned response. In: McGregor (ed) Playback and studies of animal communication. Plenum Press, New York, p 97

Dabelsteen T, Pedersen SB (1991) A portable digital sound emitter for interactive playback of animal vocalizations. Bioacoustics 3: 193–206

Davis WJ (1986) Acoustic recognition in the belted kingfisher: cardiac response to playback vocalizations. Condor 88: 505–512

Diehl P (1992) Radiotelemetric measurements of heart rate in singing blackbirds (*Turdus merula*). J Ornithol 133: 181–195

Dooling RJ (1982) Auditory perception in birds. In: Kroodsma DE, Miller EH, Ouellet H (eds) Acoustic communication in birds, vol 1. Academic Press, New York, p 95

Duellman WE, Pyles RA (1983) Acoustic resource partitioning in anuran communities. Copeia 1983: 639–649

Emlen ST (1972) An experimental analysis of the parameters of bird song eliciting species recognition. Behaviour 41: 130–171

Evans CS (1993) Recognition of the spectral characteristics of conspecific signals in ducklings: evidence for a simple perceptual process. Anim Behav 45: 1071–1082

Evans CS, Marler P (1991) On the use of video images as social stimuli in birds: audience effects on alarm calling. Anim Behav 41: 17–26

Falls JB (1981) Mapping territories with playback: an accurate census method for songbirds. Stud Avian Biol 6: 86–91

Falls JB (1992) Playback: a historical perspective. In: McGregor PK (ed) Playback and studies of animal communication. Plenum Press, New York, p 11

Falls JB, Horn AG, Dickinson TE (1988) How western meadowlarks classify their songs: evidence from song matching. Anim Behav 36: 579–585

Farabaugh SM (1982) The ecological and social significance of duetting. In: Kroodsma DE, Miller EH, Ouellet H (eds) Acoustic communication in birds, Vol 2. Academic Press, New York, p 85

Ficken RW, Ficken MS, Hailman JP (1974) Temporal pattern shifts to avoid acoustic interference in singing birds. Science 183: 762–763

Fisher J (1954) Evolution and bird sociality. In: Huxley J, Hardy AC, Ford EB (eds) Evolution as a process. George Allen and Unwin, London, p 71

Fletcher H (1940) Auditory patterns. Rev Mod Phys 12: 47–65

Fletcher H, Munson WA (1937) Relation between loudness and masking. J Acoust Soc Am 9: 1–10

Garner RL (1892) Speech of monkeys. Webster, New York

Gerhardt HC (1991) Female mate choice in tree frogs: static and dynamic acoustic criteria. Anim Behav 42: 615–635

Gerhardt HC (1992) Conducting playback experiments and interpreting their results. In: McGregor PK (ed) Playback and studies of animal communication. Plenum Press, New York, p 59

Gerhardt HC, Diekamp B, Ptacek M (1989) Inter-male spacing in choruses of the spring peeper, *Pseudacris (Hyla) crucifer*. Anim Behav 38: 1012–1024

Gilbert G, McGregor PK, Tyler G (1994) Vocal individuality as a census tool: practical considerations illustrated by a study of two rare species. J. Field Ornithol 65: 335–348

Gish SL, Morton ES (1981) Structural adaptations to local habitat acoustics in Carolina wren songs. Z Tierpsychol 56: 74–84

Godard R (1991) Long-term memory of individual neighbors in a migratory songbird. Nature 350: 228–229

Godard R (1993) Tit for tat among neighboring hooded warblers. Behav Ecol Sociobiol 33: 45–50

Gottlander K (1987) Variation in the song rate of the male pied flycatcher (*Ficedula hypoleuca*): causes and consequences. Anim Behav 35: 1037–1043

Greenberg R, Niven DK, Hopp SL, Boone C (1993) Frugivory and coexistence in a resident and a migratory vireo on the Yucatan Peninsula. Condor 95: 990–999

Grinnell J, Packer C, Pusey AE (1995) Cooperation in male lions: kinship, reciprocity or mutualism? Anim Behav 49: 95–105

Hambler C, Newing J, Hambler K (1994) Population monitoring for the flightless rail *Dryolimnas cuvieri aldabranus*. Bird Conserv Int 3: 307–318

Hansen P (1979) Vocal learning: its role in adapting sound structures to long distance propagation, and a hypothesis on its evolution. Anim Behav 27: 1270–1271

Hawkins AFA (1994) Conservation status and regional population estimates of the white-breasted mesite *Mesitornis variegata*, a rare Malagasy endemic. Bird Conserv Int 4: 279–303

Hodl W (1977) Call differences and calling site segregation in anuran species from central Amazonian floating meadows. Oecologia 28: 351–363

Hopp SL, Owren MJ, Marion J (1985) Olfactory discrimination of littermates in rats (*Rattus norvegicus*). J Comp Psychol 99:248–251

Horn AG (1992) Field experiments on the perception of song types by birds. In: McGregor PK (ed) Playback and studies of animal communication. Plenum Press, New York, p 191

Hurlbert SH (1984) Psuedoreplication and the design of ecological field experiments. Ecol Monogr 54: 187–211

Irwin FW (1958) An analysis of the concepts of discrimination and preference. Am J Psychol 71: 152–163

Jackson JE (1991) A users guide to principal components. Wiley, New York

James FC, McCulloch CE (1985) Data analysis and the design of experiments in ornithology. Curr Ornithol 2: 1–63

Klump GM, Gerhardt HC (1992) Mechanisms and function of call timing in male-male interactions in frogs. In: McGregor PK (ed) Playback and studies of animal communication. Plenum Press, New York, p 153

Krebs JR, Ashcroft R, Van Orsdol K (1981) Song matching in the great tit *Parus major*. Anim Behav 29: 918–923

Kroodsma DE (1984) Songs of the alder flycatcher (*Empidonax alnorum*) and willow flycatcher (*Empidonax traillii*) are innate. Auk 101: 13–24

Kroodsma DE (1985) Development and use of two song forms by the eastern phoebe. Wilson Bull 97: 21–29

Kroodsma DE (1986) Design of song playback experiments. Auk 103: 640–642

Kroodsma DE (1989a) Suggested experimental designs for song play-backs. Anim Behav 37: 600–609

Kroodsma DE (1989b) Inappropriate experimental designs impede progress in bioacoustic research: a reply. Anim Behav 38: 717–719

Kroodsma DE (1989c) Male eastern phoebes (*Sayornis phoebe; Tyrannidae*, Passeriformes) fail to imitate songs. J Comp Psychol 103: 227–232

Kroodsma DE (1990) Using appropriate experimental designs for intended hypotheses in 'song' playbacks, with examples for testing effects of song repertoire sizes. Anim Behav 40: 1138–1150

Kroodsma DE, Konishi M (1991) A suboscine bird (Eastern Phoebe, *Sayornis phoebe*) develops normal song without auditory feedback Anim Behav 42:477–487

Lambrechts MM, Dhondt AA (1995) Individual voice discrimination in birds. In: Power DM (ed) Current ornithology, vol 12. Plenum Press, New York, p 115

Langbauer WR Jr, Payne KB, Charif RA, Rapaport L, Osborn F (1991) African elephants respond to distant playbacks of low-frequency conspecific calls. J Exp Biol 157: 35–46

Lehner PN (1979) Handbook of ethological methods. Garland Press, New York

Lemon RE, Monette S, Roff D (1987) Song repertoires of American warblers (Parulinae): honest advertisement or assessment? Ethology 74: 265–284

Lewis C 1993) Bayesian methods for the analysis of variance. In: Keren G, Lewis C (eds) A handbook for data analysis in the behavioral sciences. Earlbaum, Hillsdale, New Jersey, p 233

Lynch JF (1989) Distributions of overwintering Nearctic migrants in the Yucatan Peninsula, I: general patterns of occurrence. Condor 91: 515–544

Lynch JF, Morton ES, Van der Voort ME (1985) Habitat segregation between the sexes of wintering hooded warblers (*Wilsonia citrina*). Auk 102: 714–721

Mabey SE, Morton ES (1990) Demography and territorial behavior of wintering Kentucky warblers in Panama. In: Hagan JM, Johnston DW (eds) Ecology and conservation of Neotropical migrant landbirds. Smithsonian Press, Washington DC, p 329

Machlis L, Dodd PWD, Fentress JC (1985) The pooling fallacy: problems arising when individuals contribute more than one observation to the data set. Z Tierpsychol 68: 201–214

Macedonia JM, Evans CS (1993) Variation among mammalian alarm call systems and the problem of meaning in animal signals. Ethology 93: 177–197

Marler P, Mitani J (1988) Vocal communication in primates and birds: Parallels and contrasts. In: Todt D, Goedeking P, Symmes D (eds) Primate vocal communication. Springer, Berlin Heidelberg New York, p 3

Marler P, Pichert R (1984) species-universal microstructure in the learned song of the swamp sparrow (*Melospiza georgiana*). Anim Behav 32: 673–689

Marler P, Evans CS, Hauser M (1992) Animal signals: motivation, referential, or both? In: Papoušek H, Jürgens U, Papoušek M (eds) Nonverbal vocal communication: comparative and developmental approaches. Cambridge University Press, Cambridge, p 66

Marten K, Marler P (1977) Sound transmission and its significance for animal communication. I. Temperate habitats. Behav Ecol Sociobiol 2: 271–90

Martin G (1981) Avian vocalizations and the sound interference model of Roberts et al.. Anim Behav 29: 632–633

Martin P, Bateson P (1986) Measuring behavior: an introductory guide. Cambridge University Press, New York

McComb K, Packer C, Pusey A (1994) Roaring and numerical assessment in contests between groups of female lions, *Panthera leo*. Anim Behav 47: 379–387

McGregor PK (ed) (1992a) Playback and Studies of Animal Communication. Plenum Press, New York

McGregor PK (1992b) Quantifying responses to playback: one, many, or composite multivariate measures? In: McGregor PK (ed) Playback and studies of animal communication. Plenum Press, New York, p 79

McGregor PK, Avery MI (1986) The unsung songs of great tits (*Parus major*): learning neighbor's songs for discrimination. Behav Ecol Sociobiol 18: 311–316

McGregor PK, Falls RB (1984) The response of western meadowlarks (*Sturnella neglecta*) to the playback of undegraded and degraded calls. Can J Zool 62: 2125–2128

McGregor PK, Horn AG (1992) Strophe length and response to playback in great tits. Anim Behav 43: 667–676

McGregor PK, Krebs JR (1984) Sound degradation as a distance cue in great tit (*Parus major*) song. Behav Ecol Sociobiol 16: 49–56

McGregor PK, Ranft RD (1994) Equipment for sound analysis and playback: a survey. Bioacoustics 6: 83–86

McGregor PK, Krebs JR, Ratcliffe LM (1983) The reaction of great tits (*Parus major*) to playback of degraded and undegraded songs: the effect of familiarity with the stimulus song type. Auk 100: 898–906

McGregor PK, Catchpole CK, Dabelsteen T, et al. (1992) Design of playback experiments: the Thornbridge Hall NATO ARW consensus. In: McGregor PK (ed) Playback and studies of animal communication. Plenum Press, New York, p 1

Michelsen A (1978) Sound reception in different environments. In: Ali MA (ed) Sensory ecology, review and perspectives. Plenum Press, New York, p 345

Morton ES (1975) Ecological sources of selection on avian sounds. Am Nat 109: 17–34

Morton ES (1982) Grading, discreetness, redundancy, and motivation-structural rules. In: Kroodsma DK, Miller EH, Ouellet H (eds) Acoustic communication in birds, vol I. Academic Press, New York, p 183

Morton ES (1986) Predictions from the ranging hypothesis for the evolution of long distance signals in birds. Behaviour 99: 65–86

Morton ES (1990) Habitat segregation by sex in the hooded warbler: experiments on proximate causation and discussion of its evolution. Am Nat 135: 319–333

Morton ES (1996) A comparison of vocal behavior among tropical and temperate passerine birds. In: Kroodsma DE, Miller EH (eds) Ecology and evolution of acoustic communication in birds. Cornell University Press, Ithaca, p 258

Morton ES, Young K (1986) A previously undescribed method of song matching in a species with a single song "type", the Kentucky warbler (*Oporornis formosus*). Ethology 73: 334–342

Morton ES, Gish SL, Van der Voort M (1986) On the learning of degraded and undegraded songs in the Carolina wren. Anim Behav 34: 815–820

Morton ES, Van der Voort M, Greenberg R (1993) How a warbler chooses its habitat: field support for laboratory experiments. Anim Behav 46: 47–53

Narins PM, Capranica RR (1976) Sexual differences in the auditory system of the treefrog *Eleutherodactylus coqui*. Science 192: 378–380

Narins PM, Smith SL (1986) Clinal variation in anuran advertisement calls: basis for acoustic isolation? Behav Ecol Sociobiol 19: 135–41

Nelson DA (1989) Song frequency as a cue for recognition of species and individuals in the field sparrow (*Spizella pusilla*). J Comp Psychol 103: 171–176

Nelson DA, Marler P (1990) The perception of birdsong and an ecological concept of signal space. In: Stebbins WC, Berkeley MA (eds) Comparative perception vol II. Complex signals. Wiley, New York, p 443

Nelson DA, Marler P (1994) Selection-based learning in bird song development. Proc Natl Acad Sci USA 91: 10498–10501

Nowicki S, Mitani JC, Nelson DA, Marler P (1989) The communicative significance of tonality in birdsong: responses to songs produced in helium. Bioacoustics 2: 35–46

Otter K, Njegovan M, Naugler C, Fotheringham J, Ratcliffe L (1993) A simple technique for interactive playback experiments using a Macintosh Powerbook computer. Bioacoustics 5: 303-307

Owren MJ, Seyfarth RM, Hopp SL (1992) Categorical vocal signaling in nonhuman primates. In: Papoušek H, Jürgens U, Papoušek M (eds) Nonverbal vocal communication: comparative and developmental approaches. Cambridge University Press, Cambridge, p 102

Pallett JR, Passmore NI (1988) The significance of multi-note advertisement calls in the reed frog, *Hyperolius tuberilinguis*. Bioacoustics 1: 13–24

Payne RB (1986) Bird songs and avian systematics. Curr Ornithol 3: 87–126

Petrinovich L, Patterson TL (1980) Field studies of habituation. III. Playback contingent on the response of the white-crowned sparrow. Anim Behav 28: 742–751

Popp JW, Ficken RW (1987) Effects of non-specific singing on the song of the ovenbird. Bird Behav 7: 22–26

Price K (1994) Center-edge effect in red squirrels: evidence from playback experiments. J Mammal 75: 545–548

Pridmore-Brown DC, Ingard U (1955) Sound propagation into the shadow zone in a temperature-stratified atmosphere above a plane boundary. J Acoust Soc Am 27: 36–42

Radesäter T, Jakobsson S, Andbjer N, Bylin A, Nyström K (1987) Song rate and pair formation in the willow warbler, *Phylloscopus trochilus*. Anim Behav 35: 1645–1651

Reid ML (1987) Costliness and reliability in the singing vigour of Ipswich sparrows. Anim Behav 35: 1735–1743

Rendall D, Rodman PS, Emond RE (1996) Vocal recognition of individuals and kin in free-ranging rhesus monkeys. Anim Behav 51: 1007–1015

Richards DG (1981) Estimation of distance of singing conspecifics by the Carolina wren. Auk 98: 127–133

Richards DG, Wiley RH (1980) Reverberations and amplitude fluctuations in the propagation of sound in a forest: implications for animal communication. Am Nat 115: 381–399

Roberts J, Hunter ML, Kacelnik A (1981) The ground effect and acoustic communication. Anim Behav 29: 633–634

Robinson SR, Terborgh J (1995) Interspecific aggression and habitat selection by Amazonian birds. J Anim Ecol 64: 1–11

Rowland WJ, Bolyard KJ, Jenkins JJ, Fowler J (1995) Video playback experiments on stickleback mate choice: female motivation and attentiveness to male color cues. Anim Behav 49: 1559–1567

Rowley I, Russell E (1993) The purple-crowned fairy-wren *Malurus coronatus*: II. Breeding biology, social organization, demography and management. Emu 93: 235–250

Rushe RH, Gottman JM (1993) Essentials in the design and analysis of time-series experiments In: Keren G, Lewis C (eds) A handbook for data analysis in the behavioral sciences: statistical issues. Erlbaum, Hillsdale, New Jersey, p 493

Ryan MJ, Brenowitz EA (1985) The role of body size, phylogeny, and ambient noise in the evolution of bird song. Am Nat 126: 87–100

Schwartz JJ, Wells KD (1983a) An experimental study of acoustic interference between two species of neotropical treefrogs. Anim Behav 31: 181–190

Schwartz JJ, Wells KD (1983b) The influence of background noise on the behavior of a neotropical treefrog, *Hyla ebraccata*. Herpetologia 39: 121–125

Schwartz JJ, Wells KD (1984a) Vocal behavior of the neotropical treefrog *Hyla phlebodes*. Herpetologia 40: 452–463

Schwartz JJ, Wells KD (1984b) Interspecific acoustic interactions of the neotropical treefrog *Hyla ebraccata*. Behav Ecol Sociobiol 14: 211–224

Schwartz JJ, Wells KD (1985) Intra- and interspecific vocal behavior of the neotropical treefrog *Hyla microcephala*. Copeia 1985: 27–38

Searcy WA (1989) Pseudoreplication, external validity and the design of playback experiments. Anim Behav 38: 715–717

Searcy WA (1992) Measuring responses of female birds to male song. In: McGregor PK (ed) Playback and studies of animal communication. Plenum Press, New York, p 175

Searcy WA, Yasukawa K (1996) Song and female choice. In: Kroodsma DE, Miller EH (eds) Ecology and evolution of acoustic communication in birds. Cornell University Press, Ithaca, p 454

Searcy WA, Coffman S, Raikow DF (1994) Habituation, recovery and the similarity of song types within repertoires in red-winged blackbirds (*Agelaius phoeniceus*) (Aves, Emberizidae). Ethology 98: 38–49

Searcy WA, Podos J, Peters S, Nowicki S (1995) Discrimination of song types and variants in song sparrows. Anim Behav 49: 1219–1226

Seyfarth RM, Cheney DL (1990) The assessment by vervet monkeys of their own and another species' alarm calls. Anim Behav 40: 754–764

Seyfarth RM, Cheney DL, Marler P (1980) Monkey responses to three different alarm calls: Evidence for predator classification and semantic communication. Science 210: 801–803

Shy E, Morton ES (1986a) Adaptation of amplitude structure of songs to propagation in field habitat in song sparrows. Ethology 72: 77–84

Shy E, Morton ES (1986b) The role of distance, familiarity, and time of day in Carolina wrens responses to conspecific songs. Behav Ecol Sociobiol 19: 393–400

Sliwa A, Sherry TW (1992) Surveying winter warbler populations in Jamaica: point counts with and without broadcast vocalizations. Condor 94: 924–936

Smith WJ (1996) Using interactive playback to study how songs and singing contribute to communication about behavior. In: Kroodsma DE, Miller EH (eds) Ecology and evolution of acoustic communication in birds. Cornell University Press, Ithaca, NY, p 377

Smith WJ, Smith AM (1992) Behavioral information provided by two song forms of the eastern kingbird, *Tyrannus tyrannus*. Behaviour 120: 90–102

Snowdon C, Hodun A, Rosenberger A, Coimbra-Filho A (1986) Long-call structure and its relation to taxonomy in lion tamarins. Am J Primatol 11: 253–262

Sorjonen J (1983) Transmission of the two most characteristic phrases of the song of the thrush nightingale (*Lucinia lucinia*) in different environmental conditions. Ornis Scand 14: 278–288

Stevens DA (1975) Laboratory methods for obtaining olfactory discrimination in rodents. In: Moulton DG, Turk A, Johnston JW Jr (eds) Methods in olfactory research. Academic Press, New York, p 375

Stoddard PK (1990) Audio computers: theory of operation and guidelines for selection of systems and components. Bioacoustics 2: 217–239

Stoddard PK (1996) Vocal recognition of neighbors by territorial passerines. In: Kroodsma DE, Miller EH (eds) Ecology and evolution of acoustic communication in birds. Cornell University Press, Ithaca p 356

Stoddard PK, Beecher MD, Campbell SE, Horning CL (1992) Song-type matching in the song sparrow. Can J Zool 70: 1440–1444

Strain JG, Mumme RL (1988) Effects of food supplementation, song playback, and temperature on vocal territorial behavior of Carolina wrens. Auk 105: 11–16

Struhsaker T (1967) Auditory communication among vervet monkeys (*Cercopithecus aethiops*). In: Altmann S (ed) Social communication among primates. Chicago University Press, Chicago, p 281

Symmes D, Biben M (1986) Maternal recognition of individual infant squirrel monkeys from isolation call playbacks. Am J Primatol 9: 39–46

Thompson R, Spencer D (1966) Habituation: a model phenomenon for the study of neuronal substrates of behavior. Psychol Rev 73:16–43

Timm NM (1993) MANOVA and MANCOVA: an overview. In: Keren G, Lewis C (eds) A handbook for data analysis in the behavioral sciences. Erlbaum, Hillsdale, New Jersey, p 129

Vallet E, Kreutzer M (1995) Female canaries are sexually responsive to special song phrases. Anim Behav 49: 1603–1610

Villard MA, Merriam G, Maurer BA (1995) Dynamics in subdivided populations of neotropical migratory birds in a fragmented temperate forest. Ecology 76: 27–40

Waser PM, Brown CH (1984) Is there a "sound window" for primate communications? Behav Ecol Sociobiol 15: 73–76

Waser PM, Brown CH (1986) Habitat acoustics and primate communication. Am J Primatol 10: 135–54

Waser PM, Waser MS (1977) Experimental studies of primate vocalization: specializations for long-distance propagation. Z Tierpsychol 43: 239–263

Wasserman FE (1977) Intraspecific acoustical interference in the white-throated sparrow, (*Zonotrichia albicollis*). Anim Behav 25: 949–952

Weary DM, Kramer DL (1995) Response of eastern chipmunks to conspecific alarm calls. Anim Behav 49: 81–93

Weeden JS, Falls JB (1959) Differential responses of male ovenbirds to recorded songs of neighboring and more distant individuals. Auk 76: 343–351

Whitehead JM (1987) Vocally mediated reciprocity between neighboring groups of mantled howling monkeys, *Aloutta palliata palliata*. Anim Behav 35: 1615–1627

Wilson EO (1975) Sociobiology: the new synthesis. Belknap Press of Harvard University Press, Cambridge

Winkler RL (1993) Bayesian statistics: an overview. In: Keren G, Lewis C (eds) A handbook for data analysis in the behavioral sciences. Erlbaum, Hillsdale, New Jersey, p 201

Wooller RD (1978) Individual vocal recognition in the kittiwake gull, *Rissa tridactyla*. Z Tierpsychol 48: 68–86

The Laboratory Use of Conditional and Natural Responses in the Study of Avian Auditory Perception

J. Cynx and S. J. Clark

1
Introduction

This chapter provides an introduction to the use of both conditional and natural response methods for studying animal auditory perception. Here we outline the history and use of these two areas in their application to understanding auditory perception. The context of this chapter adresses songbirds, although with appropriate modification the techniques can be easily adapted for use with other taxonomic groups. These techniques have been widely used for addressing questions about perception of both basic and complex auditory stimuli, and several reference sources are available (e.g., Berkley and Stebbins 1990; Burdick 1990; Stebbins and Berkley 1990; Klump et al 1995; see Tyack, this Volume, a for review of auditory perception by organisms in aquatic environments). In addition to the use of these procedures, we outline some considerations of their underlying assumptions, and examine issues of internal and external validity.

2
Conditional and Natural Responses

It is worthwhile defining some of the words and phrases in the chapter title. The phrase, *conditioned response*, is— as most first year psychology students know— a mistranslation from the Russian of what should be *conditional* (Gantt 1966). Pavlov, a student of the autonomic nervous system and its digestive sequelae, had begun to study more cerebral events. By referring to conditional reflexes or responses, he was dividing Sherrington's reflexes (1906) into those that, like the ones he had studied, occurred *unconditionally* (automatically) and those that, like the ones he was now interested in, occurred given certain environmental contingencies or conditions. He used the following metaphor to point out the difference between the two:

"My residence may be connected directly with the laboratory by a private line, and I may call up the laboratory whenever it pleases me to do so; or on the other hand, a connection may have to be made through the central exchange. But the result in both cases is the same. The only point of distinction between the methods is that the private line provides a permanent and readily available cable, while the other line necessitates a preliminary central connection being established" (Pavlov 1960, p. 25).

In other words, Pavlov used the terms to reflect the conditional or adaptive behavior of the animal, not the conditions of the behavioral laboratory. Methods have been developed to exploit conditional responses as behavioral readouts of animal perception. Other methods have also been developed which do not use conditioning techniques. These methods use naturally occurring behavioral responses to the stimuli as their assay. These methods are sometimes referred to as using unconditioned reflexes but this is potentially confusing terminology given Pavlov's view that unconditioned reflexes are "hard-wired". The behavioral reflexes used in these methods may indeed be innate or they may have been learned by the animal prior to being included in the experiment. The crucial distinction— at least as regards this chapter— is not between learned and unlearned but between behavioral responses that have been conditioned by the experimenter and naturally occurring behavioral reflexes. To avoid this confusion we will refer to these latter behavioral reflexes as *natural responses*.

2.1
Animal Perception

Animal perception is covert, and must be inferred from overt behaviors. No one procedure, statement, or equation will completely capture what students of perception mean by the term (Stebbins 1990). Perhaps the most specific one can be is to say that an animal's auditory perception is determined by: (1) what the animal can hear and (2) what it extracts as meaningful information. What we know about the animal's perception is determined by (3) what we ask the animal and (4) how we ask the animal to indicate its answer. If this loose operational definition of auditory perception seems reasonable, it can not be said that it was always obvious. There are venerable experimental histories addressing some of these facets to the mutual exclusion of others.

Classical *psychophysical* studies, beginning with Weber (1834) and Fechner (1860), focused on the first question, what can be heard. The formal relation between the physical (acoustical in this chapter) and the inner (psychological) world was the playing field on which psychophysical experiments were conducted. These two worlds were held as constant (some would say static) as possible, so precise measurements could be made to reduce perception to its essential structures, such as absolute thresholds and difference limens. The other questions then could be answered by building on these essences, or dismissed as not the prerogative of someone interested in perception.

Ethological studies, beginning with von Uexküll (1909), were concerned primarily (though not exclusively) with the next two questions, what was meaningful to the animal in its unique sensory-perceptual world, and who was defining the playing field. Classical ethology disavowed the form of the psychophysical games and argued that the animal in its own world (*Umwelt*) behaved in unique ways that could not be captured if the experimenters were allowed to determine the rules. By this view, psychophysicists not only did not give animal perceptions a sporting chance, but domesticated them until they little resembled the natural phenomena.

Finally, classic animal conditioning studies, beginning with Pavlov (1960) and Thorndike (1911), were tied to the last question, how an animal performed. The philosophy here, as with psychophysics, was structural. But, it was intent on reducing behavior, not perception. It asserted that behavior was composed of associations and stimulus-response contingencies, regardless of the context. An acoustic phenomenon was simply another stimulus in stimulus-response conditioning.

These rather ideological differences colored the study of animal perception through much of the first half of this century. The philosophical myopia resulted in separate vocabularies for studying perception. Only ethologists had a clear mandate to study perception as a part of animal communication. They provided some of the most profound findings in animal perception and communication; for example, echolocation in bats (Griffin 1974), song learning in birds (Thorpe 1961), and the dances of bees (von Frisch 1965). The ethologists were pluralists, and made fruitful discoveries because evolution and resultant animal behaviors are also pluralistic (Marler 1961).

2.2
Conditional Responses

However, one should not dismiss laboratory studies using animal conditioning. Their validity can be partially illustrated by comparison to another laboratory technique, electrophysiology. In studies of the birdsong system, anesthetized birds have been placed in stereotaxic devices, then wired for electrophysiological monitoring (e.g., Katz and Gurney 1979; Margoliash 1983, 1986; Williams and Nottebohm 1985). We know of no objections against this providing useful proximal data on song behavior. Yet, it is a highly artificial preparation. Part of the reason why it raises hackles less than those using animal conditioning may be due to the separate language developed for studying animal conditioning in this century. An argument strictly against animal conditioning studies reflects a narrow sensibility of how one does science rather than some natural division in the world like the sound barrier. In what immediately follows, we define Pavlovian (classical) and operant (instrumental) conditioning, provide a short history of how they have been used to study auditory perception in birds, and discuss some recent advances.

Pavlov developed a procedure in which an arbitrary environmental event, for example a bell, signaled the imminent arrival of a second environmental event, for example, a piece of meat. This second event was chosen because it unconditionally led to a third, physiological event, for example the animal salivating. We might as well call it transitive conditioning as Pavlovian conditioning. Because the first two environmental events were linked by the experimenter, and the last two by the animal's physiology, the first and the third events then became *conditionally* linked. The experimenter could now algebraically manipulate the equations, and then experimentally solve them. For example, he might vary the number of times a bell was paired with meat, and record a dependent variable, for example, the amount of saliva produced after five pairings. This then provided a window into how the animal associated these events, how long the association lasted, whether other associations could interfere with it, etc. Pavlov recognized that the procedure provided a window into animal perception:

"... if a tone of 1000 d.v. [Hz] is established as a conditioned stimulus, many other tones spontaneously acquire similar properties, such properties diminishing proportionally to the intervals of these tones from the one of 1000 d.v. Similarly, if a tactile stimulation of a definite circumscribed area of skin is made into a conditioned stimulus, tactile stimulation of other skin areas will also elicit some conditioned reaction, the effect diminishing with increasing distance of these areas from the one for which the conditioned reflex was originally established. The same is observed with stimulation of other receptor organs. This spontaneous development of accessory reflexes, or, as we have termed it, generalization of stimuli, can be interpreted from a biological point of view by reference to the fact that natural stimuli are in most cases not rigidly constant but range around a particular strength and quality of stimulus in a common group. For example, the hostile sound of any beast of prey serves as a conditioned stimulus to a defence reflex in the animals which it hunts. The defence reflex is brought about independently of variations in pitch, strength, and timbre of the sound produced by the animal according to its distance, the tension of its vocal cords and similar factors" (Pavlov 1960, p.113).

Pavlov explicated the use of conditional responses in studying perception with this tripartite statement. First, he noted perception of a tone is not absolute. Its frequency can be confused with other frequencies. Second, he showed that this confusion of stimuli can be scaled. The perception of a frequency generalizes in an orderly fashion to other frequencies. Of course, these two statements apply to perception regardless of whether the response is conditional. Third, he reasoned that this generalization was a biologically relevant perceptual adaptation. A conditional response was not an artifact of a laboratory procedure.

There are two corollaries here that become more important in the light of intervening research. First, generalization is the keystone to understanding perception. If there are other ways to study perception, they haven't appeared. Generalization models have become highly sophisticated (Mostofsky 1965; Shepard 1987), and applied to field studies as well as laboratory studies (Nelson 1988; Nelson and Marler 1990). Second, conditional response methods are built on animal learning theory. One cannot study animal perception with these methods without studying animal learning. This is not an astounding insight into natural animal behavior. Perception and learning are interwoven throughout the lives of animals. However, the connection is not always easily gleaned from studies on perception that use conditioning. As will become evident, perceptual findings may sink or swim given how well the investigator understands the effects of the conditioning paradigm on the subjects (e.g., see Sect. 3.3).

2.3
Operant Conditioning

"The essence of [respondent or classical] conditioning is the substitution of one stimulus for another, or, as Pavlov put it, signalization... In [operant conditioning] there is no...signalization... the organism selects from a large repertory of unconditioned movements" (Skinner 1938, p. 111).

The development of operant conditioning can be traced to Thorndike, whose studies of how animals learned to escape from puzzle boxes led to his formulation of the law of effect (1911). Thorndike arranged experiments so that operations on a mechanism (for example, a latch) would have a certain effect (for example, freedom). An animal that discovered this relation then increased the behavior that exploited it. Because in this form of conditioning the animal operates the instrument of conditioning, it is called either *operant* or *instrumental conditioning*.

Although it may or may not be theoretically possible to dissociate operant and respondent conditioning paradigms (Mackintosh 1983), most, if not all conditioning procedures include both signaling and the animal selecting a behavior. Conditioning studies of avian auditory perception began with paradigms that were heavily Pavlovian and then moved to more operant paradigms. There were considerable empirical, but not theoretical reasons for this. Experimenters in animal perception chose the behavioral paradigms that worked best. Further, the ease of data collection was facilitated by the introduction of automated and computerized devices into behavioral laboratories.

3
A History Specific to Songbirds

3.1
Sensory Measurements

Methods were initially developed to measure sensory processes, and then were adapted to study more cognitive aspects. Early experiments focused on determining the frequency range across which birds could hear sounds. Audiograms were obtained for a number of species of birds, including one species of songbird, using aversive or avoidance conditioning procedures (e.g., Brand and Kellogg 1939a,b; Edwards 1943; Trainer 1946; Schwartzkopff 1949; Dooling et al. 1971; Dooling and Saunders 1975). In aversive conditioning, the presence of a tone predicted a shock. The experimenter watched for a startle reflex as the intensity of a tone reached threshold. In avoidance conditioning, the animal was taught to avoid shock by moving to another part of the cage when a tone sounded. The experimenter then observed at what tone intensity the animal began to move. Sound levels were varied according to various psychophysical methods so that thresholds could be determined. There were reports that many of these birds, regardless of whether they were wild-caught or not, were poor laboratory subjects. There were also some reports using operant conditioning for determining auditory acuity in pigeons (e.g., Heise 1953), but not songbirds. Problems of behavioral control often seemed to deter examining auditory discrimination in any systematic way. This can be contrasted with studies of visual sensory processing in pigeons. Pigeons were used in classic work determining generalization gradients for the visible light spectrum (Guttman and Kalish 1956; Blough 1966; Honig and Urcuioli 1981). There are a number of reasons for the differences between visual and auditory studies. First, the auditory studies were conducted with a variety of birds. This required more flexibility in training and testing procedures. Second, as stated above, much of the auditory work was Pavlovian, while the visual work used operant paradigms with lit keys. The instrumental pecking at a localized light source probably exerted better experimental control than a response to diffuse sine waves.

Up to the mid-1970s, the precise psychophysical functions of avian hearing, especially in songbirds, were unclear. For example, what was the smallest frequency difference that could be detected? What was the smallest time difference that could be detected? Further, it was unknown whether these parameters were similar across species, and how they compared with mammalian hearing. Answers to these questions were seen as able to provide a window on the evolution of vertebrate hearing. More important to the topic at hand, specific hypotheses had been raised concerning the relation of hearing and song production in songbirds. One hypothesis was that the avian auditory system was capable of finer temporal and frequency resolution than humans (Pumphrey 1948; Greenwalt 1968). This presumably allowed for perception of the rapidly modulated features in much of birdsong. A second hypothesis concerning communication was based on the knowledge that each species shows selective biases to its own song. This suggested that each species' peripheral auditory system might be "tuned" to species-specific settings (Schwartzkopff 1955). The sensory adaptations could then bias song learning. Beginning in the 1970s the universality of both these hypotheses was either thrown into doubt or falsified. In general, peripheral auditory resolving power for song birds was shown to be no greater than for humans. In some instances it was worse (Fay 1988). And there are only a few species differences in the audiograms of songbirds tested (Dooling 1982). These psychophysical findings have proven valuable in documenting the conservative nature of the peripheral hearing system in birds and mammals. Regardless of the morphological differences, the auditory processing remains strikingly similar.

3.2
Ergonomics for Birds

Testing these hypotheses required the development of new behavioral technologies. Stebbins had developed an operant conditioning technique for determining audiograms in monkeys that he then modified to work with pigeons (Stebbins 1970). Following on this, Hienz et al. (1977) trained two species of wild caught songbirds to work for seed on different key pecking operants (*go-right/go-left* and *go/nogo*). In a go-right/go-left procedure, the bird begins a trial by pecking a lit center key. This usually darkens the center key and causes the two side keys to light up. It simultaneously produces one of two single (or sets of) auditory stimuli. One stimulus is associated with a key to the right of the center key; the second stimulus is associated with a key to the left of the center key. Picking the correct key, based on the stimulus, turns off the key lights, and produces a seed reward. Pecking the wrong key darkens the chamber for a few seconds. In a go/nogo procedure, a bird again begins a trial by pecking a lit key, turning off the key light, and turning on the light on a second key, producing one of the two stimuli. If one stimulus occurs, the bird refrains from pecking a second key until the trials ends (e.g., 4 s) and the key light goes off. If the bird pecks the key, the chamber is darkened for a few seconds. If the other stimulus occurs, the subject pecks the second key, which turns off the key light and produces a food reward. These procedures have been varied and elaborated and refined in a number of ways, some of which are discussed below. Recent detailed methodological reviews concerning go-right/go-left and go/nogo methods can be found in Dooling and Okanoya (1995a,b), and Hulse (1995). These procedures may seem somewhat arbitrary, but there is a considerable literature concerning how they affect performance (Konorski 1967; Burdick 1990). The problem is analogous to making user friendly devices for humans.

3.3
Overshadowing

The use of key lights in these operant procedures is partly an artifact of the training that occurs before sound stimuli are introduced. A bird is trained initially by *autoshaping*, in which its predisposition to peck at bright objects is exploited by using lit keys (Brown and Jenkins 1968). The bird then can be trained to peck the keys in the proper order by sequentially lighting them. However, a lit key may produce curiously confusing results in auditory perception experiments (Cynx 1993). Trainer (1946), using a Pavlovian aversive conditioning procedure, noticed that starlings (*Sturnus vulgaris*) , unlike other birds, had to be retrained at each frequency to construct an audiogram (all the other species were non-songbirds). If he trained a starling that a 1-kHz tone predicted a shock, the starling failed to generalize this conditional response to a 2-kHz tone. This seems remarkably maladaptive given the aversiveness of a shock. Trainer had no ready explanation, but noted that the phenomenon might indicate an "acute tonal discriminative ability (p. 77)." Dooling and his colleagues reported the same phenomenon with canaries (*Serinus canarius*; Dooling et al. 1971), but not with a parakeet (*Melopsittacus undulatus*), a non-songbird (Dooling 1973). These findings, both discovered in the course of sensory studies, suggested that there may be something special about pitch perception or stimulus control in songbirds. This was further indicated in studies with starlings, and other songbirds discriminating pitch sequences (Hulse and Cynx 1985). As stated below, these songbirds treated frequencies in an absolute fashion rather than along an ordinal or ratio scale. It seemed that, regardless whether one used Pavlovian or operant procedures, songbirds at times showed no or little frequency generalization.

Cynx replicated Trainer's findings using an go/nogo procedure (Cynx 1993). Starlings were trained to peck a lit key. If a 2-kHz tone sounded, they then pecked a second lit key for a food reward. After training, a generalization gradient from 0.5 to 8 kHz was obtained. One would normally expect more go responses when tones were closer to 2 kHz. However, the starlings produced the same number of go responses regardless of the frequency. This finding— that a tone acquires relatively little control over responding in the presence of a key light— had already been reported for pigeons (Jenkins and Harrison 1960; Rudolph and Van Houten 1977). It seems that light from the key interferes with or *overshadows* sound in controlling responses in both songbirds and non-songbirds. The bird attends so strongly to the light source that it fails to pay attention to the frequency of the stimulus tone. When another set of starlings were trained with unlit keys, the starlings produced the expected generalization gradient, with the number of go responses decreasing as the stimulus frequency deviated further from 2 kHz.

These results across 40 years of studies are not completely consistent, especially concerning differences between song and non-songbirds. This may in part be due to our lack of data concerning stimulus control across species and pitch perception in non-songbirds. More important, the use of lit and unlit keys may bias perceptual studies. Presumably, researchers should be aware of the possibility of overshadowing. Yet, lit keys are, to our knowledge, used in almost all operant studies of bird auditory perception. Besides turning off the key lights, one way to avoid the problem is to use another operant, such as perch-hopping (Weary 1989; Cynx et al. 1990).

3.4
Cognitive Measurements

Technical advances in conditioning techniques resulted in considerable change in the study of auditory perception in songbirds. Several studies showed that both cage-raised and wild-caught animals could be tested, and that operant techniques could be valid, reliable, and rapid. Animals could be tested over long periods of time without developing the behavioral problems known to occur with aversive conditioning (Hienz et al. 1977). They allowed more precise measurements of a dependent variable. In the earlier work, the investigators often reported either the presence or absence of a single startle response. In the new procedures, more subtle measures could be taken, such as reaction time to peck. However, as stated above, sensory studies gave little insight into anything special about bird song and hearing. This awaited the asking of cognitive questions about how songbirds discriminate, remember, and attend to sounds.

Early cognitive work examined whether songbirds could discriminate between tone sequences, based on abstract rules of pitch and rhythm (Hulse et al. 1984a,b). This presumed that songbirds were engaged in higher-level auditory information processing in dealing with complex sounds in the environment. Starlings were trained, using both go-right/go-left and go/nogo procedures to discriminate between either pitch or rhythmic structures. The birds acquired the pitch discrimination, but only when on the go/nogo procedure. The birds trained on rhythm acquired the discriminations with either procedure. The studies on pitch perception, using a combination of probe and transfer procedures (see Hulse 1995 for procedural details) indicated that songbirds possessed a form of absolute pitch perception (Hulse et al. 1986). This finding was given some ecologically validity by field studies, which showed that some species of birds use absolute pitch perception in discriminating between songs (Nelson 1988). Later research has shown that avian absolute pitch perception is not limited to songbirds (Cynx 1995).

3.5
Perceptual Maps

The use of the go/nogo procedure was extended by Dooling (1986) in adapting it to a *same-different judgment task*. Dooling and his colleagues provided evidence for how the budgerigars, a non-songbird, and zebra finches (*Taeniopygia guttata*) sort arbitrary sounds and calls. They trained birds to sort two sounds based on whether they were the same or different, tested them on a range of natural sounds, then used multidimensional scaling to build perceptual maps (Dooling 1986; Dooling et al. 1987, 1990). This allows the investigator to visualize the dimensions along which the birds sort the sounds. The same sort of procedure has been used with cliff swallows (*Hirundo pyrrhonota*) and starlings (Loesche et al. 1992). Using bird vocalizations, synthetic stimuli, and human speech sounds, researchers have been able to show the acoustic perceptual categories that exist in songbirds and non songbirds, and that there is a correspondence between the functional and acoustic classes of vocal signals. They have been able to show how learning is involved in the formation of these categories, and that certain speech sounds probably sound the same to budgerigars as to humans. A methodological review is given by Dooling and Okanoya (1995b).

Dooling and his colleagues have also expanded the go/nogo procedure to include an adaptive-tracking procedure and the method of constant stimuli to measure auditory sensitivity (Dooling and Okanoya 1995a; Okanoya 1995). These procedures require much more training than the simple go/nogo methods, often entailing months of daily training sessions. This requires considerable patience on the part of the experimenter. It remains an open question as to how the use of such extensive regimens affects the results. The difficulty of learning seems at odds with the adaptive behavior of animals in the wild.

3.6
Studies of Birdsong Perception

Whatever song conveys in the wild, there is no a priori reason to assume that a bird can recognize it in the laboratory. However, there are a number of studies that provide this kind of external validity for conditioning experiments with songs as stimuli (see the second half of this chapter for a more extended discussion of different types of validity). Sinnott (1980) trained cowbirds (*Molothrus ater*) and red-winged blackbirds (*Agelaius phoeniceus*) with a go/nogo procedure on two discriminations. Each bird learned to discriminate between two redwing themes and between two cowbird themes. After the birds were trained, songs was separated into their initial and terminal portions, and the birds were tested on their identification of these isolated elements. Results revealed no species differences in discriminating initial portions. However, each species was better at discriminating terminal portions of its conspecific song. Cynx and Nottebohm (1992a) trained zebra finches on a go/nogo operant to discriminate between two conspecific songs. Males discriminating between their own songs and other songs reached their criteria in the fewest number of trials, followed by males discriminating between two familiar songs, then by males discriminating between unfamiliar songs. Females required the most trials of all birds, an amount that did not vary regardless of their familiarity with the songs.

Acquisition was relatively rapid, usually within one or two sessions and a few hundred trials. This was apparently due to the use of song stimuli, as illustrated by another experiment. Other zebra finches were trained to discriminate between two song syllables that were identical except for the presence or absence of certain harmonics. These birds required over 5000 trials to acquire the discrimination. A second set of birds was trained with the same stimuli except that they were embedded in song. Again, the only difference between stimuli was the presence or absence of certain harmonics. In this case, the birds acquired the discriminations in 1000 to 2 000 trials (Cynx 1996). It seems that discriminations between song-related sounds can be learned in fewer trials than other sounds, and that the familiarity of songs can also improve discrimination learning. Acquisition rates have also been used to measure a songbird's memory capacity for song elements (Cynx and Nottebohm 1989; Beecher and Stoddard 1990).

These and other experiments using song stimuli (Shy et al. 1986; Weary 1989) show that birdsongs can be used as stimuli without fear that motivational effects will disrupt conditioning procedures. For example,there are no reports that birds respond to song stimuli by counter-singing or approaching the speaker. We have observed only that a starling, trained and tested on rising and falling four-tone sequences a number of years ago, has incorporated these sequences into his own song (unpubl. observ.). Finally, worries that the association of song with an arbitrary reward, such as a seed reward, would invalidate the results, have not proven to be warranted.

3.7
Neuroethological Studies

Finally, acquisition rates have also been used to determine some neuroethological aspects of song perception. In these cases, the conditioning procedures described above have been combined with biological manipulations.

3.7.1
Lateralization of Song Discrimination

To test for lateralization of song discrimination in zebra finches, a technique was adapted from that used in determining laterality of vocalization discrimination in Japanese macaques (*Macaca fuscata*; Petersen et al. 1978; Heffner and Heffner 1984). Each member of a pair of male zebra finches was unilaterally deafened on the opposite side from its colleague, then trained to discriminate between their two songs. Birds A and B, for example, were discriminating songs A and B. Birds with left side lesions took at least twice as many trials as birds with right side lesions to acquire the discriminations. (Cynx et al. 1992). The same pairs of birds then were trained on a second discrimination, using artificially edited song (Cynx et al. 1990). The results were reversed from those with the birds' own song. In this case, birds with left side lesions required fewer trials than birds with right-side lesions to reach the learning criterion.

3.7.2
Hormonal Mediation of Song Discrimination

Acquisition rates in some experiments with zebra finches correlated with the time of year (Cynx and Nottebohm 1992a; Cynx et al. 1992). The amount of seasonal song production in zebra finches is known to be determined partly by levels of testosterone or its metabolites (Pröve 1974). It therefore seemed reasonable to test for hormonal mediation of song discrimination. Male zebra finches on a 12:12 light:dark cycle were castrated. This provided 14 birds with negligible endogenous levels of testosterone during the experiment. Half were implanted some weeks later with silastic tubes of testosterone. The other half were implanted with empty tubes as controls. They were divided into pairs, each pair with one testosterone-implanted bird and a control. They then were trained on a number of discriminations. Each pair was first trained on a heterospecific (canary) song discrimination, then on their own two songs, then on a heterospecific discrimination again. This design determined if any effect was due to general changes in learning or was tied to zebra finch song in particular. When the zebra finches were initially trained on the canary songs, they showed no differences in trials to a learning criterion between testosterone-implanted and control birds (Cynx and Nottebohm 1992b). When the pairs were next trained on the zebra finch songs, testosterone-implanted birds acquired the discriminations in fewer trials than the control birds in every case. Finally, the birds were trained on another set of canary song stimuli. Again, there were no consistent differences.

3.8
Summary

The combination of animal learning, cognitive science, and auditory psychophysics has allowed researchers to probe birdsong perception in highly sophisticated ways. However, all these techniques have involved some sort of animal conditioning. In the next section, we discuss some complementary techniques that take advantage of naturally occurring behavioral repertoires.

4
Use of Natural Responses to Study Perception

In the wild, animals often respond to stimuli with predictable behavioral responses. If a flock of chaffinches spies an owl, they mob it (Hinde 1958). A herring gull chick, seeing the bill of its parent, exhibits begging behavior (Tinbergen and Perdeck 1950). A vervet monkey hearing a *snake alarm call* stands on its hind legs and scans the ground around it (Seyfarth et al. 1980). The methods described in this section attempt to elicit similar natural responses to stimuli in the laboratory. We have previously stated that an animal's auditory perception is determined both by what the animal can hear and what it extracts as meaningful information. The goal of many natural response techniques, whether concerned with auditory perception in songbirds or not, is to determine the latter. Natural response techniques have frequently been used to determine which components of a signal are necessary or most effective in eliciting responses. They have also been used to infer the kind of information that the stimuli transmit to the animal, and to measure song recognition and song preferences. In this half of the chapter, we examine the analysis of complex stimuli, describe how natural responses have been used to understand birdsong perception, and discuss the rationale and limitations of these approaches.

4.1
Analyzing and Modifying Complex Stimuli

Many vertebrate communications are complex patterns involving more than one sensory modality. Natural response techniques have often been used to determine which components of a complex signal are actually acting as the stimulus. The following discussion includes natural stimuli, whether they happen to be birdsong or not.

The earliest experiments of this type were conducted with visual stimuli. A famous example is the work of ter Pelkwijk and Tinbergen (1937) who used model fish to determine which stimuli were necessary to elicit territorial attacks from male three-spined sticklebacks (*Gasterosteus aculeatus*). The investigators presented the models to the sticklebacks and counted the number of aggressive responses given. They found some characteristics of the model, such as their shape, were relatively unimportant in determining the males' responses. The most salient models possessed a red ventral side and were oriented in a vertical, head-down posture. For a review of other early experiments using models to analyze complex visual stimuli, see Marler and Hamilton (1966).

The preceding example illustrates most of the essential features of using natural responses to determine the salient properties of stimuli. The first step is to demonstrate that the natural response can be elicited in the laboratory in response to the stimulus presented by the experimenter. In ethological terms, the stimulus contains a "releaser" for the behavior. The response is quantified in some way and variation in this measure becomes the dependent variable. Relative changes in the level of response can be used to assess salience of stimuli and the disappearance of the response suggests that a necessary element has been removed from the stimulus.

Compared to studies employing visual stimuli, there was a comparative dearth of acoustic studies until the development and availability of tape recorders. The tape recorder allowed for the use of the "playback", the presentation of recorded or synthesized sounds to the test subject over a loudspeaker. The chapter by Hopp and Morton (this Volume) provides an overall discussion of playback methodology. Here, we focus on the use of playback in the laboratory combined with natural response assays. One of the first playback experiments using modified stimuli was conducted by Walker (1957) who recorded the calling songs of male tree crickets and played them to caged females (*Oecanthus* spp.). In the wild, the natural response of sexually responsive females to calling song is a movement towards the singer. The starting point for Walker's research was the fact that the calling songs of three species of tree crickets varied in the rate of pulses per second that made up the continuous trill portion of their calling songs. By playing artificial songs that varied in pulse rate to female crickets and measuring the females' movement towards and away from a speaker, Walker was able to show that pulse rate is the crucial parameter for eliciting a response from a female.

The trills that make up tree cricket calling song are produced by wing vibration and are acoustically relatively simple. Consequently, Walker was able to synthesize passable artificial songs using a signal generator. The songs of birds are often too complex to easily synthesize. One of the first studies using playbacks of modified songs to songbirds was conducted by Falls (1963). Falls was interested in determining what properties of a song are most important in species recognition. Earlier work (Weeden and Falls 1959) had shown that territorial male ovenbirds (*Seiurus aurocapillus*) responded to playback of conspecific song in much the same way as they responded to another territorial male. Because ovenbird song was too complex to synthesize with a signal generator, a limited number of manipulations were available to Falls. Simple manipulations included decreasing the loudness of the songs and playing the songs backwards and at half and double speeds. Falls also removed phrases from the song and varied the duration of the silent intervals between phrases. Today these last two manipulations could easily be done on a computer but Falls had to paint the tape with a substance that showed which portions of the tape had been magnetized and then carefully splice those sections of tape in the desired way. Falls also worked with the white-throated sparrow, *Zonotrichia albicollis*, whose whistle-like song is simple enough to be duplicated, in most but not all particulars, by an audio-oscillator. For a historical review of playback experiments with birdsong and other animals, see Falls (1992).

As computers have become more powerful, and sound analysis and editing software has become more sophisticated, the kinds of modified songs that required days of effort to produce in the 1960s can now be produced in a matter of minutes. However, it still is not possible to synthesize or modify many bird songs easily. In fact, this is true for dynamically complex signals, whether acoustic or visual, although recent progress has been made in this area (e.g., Evans and Marler 1991). Much of the work with modified song has been in the area of demonstrating whether birdsong can be used for species and individual recognition and, if so, what portions of the song convey that information.

4.2
Individual, Population, and Species Recognition

The ability to recognize the songs of one's own species, or the songs from a particular population or geographic region, or the songs of particular individuals is fundamental to the biological success of songbirds. Recognizing conspecific song is necessary for song learning, for territorial defense, and for mate choice and courtship (Searcy et al. 1981b). Recognition of geographic dialects or songs specific to a particular population may have a variety of consequences (for discussions see Payne 1981; Balaban 1986). Recognition of individual song may be important in the maintenance of territorial boundaries (Falls 1982) or for strengthening and maintaining the pair bond (Miller 1979a). Demonstrating that birds can recognize different songs and determining the parameters of song used by listeners to make these discriminations has been the subject of considerable research, most of it using natural response techniques. Many of the natural response techniques that assess the recognition of song by male birds have used territorial responses as their measure of recognition (e.g. Dilger 1956; Weeden and Falls 1959; Peters et al. 1980; Searcy et al. 1981a; Nelson 1988; Searcy et al.1994). Because these assays are usually conducted in the field, and since Hopp and Morton cover field playback techniques elsewhere in this Volume, we focus on techniques that are usually carried out in the laboratory.

4.3
Species Song Recognition

One series of laboratory studies has focused on how the brown-headed cowbird recognizes conspecific song. Cowbirds are brood parasites, laying their eggs in the nests of other species; hence young cowbirds don't normally hear conspecific song as they are growing up (King and West 1977). West, King and their colleagues (King and West 1977; West et al. 1979) developed a natural response technique for measuring song perception in female cowbirds. The subjects for this assay were hand-reared female cowbirds who had never seen or heard adult cowbirds. The birds were housed in pairs under a gradually lengthening photoperiod. When adult cowbird song was played over loudspeakers, the female cowbirds responded by adopting what King and West (1977) referred to as the "copulatory response", characterized by lowered wings, arched back and neck, and a spreading of the feathers around the cloacal region. This response is also referred to as the *copulation solicitation display* (CSD). The response measure was the presence or absence of each female's CSD during each song presentation. Their results suggested that this method could be used as an assay of the salience of species specific song.

Using modified songs as stimuli and the CSD as the assay, West et al. (1979) were able to show that a brief note between song phrases, the interphrase unit (IPU), was crucial in eliciting the female's copulatory response. Response rates for stimuli that did not contain the IPU were only about 50 % of those that did, showing that the CSD assay can also be used to analyze complex stimuli.

In many species of birds, females naturally exhibit behaviors that appear to be similar in both form and function to the CSD of female cowbirds (Searcy 1992). Although the female cowbird gives the response to playback under the conditions developed by King and West, the females of many species, including the song sparrow (*Melospiza melodia*), do not. Noting that Kern and King (1972) had found that massive doses of estradiol caused spontaneous CSDs in white-crowned sparrows (*Zonotrichia leucophrys*), Searcy and Marler (1981) were inspired to try giving song sparrows lower doses of the hormone. They found silastic implants of estradiol primed the sparrows so that they would give CSDs solely in response to playbacks. They were able to use this assay to show a differential response by female song sparrows between the songs of song sparrows and swamp sparrows (*Melospiza georgiana*). The technique developed by Searcy and Marler of using estradiol implants has enabled laboratory researchers to use the CSD assay with many species of females who do not normally give CSDs in response to playback (e.g. Searcy et al. 1981b; Searcy and Brenowitz 1988; Searcy 1990, 1992; Clark and Nottebohm 1992).

Clemmons (1995) used the gape behavior of nestling black-capped chickadees (*Parus atricapillus*) as a natural response to study the development of auditory perception. In the wild, adult chickadees make a vocalization, the "squawk," that stimulates gaping in nestlings during feeding. Using playbacks of both conspecific and heterospecific vocalizations, Clemmons demonstrated a preference for the *squawk* which appeared two to three days post-hatching and steadily increased over the next week. The auditory preferences did not appear to be modifiable by experience: Clemmons found that nestlings reinforced with food for gaping at another chickadee vocalization still gaped most frequently at the *squawk*.

4.4
Recognition of Dialects and Geographic Variation

Some of the same techniques that have been used for species recognition have also been employed to show that birds perceive the differences between song dialects from different geographic regions or from different subspecies. As with species recognition, most of the studies have used the territorial responses of males as their assay (e.g., Catchpole 1978; Baker et al. 1981; Baker 1994). It has long been thought that song may function as a mechanism for mate selection (Marler 1960). Multiple hypotheses exist to explain the functional significance, if any, of geographic variation in song (Payne 1981; Balaban 1986). One possibility is that song may act as a population marker, leading to assortative mating and reproductive isolation (King et al. 1980; Balaban 1986, 1987; Clayton 1990a). King et al. (1980) played the songs of two subspecies of brown-headed cowbird (*M a ater* and *M a obscurus*) to female cowbirds of both subspecies. Using the CSD assay, they found that both groups of females give CSDs more often to the songs of their own sub-species. Females of two subspecies of zebra finch (*T g guttata* and *T g castanotis*) also prefer songs of their own sub-species as measured by the CSD assay (Clayton and Pröve 1989; Clayton 1990b).

4.5
Individual Recognition

Some birds do not appear to discriminate the songs of specific individuals. Female brown-headed cowbirds do not respond differentially in the CSD assay to the songs of known individuals (King and West 1977; West and King 1980). There is a large body of work, mostly from field playbacks to males (but see Hooker and Hooker 1969), showing that in many species, male birds discriminate between familiar and unfamiliar songs, an ability useful for distinguishing neighbors with whom territorial boundaries have been established from new intruders (Falls 1982). Although some of these experiments do not allow one to conclude that territorial males do more than distinguish familiar from unfamiliar songs, some experiments (e.g., Brooks, cited in Falls 1969) have shown that a familiar song played from an unfamiliar location provokes as strong a response as an unfamiliar song. Here, we survey some of the other natural response techniques that have been developed to investigate this question.

    The stimuli normally consist of recordings of different individuals. The major differences between methods are in whether the stimuli are presented simultaneously or successively (Beer 1970) and in the type of natural response employed. Methods using simultaneous presentation of stimuli often use a sort of "battle of the bands approach", two speakers placed at opposite ends of an arena or cage and the use of approach or amount of time spent in proximity to one speaker or another as the assay. Tschanz (1965) found that young guillemots (*Uria aalbe*) that were exposed to recordings of the

"*Lockrufe*" call of a specific adult during the last 3-4 days pre-hatching would approach the loudspeaker playing that adult's call, whereas chicks not exposed to any calls would not express a preference for one speaker over another.

Miller (1979a) used the two-speaker approach method to test whether female zebra finches recognize the songs of their mates. Females were separated from their mates for two 2-3 days and then simultaneously presented with the song of their mate and the song of a neighbor. Females demonstrated a significant preference for the song of their mate. Using the same methodology, Miller (1979b) showed that female zebra finches who had been separated from their father for at least two months expressed a preference for the father's song over both similar and dissimilar songs.

Sequential presentation of song stimuli has also been used to measure individual recognition. Beer (1970) used a variety of natural responses including vocalizations, orientation, and approach to show that gull chicks recognized the calls of their parents. In contrast, the calls of a gull from a different part of the gullery caused withdrawal and silent crouching by the chick. Approach methods and sequential presentations have also been used in field settings, e.g. Nemeth (1994) showed that female reed buntings (*Emberiza schoeniclus*) will approach the songs of their mates but not other males.

Several studies besides Beer's have used vocalizations to measure whether birds distinguish between familiar and unfamiliar songs, regardless of whether there is true individual recognition. The earliest example comes from the work of Hinde (1958) who played a variety of songs to male chaffinches and used the number of songs given in reply as the natural response. Response was greatest to songs similar to the bird's own song. Dietrich (1981) found that female Bengalese finches (*Lonchura striata*) call more frequently to the songs of familiar males than to those of strange males.

4.6
Sexual Selection and Song Preferences

A number of studies have used natural response methods, mostly the CSD techniques pioneered by King and West (1977) and Searcy and Marler (1981), to measure female song preferences, with the implication that these song preferences reflect mating preferences as well. Many of these studies have recently been reviewed by Searcy (1992). Still more recently, Vallet and Kreutzer (1995) found that certain phrases of male canary song (albeit based on the song of one particular male) were more effective in eliciting CSDs from female canaries than other parts. The authors argued that the differential response in the CSD assay reflects a differential sexual response and that these particular song features may act as "sexual releasers". Another set of natural responses was used by Kroodsma (1976) to measure the effects of song on female canaries. He found that the rates with which females built nests and laid eggs were greater for large song repertoires than for small repertoires.

5
General Considerations
When Using Natural Response Methods

We hope that this brief overview shows that natural response methods are powerful techniques for studying avian auditory perception. They are widely used, and the sound playback is considered a standard ethological method (e.g., Lehner 1979). However, the literature contains little discussion of the rationale of these methods and the conditions when the use of these methods is appropriate, although see Balaban (1986) and Searcy (1992). Here, we will attempt to expand on the discussion begun by these authors.

We begin by considering what conclusions can be drawn when a subject shows a response or fails to make a response to the presentation of a stimulus. A first demonstration in almost all natural response methods is to show that the animal responds to the wild-type stimulus, for example, to demonstrate that a female bird gives a copulation solicitation display in response to male song. What can we conclude from this response? There are two possible explanations that are undesirable if we seek to use natural response methods as assays of perception. The first is that the behavior is spontaneous or elicited in response to internal or external stimuli which have nothing to do with the test situation. A second possibility is that the bird is responding to some aspect of the testing protocol that has nothing directly to do with the stimulus being presented— for example, some aspect of the testing chamber. It is relatively easy to eliminate both of these possibilities by the use of control stimuli. The third (desirable) possibility is to conclude that an auditory stimulus is sufficient to elicit the natural response under the test conditions.

Having demonstrated that a bird will give the desired response under test conditions, the experimenter presents the bird with modified stimuli: they may have been modified by a computer, synthesized, be from a different geographic region, etc. What can we conclude if the response to our modified stimulus is not different than the response to the original? Again, there are three possibilities: the subject does not perceive the differences between the two stimuli, the subject perceives the differences but they are functionally equivalent, or the subject perceives the differences, the stimuli are not functionally equivalent but the particular assay is inappropriate for revealing that differential salience (Balaban 1986).

How do we interpret the results if the response to the modified stimulus differs from the response to the original stimulus? In this case we can cautiously conclude that the differences between stimuli are both perceptible and salient to the animal. Further interpretation of the results is an exercise fraught with peril because there are at least three different ways that the results could be misinterpreted. These three ways correspond to the classic problems in experimental design of internal, external and construct validity. We will examine what is meant by these concepts and how they apply to the use of natural response methods.

5.1
Internal Validity

Internal validity is the extent to which the conclusions drawn from the experimental results may not reflect what actually transpired in the experiment (Campbell and Stanley 1963). High internal validity means that variation in the dependent variable is due to manipulations of the independent variable. However, in any study, it is possibile that the results are due to another variable that covaries with the independent variable. Balaban (1986) has discussed one possible challenge to internal validity: the problem of differential familiarity. To learn whether birds perceive geographic variation in song, one must take care that we are not giving birds a choice between a strange song from a different geographic area and the song of singer individually known to the test subject. Similar problems crop up in attempting to measure individual recognition (Beer 1970) or which components of a stimulus are most salient to the listener.

Another challenge to internal validity in natural response methods are *stage-of-practice effects*. In within-subject designs, where each test subject is tested with multiple levels of the independent variable, care must be taken to balance the presentation of stimuli. Repeated exposure to the test situation can change the responsiveness of a subject in several ways that have nothing to do with the particular stimulus being presented. For example, repeated exposures may lessen the stressfulness of the procedure for the animal and therefore make it more responsive. Or repeated exposure may habituate the animal. There may be other consequences to repeated exposures such as refractory periods. Most scientists are aware of the need to balance stage-of-practice effects in within-subjects designs and we do not wish to belabor this point. We do wish to point out, however, that for some irreversible procedures— for example, combining brain lesions with natural response methods— it is not possible to balance practice effects and in these cases investigators would do well to consider between-subjects designs.

5.2
External Validity

External validity is the extent to which the findings of an experiment generalize to other individuals, settings and conditions beyond the scope of the present experiment. In other words, do the laboratory results increase our understanding of what is happening in the field? Evidence exists in the literature that the results of a particular natural response method may not always generalize to other conditions. For example, ring-billed chicks can recognize their parents when stimuli are presented simultaneously but not when they are presented sequentially (Evans, cited in Beer 1970). Evans argues that the behavior of a chick may differ in a recognition task depending on whether it has lost its parents or not. If an orphaned chick is to survive, it must be adopted by other adults. Consequently there should be selection against the withdrawal behavior normally shown by chicks to calls from adults other than their parents. Beer (1970) found that the natural responses of laughing gull chicks in recognition tasks varied with their age.

More recently, Johnsrude et al. (1994) found that some features of songbird perception by female cowbirds differed depending on whether the females were tested with the CSD assay or by operant methods.

Another challenge to external validity is the problem of *pseudoreplication* (Kroodsma 1986). As this issue is discussed elsewhere in this Volume (see Hopp and Morton, this Volume) and has been the subject of recent works (e.g., McGregor 1992), we will not discuss it here.

External validity means that we cannot stop thinking like ethologists simply because we are doing our testing in the laboratory. As the name implies, natural response methods attempt to elicit normal behavior in a lab setting. In the wild we expect individual variation which is dependent on a multitude of factors including the context in which a stimulus is perceived and the subject's own unique history. These same factors must also be considered when carrying out natural response assays in the laboratory.

5.3
Construct Validity

Construct validity refers to the extent the variable actually measured relates to the conceptual variable we are interested in measuring (Carmines and Zeller 1979). Construct validity often fails when the experimenter cannot directly measure the quantity in which he or she is interested. A taxonomist interested in the length of a bird's beak can measure that quantity directly; an ethologist interested in a bird's mating preferences must find an indirect measure. In the case of natural responses we are seldom content to simply report a differential response to test stimuli. More often we would like to go beyond this level of analysis and conclude that a difference in responsiveness as measured by the assay indicates a difference in ecological significance: more CSDs given to a stimulus indicates a higher degree of sexual responsiveness, closer approach to a speaker in a field playback indicates a higher degree of aggressiveness and so on. When we do this, construct validity becomes a problem which must be addressed.

There are several assumptions implicit in using the natural response methods that relate to the question of construct validity. The first is that there is a logical relation between the variable measured and the variable of interest. This relationship seems clear for some natural responses. For example, in the wild when a female gives a copulation solicitation display it is usually followed by the female mating with the male. Thus, it is believable that the CSD is an assay of sexual acceptance, yet even this apparently straightforward conclusion can be problematic. Lambrechts et al. (1993) suggest, for example, that wing quivering in the black-capped chickadee may be functioning more to reduce aggression than as a sexual invitation, especially during the nestling period. Furthermore, this conclusion assumes that a CSD given under the experimental conditions "means" the same thing to the bird as one given in the wild. This may not be such a good assumption if the manipulations necessary to produce the response (such as estradiol implantation) have changed the subject's threshold of response. There have been a few studies which attempt to demonstrate the construct validity of natural response methods to preferences found in the wild (e.g., West et al. 1981; see Searcy 1992 for others). More such studies are needed.

If even the apparently straightforward CSD posture is subject to alternate interpretation, it may be more difficult to make sense of other response measures, such as speaker approach or changes in heart rate. Searcy (1992) has pointed out that, depending on the context, phonotaxis may be aggressive, affiliative or indicative of mating preferences. As with external validity, understanding the ecological significance of a natural response is essential if we are to use that response to infer the kind of information the stimulus transmits to the subject (Marler 1961).

The more we attempt to interpret the meaning of natural response experiments, the larger the problem of construct validity becomes. One such problem when measuring female song preferences has been pointed out by William Searcy (pers comm 1988). The advantage of the playback technique is that we can assess female preferences to song alone without other features of the male confounding her behavior. As Searcy points out, however, in the wild, females would be presented not only with physical aspects of the male but also with such variables as the quality of the male's territory. In many species where territory quality is more important than male quality in a female's decision to nest and mate, natural response methods that assay song preferences may not be good assays of mate choice.

A final issue that relates to the construct validity of natural response methods is what to make of quantitative variation in an assay. It may be that a given natural response indirectly measures something else, e.g., the CSD measures sexual preference, to stay with the example we have been using. Yet there is no a priori reason to assume that variation in the assay, as measured in the lab, has ecological significance. Consider that two CSDs given in response to a playback may not indicate any more sexual preference than one CSD. Given that, in the wild, a courting male immediately mounts a female when she performs a CSD, both cases indicate the same thing: sexual acceptance. Perhaps the number of CSDs given is a measure of how quickly the female habituates to the stimulus when the unexpected happens (no male mounts her). This may or may not be related to her preference. Perhaps latency to the first CSD is a better measure. A similar problem occurs when we attempt to combine various kinds of responses into some sort of composite measure. A common practice when using the CSD assay, for example, is to grade the degree of response from simple crouching to the complete CSD. This may be an appropriate procedure but it assumes that the gradations represent an equal-interval scale. If simple crouching is scored as "one" and a complete CSD as "three" then three crouches generate the same score as one CSD. They may not, however, have the same ecological significance. It may be that a female who is only crouching will rebuff a male who tries to mount her in which case no number of crouches is equal to a CSD. On the other hand, perhaps a crouching female will accept a male as readily as one giving a CSD. In the latter case it might be best to give crouches and CSDs the same score.

In this section discussion has focused on the CSD assay. This was only to provide a concrete example upon which to base the discussion. The issues apply to all natural response methods. Moreover, we do not necessarily subscribe to many of the issues raised, that two CSDs are not more than one, for example. These are issues that must be considered by researchers in this area, however, even if they are ultimately rejected.

6
Summary

Fifty years ago, the question of how birds perceived sounds was given wide berth by competent experimenters. Rather, inferences were made from what birds sang. Few if any tools had been developed with which to investigate this aspect of animal behavior. Now this has been partially remedied. Precise questions can be asked as to what a bird can hear, what it listens to, and what it does with the information. Part of this is due to the application of behavioral field and laboratory techniques. But, just as important, students of animal behavior have been educated in how to ask precise questions and interpret the results. Auditory perception is not some monolithic event that offers a single answer. Rather, it comprises an integrated part of an animal's behavior that has both proximate and ultimate causes (Tinbergen 1951). It's worth contemplating what will be said of studies of animal communication in another half century. One would hope researchers embrace rather than avoid the pluralism inherent in nature.

References

Balaban ES (1986) Cultural and genetic variation in swamp sparrows (*Melospiza georgiana*). Doctoral Diss, The Rockefeller University, New York
Balaban E (1987) Behavioural salience of geographical song variants. Behaviour 105: 297-32
Baker MC (1994) Loss of function in territorial song: comparison of island and mainland populations of the single honeyeater (*Meliphaca virescens*). Auk 111: 178-184
Baker MC, Thompson D, Sherman G, Cunningham M (1981) The role of male vs. male interactions in maintaining population dialect structure. Behav Ecol Sociobiol 8: 65-69
Beecher MD, Stoddard PK (1990) The role of birdsong and calls in individual recognition: contrasting field and laboratory results. In: Stebbins WC, Berkley MA (eds) Comparative perception, vol. II. Complex signals. Wiley, New York, p 375
Beer CG (1970) Individual recognition of voice in the social behavior of birds. In: Lehrman DS, Hinde RA, Shaw E (eds) Advances in the study of behavior, vol 3. Academic Press, New York, p 27
Berkley MA, Stebbins WC (eds) (1990) Comparative perception, vol I. Basic Mechanisms. Wiley, New York
Blough DS (1966) The study of animal sensory processes by operant method. In: Honig WK (ed) Operant behavior: areas of research and application. Prentice-Hall, Englewood Cliffs, p 345
Brand AR, Kellogg PP (1939a) Auditory responses of starlings, English sparrow, and domestic pigeons. Wilson Bull 51: 38-41
Brand AR, Kellogg PP (1939b) The range of hearing of canaries. Science 90: 354
Brown P, Jenkins HM (1968) Autoshaping of the pigeon's keypeck. J Exp Anal Behav 11: 1-8
Burdick CK (1990) The effect of behavioral paradigm on auditory discrimination learning: a literature review. J Aud Res 19: 59-82
Campbell DT, Stanley JC (1963) Experimental and quasi-experimental designs for research. Rand McNally, Chicago
Carmines EG, Zeller RA (1979) Reliability and validity assessment. Sage, Beverly Hills
Catchpole CK (1978) Intraspecific territorialism and competition in Acrocephalus warblers as revealed by playback experiments in areas of sympartry and allopatry. Anim Behav 26: 1072-1080
Clark SJ, Nottebohm F (1992) Perception of birdsong by female zebra finches and canaries. Soc Neurosci Abstr 16: 1100
Clayton NS (1988) Song discrimination learning in zebra finches. Anim Behav 36: 1016-1024
Clayton NS (1990a) Assortative mating in zebra finch subspecies, *Taeniopygia guttata guttata* and *T. G. castanotis*. Philos Trans R Soc Lond Biol Sci 330: 351-370
Clayton NS (1990b) Subspecies recognition and song learning in zebra finches. Anim Behav 40: 1009-1017
Clayton NS, Pröve E (1989) Song discrimination in female zebra finches and Bengalese finches. Anim Behav 38: 352-354

Clemmons JR(1995) Development of a selective response to an adult vocalization in nestling black-cap-
    ped chickadees. Behaviour 132: 1-20
Curtis H (1983) Biology, 4th edn. Worth, New York
Cynx J (1993) Auditory frequency generalization and a failure to find octave generalization in a songbird,
    the European starling (*Sturnus vulgaris*). J Comp Psychol 107: 140-146
Cynx J (1995) Similarities in absolute and relative pitch perception in song birds (starling and zebra
    finch) and a non-song bird (pigeon). J Comp Psychol 109: 261-267
Cynx J (1996) Neuroethological studies on how birds discriminate song. In: Moss CF, Shettleworth SJ
    (eds) Neuroethology of cognitive and perceptual processes. Westview Press, Boulder, p 63
Cynx J, Hulse SH, Polyzois S (1986) A psychophysical measure of loss of pitch discrimination resulting
    from a frequency range constraint in European starlings (*Sturnus vulgaris*). J Exp Psychol: Anim
    Behav Processes 12: 394-402Cynx J, Nottebohm F (1989) A songbird's memory capacity for song
    elements. Proc East Psychol Assoc 60, Boston
Cynx J, Nottebohm F (1992a) Role of gender, season, and familiarity in discrimination of conspecific
    song by zebra finches (*Taeniopygia guttata*). PNAS USA 89: 1368-1371
Cynx J, Nottebohm F (1992b) Testosterone facilitates some conspecific song discriminations in castrated
    zebra finches (*Taeniopygia guttata*). PNAS USA 89: 1376-1378
Cynx J, Williams H, Nottebohm F (1992) Hemispheric differences in avian song discrimination. PNAS
    USA 89: 1372-1375
Cynx J, Williams H, Nottebohm, F (1990) Timbre discrimination in zebra finch (*Taeniopygia guttata*)
    song syllables. J Comp Psychol 104: 303-308
Dietrich K (1981) Unterschiedliche Reaktionen von Weibchen des Japanischen Möwchens (*Lonchura
    striata var. domestica*) auf Gesänge verwandter und nicht verwandter Artgenossen. Z Tierpsychol
    57: 235-245
Dilger WC (1956) Hostile behavior and reproductive isolating mechanisms in the avian genera *Catharus*
    and *Hylocichla*. Auk 73: 313-353
Dooling RJ (1973) Behavioral audiometry with the parakeet *Melopsittacus undulatus*. J Acoust Soc Am
    53: 1757-1758
Dooling RJ (1982) Auditory perception in birds. In: Kroodsma DE, Miller EH, Ouellet H (eds) Acoustic
    communication in birds, vol 1. Academic Press, New York, p 95
Dooling RJ (1986) Perception of vocal signals by budgerigars (*Melopsittacus undulatus*). Exp Biol 45:
    195-218
Dooling RJ, Okanoya K (1995a) The method of constant stimuli in testing auditory sensitivity in small
    birds. In: Klump GM, Dooling RJ, Fay RR, Stebbins WC (eds) Methods in comparative psychoa-
    coustics. Birkhuser, Basel, p161
Dooling RJ, Okanoya K (1995b) Psychophysical methods for assessing perceptual categories. In Klump,
    GM, Dooling, RJ, Fay RR, Stebbins WC (eds) Methods in comparative psychoacoustics. Birkhuser,
    Basel, p 307
Dooling RJ, Saunders JC (1975) Hearing in the parakeet (*Melopsittacus undulatus*): absolute thresholds,
    critical ratios, frequency difference limens, and vocalizations. J Comp Physiol Psychol 88: 1-20
Dooling RJ, Searcy M (1980) Early perceptual selectivity in the swamp sparrow. Dev Psychobiol 13:
    499-506
Dooling RJ, Mulligan JA, Miller JD (1971) Auditory sensitivity and song spectrum of the common canary
    (*Serinus canarius*). J Acoust Soc Am 50: 700-709
Dooling RJ, Park TJ, Brown SD, Okanoya K, Soli SD (1987) Perceptual organization of acoustic stimuli
    by budgerigars (*Melopsittacus undulatus*), II. Vocal signals. J Comp Psychol 101: 367-381
Dooling RJ, Brown SD, Park TJ, Okanoya K (1990) Natural perceptual categories for vocal signals in
    budgerigars (*Melopsittacus undulatus*). In: Stebbins WC, Berkley MA (eds) Comparative percep-
    tion: vol II. cCmplex signals. Wiley, New York, p 4
Edwards EP (1943) Hearing ranges of four species of birds. Auk 60: 239-241
Evans CS, Marler P (1991) On the use of video images as social stimuli in birds: audience effects on alarm
    calling. Anim Behav 41: 17-26
Falls JB (1963) Properties of bird song eliciting responses from territorial males. Proc 13th Int Ornithol
    Congr 1: 259-271
Falls JB (1969) Functions of territorial song in the white-crowned sparrow. In: Hinde RA (ed) Bird
    vocalizations. Cambridge University Press, London, p 207
Falls JB (1982) Individual recognition by sound in birds. In: Kroodsma, DE, Miller EH, Ouellet H (eds)
    Acoustic communication in birds, vol 2. Academic Press, New York, p 237
Falls JB (1992) Playback: a historical perspective. In: McGregor PK (ed) Playback and studies of animal
    communication. Plenum Press, New York, p 11
Fay RR (1988) Hearing in vertebrates: a psychophysics databook. Hill-Fay Associates, Chicago

Fechner GT (1860) Element der Psychophysik. Breitkopf and Härterl, Leipzig

Frisch K von (1965) Tanzsprache und Orientierung der Bienen. Springer, Berlin Heidelberg New York

Gantt WH (1966) Conditional or conditioned, reflex or response? Cond Reflex 1: 69-74

Greenewalt CH (1968) Bird song: acoustics and physiology. Smithsonian Institution Press, Washington, DC

Griffin DR (1974) Listening in the dark. Dover Publications, New York (Original work published 1958)

Guttman N, Kalish H (1956) Discriminability and stimulus generalization. J Exp Psychol 29: 390-400

Heffner HE, Heffner RS (1984) Temporal lobe lesions and perception of species-specific vocalizations by macaques. Science 226: 75-76

Heise GA (1953) Auditory thresholds in the pigeon. Am J Psychol 66: 1-19

Hienz RD, Sinnott JM, Sachs MB (1977) Auditory sensitivity of the red wing blackbird (*Agelaius phoeniceus*) and brown-headed cowbird (*Molothrus ater*). J Comp Physiol Psychol 91: 1365-1376

Hinde RA (1958) Alternative motor patterns in chaffinch song. Br J Anim Behav 6: 211-218

Honig W, Urcuioli PJ (1981) The legacy of Guttman and Kalish (1956): 25 years of research on stimulus generalization. J Exp Anal Behav 36: 405-445

Hooker T, Hooker BI (1969) Duetting. In: Hinde, RA (ed) Bird vocalizations. Cambridge University Press, London, p 185

Hulse SH (1995) The discrimination-transfer procedure for studying auditory perception and perceptual invariance in animals. In: Klump GM, Dooling RJ, Fay RR, Stebbins WC (eds) Methods in comparative psychoacoustics. Birkhäuser, Basel, p 319

Hulse SH, Cynx J (1985) Relative pitch perception is constrained by absolute pitch in songbirds (*Mimus, Molothrus*, and *Sturnus*). J Comp Psychol 99: 176-196

Hulse SH, Cynx J, Humpal J (1984a) Absolute and relative discrimination in serial pitch perception by birds. J Exp Psychol: Gen 113: 38-54

Hulse SH, Humpal J, Cynx J (1984b) Discrimination and generalization of rhythmic and arrhythmic sound patterns by European starlings (*Sturnus vulgaris*). Music Percept 1: 442-464

Jenkins HM, Harrison RH (1960) Effect of discrimination training on auditory generalization. J Exp Psychol 59: 246-253

Johnsrude IS, Weary DM, Ratcliffe LM, Weisman RG (1994) Effect of motivational context on conspecific song discrimination by brown-headed cowbirds (*Molothrus ater*). J Comp Psychol 108: 172-178

Katz LC, Gurney ME (1979) Auditory responses in the zebra finch's motor system for song. Brain Res 221: 192-197

Kern MD, King JR (1972) Testosterone-induced singing in female white-crowed sparrows. Condor 14: 204-209

King AP, West MJ (1977) Species identification in the North American cowbird: appropriate responses to abnormal song. Science 195: 1002-1004

King AP, West MJ, Eastzer DH (1980) Song structure and song development as potential contributors to reproductive isolation in cowbirds (*Molothrus ater*). J Comp Physiol Psychol 94: 1028-1039

Klump GM, Dooling RJ, Fay RR, Stebbins WC (eds) (1995) Methods in comparative psychoacoustics. Birkhuser, Basel

Konorski J (1967) Integrative activity of the brain: an interdisciplinary approach. University Chicago Press, Chicago

Kroodsma DE (1976) Reproductive development in a female songbird: differential stimulation by quality of male song. Science 192: 574-575

Kroodsma DE (1986) Design of song playback experiments. Auk 103: 640-642

Lambrechts MM, Clemmons JR, Hailman JP (1993) Wing quivering of black-capped chickadees with nestlings: invitation or appeasement? Anim Behav 46: 397-399

Lehner PN (1979) Handbook of ethological methods. Garland STMP Press, New York

Lemon R (1967) The response of cardinal to songs of different dialects. Can J Zool 44: 413-428

Lissman HW (1932) Die Umwelt des Kampffisches (*Betta splendens* Regan). Z vgl Physiol 18: 65-111

Loesche P, Beecher MD, Stoddard PK (1992) Perception of cliff swallow calls by birds (*Hirundo pyrrhonota* and *Sturnus vulgaris*) and humans (*Homo sapiens*). J Comp Psychol 106: 239-247

Mackintosh NJ (1983) Conditioning and associative learning. Oxford University Press, New York

Margoliash D (1983) Acoustic parameters underlying the responses of song-specific neurons in the white-crowned sparrow. J Neurosci 3: 1039-1057

Margoliash D (1986) Preference for autogenous song by auditory neurons in a song system nucleus of the white-crowned sparrow. J Neurosci 6: 1643-1661

Marler P (1960) Bird songs and mate selection. In: Lanyon WE, Tavolga WN (eds) Animal sounds and communication. Am Inst Biol Science, Washington, DC

Marler P (1961) The logical analysis of animal communication. J Theor Biol 1: 295-317

Marler P, Hamilton WJ III (1966) Mechanisms of animal behavior. Wiley, New York

McGregor PK (ed) (1992) Playback and studies of animal communication. Plenum Press, New York

Miller DB (1979a) The acoustic basis of mate recognition by female zebra finches (*Taeniopygia guttata*). Anim Behav 27: 376-380

Miller DB (1979b) Long-term recognition of father's song by female zebra finches. Nature 280: 389-391

Mostofsky DI (1965) Stimulus generalization. Stanford University Press, Stanford

Nelson DA (1988) Feature weighting in species song recognition by the field sparrow (*Spizella pusilla*). Behaviour 106: 158-182

Nelson DA, Marler P (1990) The perception of birdsong and an ecological concept of signal space. In: Stebbins WC, Berkeley MA (eds) Comparative perception, vol 2. Wiley New York, p 443

Nemeth E (1994) Individual recognition of song by the female and song activity of the male in the Reed Bunting (*Emberiza schoeniculus*). J Ornithol 135: 217-222

Okanoya K (1995) Adaptive tracking procedures to measure auditory sensitivity in songbirds. In: Klump GM, Dooling RJ, Fay RR, Stebbins WC (eds) Methods in comparative psychoacoustics. Birkhuser, Basel, p 149

Pavlov IP (1960) Conditioned reflexes: An investigation of the physiological activity of the cerebral cortex. Dover Press, New York (Original work published 1927)

Payne R (1981) Population structure and social behavior: models for testing ecological significance of song dialects in birds. In: Alexander RD, Tinkle D (eds) Natural selection and social behavior: recent research and new theory. Chiron, Wilmette, Illinois

Pelkwijk JJ ter, Tinbergen N (1937) Eine reizbiologische Analyse einiger Verhaltensweisen von *Gasterosteus aculeatus* L. Z Tierpsychol 1: 193-200

Peters SS, Searcy WA, Marler P (1980) Species song discrimination in choice experiments with territorial male swamp and song sparrows. Anim Behav 28: 393-404

Petersen MR, Beecher MD, Zoloth SR, Moody DB, Stebbins WC (1978) Neural lateralization of species-specific vocalizations by Japanese macaques (*Macaca fuscata*). Science 204: 324-327

Pörve E (1974) Der Einfluss von Kastration und Testosteronsubstitution auf das Sexualverhalten mannlicher Zebrafinken (*Taeniopygia guttata castonatis* Gould). J Ornithol 115: 388-347

Pumphrey RJ (1948) The sense organs of birds. Ibis 90: 171-199

Roitblat HL, Bever TG, Terrace HS (1984) Animal cognition. Erlbaum, Hillsdale, New Jersey

Rudolph RL, Van Houten R (1977) Auditory stimulus control in pigeons: Jenkins and Harrison (1960) revisited. J Exp Anal Behav 27: 327-330

Schwartzkopff J (1949) Uber sitz und Leistung von Gehor und Vibrationssinn bei Vogeln. Z vgl Physiol 31: 527-608

Schwartzkopff J (1955) On the hearing of birds. Auk 72: 340-347

Searcy WA (1990) Species recognition of song by female red-winged blackbirds. Anim Behav 40: 1119-1127

Searcy WA (1992) Measuring responses of female birds to male song. In: McGregor PK (ed) Playback and studies of animal communication. Plenum Press, New York, p 175

Searcy WA, Brenowitz EA (1988) Sexual differences in species recognition of avian song. Nature 332: 152-154

Searcy WA, Marler P (1981) A test for responsiveness to song structure and programming in female sparrows. Science 213: 926-928

Searcy WA, Balaban E, Canady RA, Clark SJ, Runfeldt S, Williams H (1981a) Responsiveness of male swamp sparrows to temporal organization of song. Auk 98: 613-615

Searcy WA, Marler P, Peters SS (1981b) Species song discrimination in adult female song and swamp sparrows. Anim Behav 29: 997-1003

Searcy WA, Marler P, Peters SS (1985) Songs of isolation-reared sparrows function in communication, but are significantly less effective than learned songs. Behav Ecol Sociobiol 17: 223-229

Searcy WA, Coffman S, Raikow DF (1994) Habituation, recovery and the similarity of song types within repertoires in red-winged blackbirds (*Agelaius phoeniceus*) (Aves, Emberizidae) Ethology 98: 38-49

Seyfarth RM, Cheney DL, Marler P (1980) Monkey responses to three different alarm calls: Evidence for predator classification and semantic communication. Science 210: 801-803

Shepard RN (1987) Toward a universal law of generalization for psychological science. Science 237: 1317-1323

Sherrington CS (1906) Integrative action of the nervous system. Yale University Press, New Haven

Shy E, McGregor PK, Krebs JR (1986) Discrimination of song types by male great tits. Behav Proc 13: 1-12

Sinnott JM (1980) Species-specific coding in bird song. J Acoust Soc Am 68: 494-497

Skinner BF (1938) The behavior of organisms: an experimental analysis. Appleton-Century-Crofts, NY

Stebbins WC (ed) (1970) Animal psychophysics. Appleton-Century-Crofts, New York

Stebbins WC (1990) Perception in animal behavior. In Berkley MA, Stebbins WC (eds) Comparative perception, vol I. Basic mechanisms. Wiley, New York, p 1

Stebbins WC, Berkley MA (eds) (1990) Comparative perception, vol II. Complex signals. Wiley, New York

Thorndike EL (1911) Animal intelligence: experimental studies. Macmillian, New York

Thorpe WH (1961) Bird-song. Cambridge University Press, London

Tinbergen N (1951) The study of instinct. Oxford University Press, Oxford

Tinbergen N, Perdeck AC (1950) On the stimulus situation releasing the begging response in the newly hatched herring gull chick (*Larus argentatus* Pont). Behaviour 3: 1-39

Trainer JE (1946) The auditory acuity of certain birds. Doctoral Diss, Cornell University, Ithaca

Tschanz B (1965) Beobachtungen und Experimente zur Entstehung der "persönlichen" Beziehung zwischen Jungvogel und Eltern bei Trottellummen. Verh Schweiz Naturforsch Geswiss Teil 1964: 211-216

Uexküll, J von (1909) Umwelt und Innenwelt der Tiere. Springer, Berlin Heidelberg New York

Vallet E, Kreutzer M (1995) Female canaries are sexually responsive to special song phrases. Anim behav 49:1603-1610

Walker T S (1957) Specificity in the response of the female tree crickets (Orthoptera, Gryllidae, Oecanthinae) to calling songs of males Ann Etmol Soc Am 50: 626-636

Weary DM (1989) Categorical perception of birdsong: how do great tits (*Parus major*) perceive temporal variation in their song? J Comp Psychol 103: 320-325

Weber EH (1834) De pulsu, resorpitione, auditu et tactu: annotationes anatomicae et physiologicae. Koehlor, Leipzig

Weeden JS, Falls JB (1959) Differential responses of male ovenbirds to recorded songs of neighboring and more distant individuals. Auk 76: 343-351

West MJ, King AP (1980) Enriching cowbird song by social deprivation. J Comp Physiol Psychol 94: 263-270

West MJ, King AP, Eastzer DH (1981) Validating the female bioassay of cowbird song: relating differences in song potency to mating success. Anim Behav 29: 490-501

West MJ, King AP, Eastzer DH, Staddon JER (1979) A bioassay of isolate cowbird song. J Comp Physiol Psychol 93: 124-133

Williams H, Nottebohm F (1985) Auditory responses in avian vocal motor neurons: a motor theory for song perception. Science 229: 279-282

CHAPTER 12

Assessing Hormonal Responses to Acoustic Stimulation

G. F. BALL AND A. M. DUFTY JR.

1
Introduction

An important component of any bioacoustic investigation is to assess the consequences that a given signal has for the receiver of the signal (Green and Marler 1979; Hauser 1996). Attaining this goal requires the answering of two related questions: (1) How is the signal actually perceived by the receiver? and (2) How does it affect the behavior and/or physiology of the receiver? As reviewed by Cynx and Clark, (this Volume), one way to answer question one is to employ conditioning methods. One can ask an animal to learn a task for a reward that requires the subject to discriminate among a variety of signals that may or may not occur in its natural environment. When applied appropriately this approach can provide insight into what in the acoustic environment an individual is physically able to perceive. Hopp and Morton, (this Volume), summarize a powerful and widely used method that partially answers question two, namely the sound playback experiment. In studies of this type, a signal is presented in a controlled fashion and the investigator is able to measure the behavioral response of a selected receiver to the signal. Typically in these studies, the target responses are overt behaviors, but one can also measure physiological responses.

The obvious aspect of physiology to measure is synaptic physiology. Acoustic signals are coded in the central nervous system by changes in the neuronal firing rate of the various nerve fibers that project along multisynaptic pathways from the receptor cells in the auditory periphery to the various auditory processing areas in the brain itself. Measuring changes in the neural firing rate in response to auditory stimulation is a rich and important source of information concerning the salience of a particular acoustic signal to an individual and the internal attributes of an organism that influence how it responds to various acoustic stimuli. The limitation of this approach is that it involves the implantation of electrodes and, in many cases, requires that the animal be anesthetized. Even when recording from fully awake animals, the type and manner of stimulus presentation is limited by the methodological constraints placed on the experimenters to ensure accurate and reliable recordings from nerve cells or fibers.

Another aspect of an animal's physiology that is known to change, in many circumstances, in response to acoustic stimulation is the endocrine physiology of the animal (Wingfield et al. 1994). Hormones do not change in response to environmental stimuli as rapidly as the neuronal firing rate of the auditory system, rather their secretion is the eventual result of such neural changes and hormones constitute one of the internal messenger systems that implement the long-term biological effects of acoustic signals. The slower time course of endocrine changes, and the relative ease of measurement of these changes, makes them a class of physiological stimuli worth considering when one is interested in assessing the consequences of acoustic stimuli.

In this chapter we review the merits and pitfalls of using the activity of the endocrine system as an indicator of the salience of an acoustic signal. Many signals have evolved to be used in the context of reproduction and aggression. Signals used in these contexts often result in the neuroendocrine physiology of the receiver being modified in some way. A bioacoustician can take advantage of this fact. By measuring hormonal modifications one can attain substantial and even unique insights into the actual consequences of a given signal. In many cases, the endocrine response will supplement what is clear from other behavioral measures. In other cases, however, the salience of a given signal may not be apparent by other measures or the endocrine response may actually contradict what one may have supposed from those indications. This chapter will not be a comprehensive discussion of the entire literature on hormones and vocal communication in vertebrates, for that the reader is referred to other reviews (e.g., Kelley and Brenowitz 1992; Wingfield et al. 1994). Rather, we will selectively review a diversity of studies with an eye toward elucidating the practicalities of the methods employed. Most of our examples will involve avian species because most of the available data concerns birds, with some notable exceptions.

The usefulness and precision of an assessment of endocrine functioning in response to acoustic stimulation are obviously functions of the methods one employs. We will select our studies to illustrate the diversity of dependent measures one can select. Early investigators were limited to rather indirect measures of endocrine activity that change over rather long periods of time, such as histological variations in an endocrine gland or changes in the size and weight of hormone dependent structures (e.g., Lehrman 1959; Marshall 1959). In the late 1960s sensitive assay methods were first developed that have provided for a more precise determination of endocrine activity by allowing one to measure plasma hormone levels. More recently, procedures that assess brain chemistry offer the prospect of measuring the regulatory neurohormones in the brain that ultimately stimulate peripheral gland activity. Attention will be given to a comparison of the relative usefulness of these different approaches. Although, in general, more advanced methods should be employed when practical, indirect measurements can still provide potentially valuable information; information that is attained with the use of relatively simple and inexpensive procedures. It should also be noted at the outset that the study of the endocrine consequences of acoustic stimulation is a special case of the general discipline of environmental endocrinology. Environmental endocrinology investigates how physical stimuli such as photoperiod, temperature and food availability modify endocrine cycles as well as how behavioral stimuli impinge on the neuroendocrine system (Silver 1992). Therefore our discussion will be framed by the more general context of how external stimuli of all types modify endocrine activity.

2
A Basic Description of the Neuroendocrine System

As stated above, the vast majority of studies that investigate acoustic effects on hormone activity concern signals that are used in the context of reproduction and/or aggression. The relevant physiological systems mediating these effects are the endocrine system and the nervous system. The nervous system and the endocrine system were in the past considered as separate physiological systems by some investigators. However, as it has become apparent that they have much in common and are functionally interrelated in important ways, most authors studying hormones in the context of behavior refer to a single neuroendocrine system. A few generalizations will be made about the neuroendocrine system to orient and introduce readers not familiar with this system at all. For a good description of the basics of neuroendocrinology as they apply to the study of hormonal responses to acoustic stimulation the reader is referred to texts by Brown (1994) and Nelson (1995).

The endocrine system consists of *duct-less* glands such as the thyroid, adrenal, pituitary, and the gonads; these glands secrete chemical messengers, known as hormones, into the blood. Hormones exert their biological effects by binding to specialized receptor proteins that occur either on the cell membrane or, in some cases, within the cell in the cytoplasm or the nucleus (notable in the case of steroid hormones and thyroid hormones). Once a hormone binds to its receptor, a biochemical cascade is initiated that, in the case of membrane-bound receptors, is induced by intracellular second messengers (such as members of the G-protein family) or, in the case of intracellular receptors is induced by the hormone-receptor complex entering the cell nucleus and acting as a transcription factor, i.e. inducing the expression of specific mRNAs that then initiate the synthesis of new proteins.

The nervous system consists of the brain, spinal cord, and peripheral nerves that are involved in rapid electrochemical communication. The building blocks of the nervous system are neurons, cells that serve as specialized information processing devices, and their associated glia cells that support and modulate neural activity. In general, neurons respond rapidly (within milliseconds) to external stimuli and secrete chemical messengers, neurotransmitters, that act on nearby cells usually by binding to membrane bound receptors. In contrast, endocrine organs respond over a longer time course (minutes to hours to weeks) and secrete hormones that generally act on distant targets. However, as one investigates these systems more closely, these distinctions become blurred. For example, neurons can synthesize and secrete hormones into the blood and endocrine organs receive neural innervation in many cases. Furthermore, the same chemical substance can act as a hormone in one part of the body but seemingly as a neurotransmitter in the brain. Because of these blurred distinctions that extend beyond even the field of neuroendocrinology, Bern (1990) has suggested that we should think of a broad field named the *Biology of Chemical Mediation* in which there are various subfields such as neuroendocrinology.

The neuroendocrine system most relevant to a consideration of hormonal responses to acoustic stimuli is the hypothalamo-pituitary-gonadal/adrenal axis (see Figure 1). The overall structure of this systems is roughly similar among all vertebrates (see Ev-

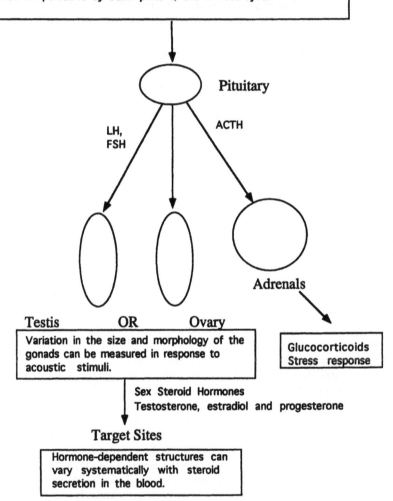

Fig. 1. A diagrammatic representation of the vertebrate hypothalamo-pituitary-gonadal axis. This axis mediates endocrine responses to environmental stimuli including acoustic stimuli. Changes in endocrine functioning in response to acoustic stimulation can, in theory, be assessed at any level along this axis. However, measuring variation in the synthesis or content of the hypothalamic releasing hormones requires that the animals be killed or that very complex surgery be performed to collect portal blood (i.e., the blood in the portal vessels linking the brain and the pituitary) so this is not typically employed. In blood samples collected from peripheral veins one can routinely measure pituitary hormones such as luteinizing hormone (*LH*), follicle-stimulating hormone (*FSH*), or adrenocorticotropic hormone (*ACTH*). Similarly, steroid hormones from either the gonads (i.e., the testis or the ovary) or the adrenal as listed in the figure can also be measured. In addition to measuring hormones, one can assess the mass or morphology of endocrine glands such as the gonads, or of hormone-dependent structures such as the oviduct, to obtain evidence for the endocrine consequences of acoustic stimuli. See text for a more detailed discussion

erett 1988 for a general review). The hypothalamus is a part of the brain (in the diencephalon) that synthesizes and secretes a variety of regulatory peptides in the context of reproduction and aggression. The most important of these is *gonadotropin-releasing hormone* (GnRH), sometimes referred to as *luteinizing hormone releasing hormone* (LHRH). All vertebrates have at least one form of this decapeptide present in the brain; in many species more than one form has been described (King and Millar 1991). However, it is generally true among vertebrates that some form of GnRH regulates the pituitary gonadotropin hormones, *luteinizing hormone* (LH) and *follicle-stimulating hormone* (FSH). Another example of a hypothalamic regulatory peptide is *corticotropin-releasing hormone* (CRH) a neuropeptide that is part of the neural system that regulates the pituitary hormone, *adrenocorticotropic hormone* (ACTH). These regulatory peptides are released into the portal blood vessels which drain into the blood vessels of the anterior pituitary gland. The gonadotropin pituitary hormones, LH and FSH, stimulate gametogenesis and stimulate the synthesis of the gonadal sex steroid hormones that can be divided into three groups: (1) *androgens*, such as testosterone; (2) *estrogens* such as 17β-estradiol, and (3) *progestins* such as progesterone. Other pituitary hormones, such as ACTH, regulate the adrenal secretion of the group of steroid hormones known as the glucocorticoids, such as cortisol or corticosterone. The different steroid hormones exert a multitude of effects both by acting on tissues subsuming reproductive functions, such as the oviduct, as well as by acting on the brain to regulate their own secretion via positive and negative feedback on the activity of the releasing hormones in the brain and by acting in other sites to influence behavior. Investigators interested in assessing the consequences of acoustic stimulation can potentially measure the effects of such stimulation at any point along this cascade of physiological events (see Figure 1). The decision as to where and what to select will greatly influence how fast and how precisely one can detect a hormone effect.

3
The Measurement of Hormone-Dependent Structures and Behaviors in Response to Acoustic Stimulation

The pioneering investigations of Rowan (1925) demonstrated that an external stimulus such as photoperiod could greatly modify endocrine activity associated with seasonal breeding. It was subsequently recognized by behavioral endocrinologists such as Lehrman (1959) that other environmental stimuli such as ongoing behavior of various sorts, including acoustic signals, could have a profound effect on endocrine functioning. Prior to the 1960s, limitations in the available methods made it difficult to directly measure hormone secretion in the blood. Indications of endocrine activity were consequently based on: (1) the assessment of histological changes in hormone secreting glands; (2) changes in the weight of hormone-dependent structures, or (3) the measurement of the frequency and intensity of hormone-dependent behaviors. The assessment of changes of these types, although in some sense providing only a crude indicator of endocrine changes, is still in many cases very useful.

3.1
Methodological Issues Related to the Measurement
of Acoustic Effects on the Size and Histology of the Gonads
and Hormone-Dependent Structures and Behaviors

Depending on the species studied, the investigation of changes in gonadal activity may or may not require a surgical manipulation of some sort. In species with internal gonads, exploratory laparotomy is often required to assess gonadal state (e.g., Bailey 1953; Risser 1971). Most vertebrates living in the temperate zone undergo extensive seasonal changes in gonad size in response to a host of environmental changes including acoustic stimulation. These can easily be ascertained by measuring either the diameter of the testis or the largest follicle in the ovary, or in the case of the testis by measuring the length and the width of the organ and calculating the volume. Testis volume in birds is often calculated with the equation $4/3\pi a^2 b$, where a is the radius of the testis at its widest point and b is half of the long axis (e.g., Dawson and Goldsmith 1983). In female birds, the most definitive measure of follicular development is egg laying itself. By looking for oviposition, one is provided with an easy to measure dependent variable. If the animal is going to be euthanized, the gonads can be dissected, weighed, and the mass determined. If the gonads are collected they can be fixed and then histologically processed to allow one to ascertain whether gametogenesis was ongoing and/or to examine the morphology of the steroid synthesizing cells. In males, one can assess spermatogenesis by visualizing the cells of Leydig that synthesize testosterone to determine if they are active or not. However, detailed information about Leydig cell function may require examination at the electron microscopic level and is therefore not particularly convenient information to collect (e.g., Lam and Farner 1976). In females, the histology of the ovary can be complex and varies to a certain extent among the different vertebrate groups. However, one can look for evidence of recently ovulated follicles and other cellular measures of the activity of the ovary at a given time. The variability of gonad morphology among species, and techniques for study among different vertebrate groups are discussed in detail by van Tienhoven (1983), Dodd (1986), and Nagahama (1986).

In addition to the activity of the gonads themselves, the effects of gonadal activity can be measured by weighing hormone-dependent structures and by studying behaviors. In some cases these are internal structures, such as the uterus and the oviduct, that require the animal to be euthanized before the structure is dissected and weighed. In other cases external organs, such as the cloacal protuberance, can be measured providing a reliable indicator of gonadal mass and sex steroid hormone secretion. For example, in Japanese quail (*Coturnix japonica*) the cloacal protuberance is tightly correlated with the size of the testis in males, and the circulating levels of testosterone in their plasma (Sachs 1967; Delville et al. 1985).

In addition to measuring changes in the size and morphology of hormone-producing and hormone-dependent structures, it is possible to use changes in reproductive behavior itself as an indicator of changes in endocrine state in response to acoustic stimulation. In canaries (*Serinus canaria*), nest building behavior is highly correlated in a positive direction with the diameter of the largest ovarian follicle, oviduct weight and plasma levels of the gonadotropin luteinizing hormone (Follett et al. 1973). The use of these various measures in relation to acoustic stimulation is illustrated below by several examples.

3.2
The Effects of Female Vocalizations on the Development of Testes and Spermatogenesis

In early studies, where testicular volume and spermatogenesis were compared in males housed alone or with females, it was found that female vocalizations stimulate male testicular activity. For example, male European starlings (*Sturnus vulgaris*) rapidly grow their gonads and initiate spermatogenesis on photoperiods of 12.5L:11.5D and longer. In the presence of females, males develop a larger testis and attain spermatogenesis sooner that males housed alone on a similar photoperiod (Burger 1953). Females have been shown to influence male testicular size and spermatogenic activity in a wide variety of vertebrate taxa including reptiles (e.g., Crews and Garrick 1980) and fish (Silverman 1978). However, in none of these studies have the acoustic signals emitted by the female per se been shown to mediate this effect. It seems highly probable, at least in certain avian species such as those that engage in vocal duets between the male and female in a pair, that acoustic signals produced by the female will facilitate testicular growth and spermatogenesis, but this has not been demonstrated explicitly to our knowledge. This is an understudied area; most research has focused on the effects of male vocalizations on female reproductive development (Inman 1986).

3.3
The Effects of Male Vocalizations on the Development
of the Ovaries and Hormone-Dependent Structures

In several vertebrate species, male vocalizations are a prominent component of courtship interactions and thus the effects of male vocalizations on the development of the ovaries have been studied in some detail. Among the first studies were investigations of budgerigars, ring doves and canaries. In budgerigars (*Melopsittacus undulatus*) initial studies demonstrated that females grew follicles of a larger size and actually laid eggs at a higher rate when they heard male vocalizations than when held under similar conditions in the absence of sounds emanating from the male (Vaugien 1951; Ficken et al. 1960; Brockway 1962). It was subsequently established that a particular class of male vocalizations, the *soft warble* that is closely associated with male precopulatory behavior, was most effective in stimulating follicular growth and oviposition (Brockway 1965).

Experimental studies of ring doves (*Streptopelia risoria*) have illustrated the context in which male vocalizations are most effective in stimulating female reproductive development. Playing colony sounds that include male and female courtship vocalizations to isolated female doves promotes ovarian growth and oviposition but visual as well as auditory stimuli must be present for maximal effect (Lott and Brody 1966; Lott et al.

1967). A clever experiment by Friedman (1977) demonstrated that vocalizations produced by the male are most effective if the female perceives a male actively courting her. Females who hear male courtship vocalizations but see the reflection of a male who is directing his courtship displays to another female do not grow their follicles, or their oviducts, or oviposit as much as females who see and hear a male who is directing his behavior at them. An added refinement to this phenomenon has been discovered by Cheng (1986, 1992) who found that females must hear themselves coo when responding to male courtship stimuli for maximal follicular growth and oviposition to occur. Muted females who were played back their own coos had a higher mean follicle size than muted females who heard other female's coos, male coos or no vocalizations at all (Cheng 1986).

In several songbird species, most notably canaries and white-crowned sparrows (*Zonotrichia leucophrys gambelii*), male song has been shown to stimulate the ovaries and/or the oviduct in conspecific females (Hinde and Steel 1978; Morton et al. 1985). These species breed seasonally and a major environmental cue that regulates the timing of reproduction is annual variation in photoperiod. Both Morton et al. (1985) and Hinde and Steel (1978) found that song was not effective in enhancing reproductive development under all photoperiodic conditions. In the case of Gambel's white-crowned sparrows there seems to be a photoperiodic threshold effect. Song played back to females on photoperiods of 11L:13D or 6L:18D did not enhance ovarian growth while it did enhance ovarian growth of females on photoperiods of 12.5L:11.5D and 14L:10D (Morton et al. 1985). In the case of canaries, Hinde and Steel (1978) reported that playing back tape-recorded male song to female canaries on a photoperiod of 11L:13D significantly increased follicular growth as compared to females on 11L:13D who did not hear male song. However, females on photoperiods of 14L:10D did not exhibit enhanced reproductive development by exposure to song. Hinde and Steel (1978) suggested that this happened because a photoperiod of 14L:10D stimulated the reproductive system to the maximal extent, and the supplementary stimulation provided by the song could not stimulate it further.

Among mammalian species, reproductive signaling frequently depends on olfactory cues (Wingfield et al. 1994). However, male red deer (*Cervus elaphus*) produce a *roaring* vocalization that has been shown to have a stimulatory effect on female reproductive development (McComb 1987). Female deer were exposed to sound playback of red deer roars for 14 days at rates chosen to simulate those normally exhibited by harem-holding stags, and then placed with fertile males. These females were shown to give birth to calves at earlier dates than females who received no conspecific stimulation prior to their being paired with fertile stags (McComb 1987). From this earlier date of birth one can infer that ovulation occurred earlier in the females who were exposed to the male vocalizations, as compared to the females who did not hear these sounds.

3.4
Effects on the Frequency and Duration of Hormone-Dependent Behaviors

As discussed previously, in canaries, nest building behavior is highly correlated in a positive direction with the diameter of the largest ovarian follicle, oviduct weight, and plasma levels of the gonadotropin luteinizing hormone (Follett et al. 1973). Kroodsma (1976) capitalized on these findings and demonstrated that females receiving sound playbacks of males with large song repertoires engaged in nest building at a far higher rate than females exposed to playbacks simulating relatively small song repertoires. Thus, previous endocrine studies show how correlating behavioral changes with hormone secretion allow one to measure the endocrine consequences of acoustic stimulation in a noninvasive manner.

4
The Measurement of Hormones in the Blood in Response to Acoustic Stimulation

Studies such as those outlined above continue to elucidate the effect of acoustic stimulation on the development of hormone-dependent structures and behaviors. However, these investigations measure events that occur with a relatively long latency (days to weeks) in response to auditory cues. Very useful additional information is provided by measuring hormone levels in the blood. Today such measurements are usually done using *radioimmunoassay* (RIA) techniques, or in some cases by *enzyme immunoassays* (EIA), although other procedures have been utilized. Before discussing the results of these studies, we first briefly review these various methods.

4.1
Methods for the Measurement of Hormones in the Blood

It is generally accepted that any endocrine assay technique must perform satisfactorily in four areas: (1) *specificity* — the method should reliably measure the hormone in question and nothing else. In practice, the problem often is largely avoided if cross-reactive substances are chromatographically isolated prior to assay; (2) *accuracy*— the total amount of hormone present in the samples should be measured. This requires correction factors for hormone loss during the various procedural steps. The best way to gauge accuracy is to take known amounts of hormone through the assay paradigm and determine the amount of hormone actually measured; (3) *precision* — the same samples, when measured repeatedly in the same assay or different assays, should give similar results (within 5-10 %); (4) *sensitivity* — the minimum amount of hormone that can be distinguished from water or plasma blanks containing no hormone reflects the sensitivity limits of the assay. There is no standard method to determine the point at which measured values are outside the assay's sensitivity. An example of one way to estimate sensitivity is to define it as the apparent blank concentration + 2.5 standard deviations (Vermeulen 1976).

4.1.1
Spectrophotometry, Fluorimetry, and Gas-Liquid Chromatography Methods

Several techniques have been used to quantify hormone titers in blood. One of the first involved spectrophotometry, both in the visible and the ultraviolet-light range. For example, steroid hormones show specific absorption spectra. Although steroids are normally colorless in solution, they will exhibit specific colorimetric changes if treated with acids (e.g., Bernstein and Lenhard 1953; Nowaczynski and Steyermark 1955). In the mid-1950s, a colorimetric technique was developed to measure estrogens in urine (Brown 1955). With subsequent modifications this procedure is the basis for the current home pregnancy tests. However, measuring hormone levels in urine is less than ideal, for often what is measured is hormone metabolites, whose levels may or may not reflect the circulating values of the target hormone. In addition, the sensitivity of colorimetric assays is only in the μg/ml range with urine or plasma (e.g., Preedy and Aitken 1961), which is much less than that of more recently developed techniques (see below).

Fluorimetric techniques were used in the 1960s to measure hormone levels (e.g., Roy and Brown 1960). Steroid hormones ordinarily do not fluoresce, but do so when heated with phosphoric or sulfuric acid (Finkelstein 1952). This technique is more sensitive than colorimetry by 10-100 times, but it suffers from technical drawbacks involving quench and self-absorption (see Goldzeiher 1963 for review). Gas-liquid chromatography (GLC) was also developed as a method to quantify steroid levels in the 1960s (Vandenheuvel et al. 1960). It is most effectively used in combination with other techniques that initially isolate and purify the sample, since interference by nontarget compounds can render the results from impure samples uninterpretable. The technique involves injection of the purified liquid sample through a microsyringe into a column (often glass or copper). A carrier gas (usually nitrogen, argon, or helium) flows through the column, whose contents (e.g., glass microbeads or inert diatomaceous earth) disperses the stationary liquid phase. The carrier gas transports the vaporized sample to a detector that is sensitive and specific for the type of compound being measured. GLC provides sensitivity to the nanogram (10^{-9}) level.

4.1.2
Competitive Protein-Binding Assays, Radioimmunoassays,
and Enzyme Immunoassays

Competitive protein-binding assays are based on the fact that steroid hormones circulate in the blood largely bound to transport proteins (e.g., Pearlman and Crepy 1967), such as the albumins, sex-hormone binding globulin (SHBG), and corticosterone-binding globulin. By creating a mixture of the appropriate binding protein, the extracted steroid from the sample, and a known amount of ^{14}C- or ^{3}H-radiolabeled steroid, a competition is established between the labeled and unlabeled hormone for binding sites on the protein. By determining the amount of radioactivity bound to the protein, and comparing the results with those from samples containing known amounts of unlabeled hormone, the quantity of hormone in the unknown sample can be calculated (Vermeulen 1976). Although competitive protein-binding assays are sensitive to the nanogram level, this technique is somewhat limited in its applicability to avian blood, because birds (unlike mammals, reptiles, or amphibians) do not have SHBG (see Wingfield 1980a for review).

The procedure most widely used today for the measurement of plasma steroid hormones is the *radioimmunoassay* (RIA). This technique has several advantages: it is relatively quick, inexpensive, and uncomplicated to perform; it is very sensitive (pg/ml); and it requires very little plasma (as little as 50 µl). This latter characteristic enables investigators to sample small animals without sacrificing them, and to sample the same animal repeatedly. The RIA is conceptually similar to competitive protein-binding assays. A protein (in this case, an antibody to the target hormone) is mixed with the extracted native hormone, plus a known quantity of radiolabeled (usually tritiated or iodinated) hormone. The labeled and unlabeled hormones compete for the antibody, and the radioactivity is measured after separation of bound from free hormone. The amount of unknown hormone present can be determined by referring to a standard curve that is generated by adding differing known amounts on unlabeled hormone to tubes containing a constant amount of labeled hormone. One can compare the amount of binding that has been inhibited in the unknown samples to the amount inhibited by the varying known quantities of unlabeled hormone in the standard curve, and ascertain the quantity of hormone that is present in the unknown sample. EIAs (enzyme immunoassays) are also conceptually related to RIAs. In this case the binding of an antigen labeled with an enzyme to an antibody is detected by a change in enzyme activity. In competitive EIAs, a labeled and unlabeled hormone can compete for the antibody, but in this case no radioactivity is involved because the bound antigen-antibody complex is detected by the change in color following the enzymatic reaction. Standard curves can also be constructed to determine the amount of hormone present.

A key prerequisite for these immunoassays is the development of an appropriate antibody. For example, steroid hormones, by themselves, usually do not induce antibody production. In order to induce antigenic activity, steroids must first be transformed into derivatives that contain a terminal carboxyl group (Erlanger et al. 1957). These hormone derivatives can then be conjugated to a protein, such as serum albumin (bovine, human, or rabbit). The final product, after treatment with an adjuvant, can be injected into the animal to be immunized. Booster injections are made every month, and weekly blood samples can be removed approximately 6 weeks after the first injection. The choice of animal (rabbit, sheep, horse, rat, etc.) depends on the amount of antisera required and the housing facilities on hand, but rabbits are the most typical species used.

Antibodies generally are much more specific than are binding proteins, so interference from non-target substances is greatly reduced. However, antibodies from different sources do tend to vary in their titer, i.e., the degree to which the antibody can be diluted and still exhibit an antibody-antigen reaction. This can cause recalibration delays when an antibody from a new source is used. The development of monoclonal antibodies (e.g., Seibert 1983) should help to standardize the production of antisera.

The specificity of antisera can also be enhanced by creating antibodies against particular sites on the hormone. This is especially useful with steroids, which all share a common four-ring structure, and differ only in the details of the side groups they contain. Despite these efforts to produce antibody specificity, cross-reactivity can be as great as 100 % for some closely related hormones (e.g., testosterone and 5α-dihydrotestosterone), and this often necessitates chromatographic separation of hormone fractions prior to conducting the assay. The first such studies were designed to measure estradiol (Abraham 1969; Midgley et al. 1969) and testosterone (Niswender and Midgley 1969), although it is now possible to quantify multiple steroids from a single plasma sample (Wingfield and Farner 1975).

4.2
Studies of Plasma Hormone Changes in Response to Social Stimuli in the Laboratory and the Field

Early studies measuring hormonal responses to social cues were confined to the laboratory, where conditions could be closely regulated. These studies, for reasons of convenience and accessibility, initially focused on domesticated species such as the Pekin duck (Garnier 1971; Jallageas et al.1974), canary (Nicholls 1974), ring dove, (Cheng and Follett 1976), and domestic chicken (Etches and Cunnigham 1977). Feral birds also were examined, but in captive conditions (Kerlan and Jaffe 1974). The first report of plasma hormone levels measured from feral birds in the wild was that of Temple (1974) in a study of starlings. However, these birds were sacrificed to collect sufficient blood, so serial samples tracking birds through different breeding stages were not obtained. The door to the field was opened by the work of Wingfield and Farner (1975, 1976, 1978a,b), who developed and applied methods that facilitated repeated sampling of even small passerines with no long-lasting adverse effects on their behavior or survival.

The emphasis of the early fieldwork was to document changes in plasma hormone levels throughout the breeding season and to correlate these with observed changes in breeding activities. It was suggested that social cues, such as auditory stimuli, play an important role in stimulating hormonal secretions of recipients and, consequently, in developing and synchronizing appropriate reproductive stages (Wingfield 1980b, 1983). However, teasing apart the various social stimuli required the control available in a laboratory. Indeed, ring doves, the test subject for so many of the pioneering studies of social effects on reproductive behavior and tissue morphology discussed above, also provided an early model for the measurement of social effects on circulating hormone levels directly. Cheng and Follett (1976) reported that LH levels increased in female ring doves during the male courtship period of the nesting cycle. Feder et al. (1977) noted that androgen levels (testosterone and 5α-DHT) increased in the blood of male ring doves within four hours after housing them with females. Similar increases in gonadotropin and sex steroid hormones have been noted by others (O'Connell et al. 1981). These studies suggest that the presence and/or performance of individual birds could influence the endocrine profile of conspecifics. Field studies of feral species support these findings, for plasma levels of reproductive hormones tend to be maximal during periods of territorial, courtship and nesting interactions with conspecifics (e.g., Wingfield and Farner 1978a,b; Dittami 1981; Dawson 1983; see Wingfield and Farner 1993 for a review).

4.3
Experimental Studies of Auditory Stimuli in Relation to Changes in Plasma Hormone Levels

The early field and laboratory studies were limited in that they could not distinguish between hormonal responses elicited by social interactions, and those resulting from environmental influences (i.e., lengthening photoperiod). Perhaps the observed changes in hormone levels were a natural response to long days and would have occurred in the absence of social stimulation. In seasonally breeding species in the temperate zone, one important consequence of lengthening photoperiods is to stimulate hormone secretions (e.g., Murton and Westwood 1977; Ball 1993). More recent fieldwork has involved experimental manipulation of social cues, and has provided stronger evidence of the sensitivity of endocrine activity to such stimuli. For instance, Moore (1982) reported that male white-crowned sparrows paired with estradiol-implanted females maintained elevated T and LH levels longer than did males paired with control females. Similar findings were found in song sparrows (*Melospiza melodia*; Runfeldt and Wingfield 1985). Further, Wingfield (1984) noted that T levels were significantly higher in male song sparrows that had been exposed to a simulated territorial intrusion than in control (foraging) birds. He also found that unmanipulated males with territories adjacent to T-implanted males had higher T levels than males whose neighbors did not receive exogenous T (Wingfield 1985). In addition, others have found relationships between such factors as pairing status (Dufty and Wingfield 1986a) or population density (Ball and Wingfield 1987) and endocrine levels.

The above experiments clearly indicate that social stimuli emanating from either same- or opposite-sexed conspecifics can have a significant effect on the pattern of hormone secretion. However, none of these studies addressed the question of what form of social stimulation is important. Birds rely largely on vocal and visual stimuli in their aggressive and reproductive activities, so a logical first step would be to experimentally tease apart the effects mediated by these two sensory modalities. Surprisingly, despite the general acceptance that social cues are important, relatively few studies have been designed to examine the direct effects of auditory (and visual) stimuli on endocrine secretions.

One of the first such studies was that of O'Connell et al. (1981), who paired female ring doves with intact or deafened males. The deafened males had significantly lower T levels than did the intact birds, indicating that auditory cues from females affect male hormone patterns. However, it is unclear what effect deafening had on the males' capacity to respond. Surgically deafened birds may suffer from a general motivational deficit as a result of the elimination of this major sensory pathway. Marler et al. (1972, p. 592) suspected "...a deficiency in the motivation of...deaf birds for singing...", and O'Connell et al. (1981) showed that deaf males exhibited several behavioral decrements relative to intact males. Motivational problems are also suspected to underlay changes in avian homing behavior that appear following the removal of another sensory modality, olfaction (e.g., Keeton and Brown discussion in Papi et al. 1978). Whether such reductions in motivation produce the observed diminution in hormone secretion remains to be determined. Additionally, as noted above, female ring doves have been shown to respond to the sound of their own vocalizations with increased ovarian development (Cheng et al. 1988); presumably, this development reflects changes in plasma hormone levels. It is not unreasonable to suggest that male ring doves may also respond hormonally to their own auditory output, and that the O'Connell et al. (1981) results may be explained, in part, by the inability of males to hear their own vocalizations.

In another study, female cockatiels (*Nymphicus hollandicus*) given only auditory contact with mates, exhibited plasma levels of LH similar to those of females given both auditory and visual contact (Shields et al. 1989). The LH levels in both groups were significantly lower than those of females allowed full contact with mates, but only in one of six samples spanning about seven weeks. This modest hormonal difference occurred despite large behavioral differences between groups in their performance of nesting behaviors: birds given the greatest degree of mate access (full contact, auditory + visual contact) were significantly more likely to inspect the nest or form a nest-bowl. The results suggest a more integrated role for vocal cues, with access to mates and to nest-boxes also contributing to LH secretion (see also Gwinner et al. 1987).

Wingfield and Wada (1989) investigated the effects of intrasexual vocal and visual cues on endocrine responses of song sparrows. They found that there was a tendency for plasma T levels to be higher in captive and free-living territorial males exposed to song alone compared to controls, but the difference was not statistically significant. The presence of a devocalized male (i.e., visual but not vocal cues) *was* stimulatory to captive male song sparrows, but not to territorial males. However, the latter birds did exhibit a hormonal response to vocal and visual cues together (i.e., tape-recording + devocalized male). Furthermore, territorial males did give strong behavioral responses to all combinations of social stimuli, that is, they aggressively approached the playback speaker, devocalized male, or both combined. Given the energetic costs of maintaining elevated T levels (e.g., Hänsler and Prinzinger 1979; Högstadt 1987), this species may have evolved a response mechanism whereby initial behavioral responses occur in the absence of endocrine changes, and T secretion increases only after an intruder is verified visually.

A similarly muted hormonal response to intrasexual auditory cues occurs in brown-headed cowbirds (*Molothrus ater*; Dufty and Wingfield 1990). Isolate-raised males tutored with conspecific vocalizations showed hormonal responses similar to those produced by non-tutored isolates. Conversely, the presence of a devocalized companion produced plasma T levels comparable to those of males housed with several other singing males, again illustrating the predominance of intrasexual visual cues in eliciting endocrine responses in males. Cowbirds do not respond to playbacks of conspecific songs (Dufty 1982), and most male-male aggression occurs over very short distances (Friedmann 1929). Hence, visual cues from males in close proximity may suffice to elicit maximal endocrine responses.

Thus far only these few studies have attempted to relate the behavioral effects of acoustic stimulation to changes in plasma hormone levels in birds. The results, as outlined above, are less than clear-cut. Acoustic stimulation, alone or in conjunction with visual cues, can elicit obvious changes in behavior. However, these changes do not always translate into parallel changes in plasma hormone levels. Vocal stimuli may simply be one component of a mosaic of cues, all of which initially can elicit a given response. The *maintenance* of the response may, indeed, require hormonal adjustments, and these physiological changes may necessitate the appearance of all (or a threshold number) of the contributing environmental stimuli. Endocrine responses can occur within minutes (Wingfield 1984; Wingfield and Wada 1989), so the time and energy lost to a bird in verifying the presence of a potential mate, intruder, etc., would be minimal. Clearly, more work is warranted on the effects of vocal (and other) stimuli on changes in plasma hormone levels.

4.4
Biochemical Measures of Brain Activation in Response to Acoustic Stimuli

The previous sections have selectively reviewed how the consequences of acoustic stimuli can be assessed by measuring changes in hormone-producing or hormone-dependent structures, and/or by measuring changes in hormone levels circulating in the plasma. The advantages of these methods is that in many cases one can assess endocrine changes relatively non-invasively, i.e., by collecting a blood sample or by measuring the size of the gonad. However, biochemical measures of brain activity are being applied to studies of the physiological consequences of acoustic stimuli. These measures have the advantage of providing information that can supplement electrophysiological studies, and allow one to identify neural pathways that mediate responses to acoustic signals. They have the obvious disadvantage that the animal must be sacrificed for the brain to be collected and investigated. In this section we will briefly highlight some recent approaches to the study of changes in brain activity in response to acoustic stimulation.

Two markers of metabolic activity in the brain that have been used to identify auditory pathways include the assessment of changes in cytochrome oxidase histochemistry (CO) and in the activity of 2-deoxy-d-glucose (2-DG). These markers provide relatively long-term measures of cell activity that require several hours to days to be detected by these methods. Both of these methods allow one to selectively identify central auditory pathways after acoustic stimulation (e.g., Muller and Scheich 1985; Brauth 1990). In tree frog species such as *Hyla cinerea* and *H. versicolor,* variation in CO activity in response to auditory stimulation distinguishes between the central and anterior thalamic auditory circuits. These circuits are part of the pathway in the female brain that carries information from the auditory periphery to areas in the hypothalamus and preoptic region that stimulate endocrine secretions (Wilczynski et al. 1993).

An exciting development relevant to the assessment of neural activation in response to auditory stimuli has involved the application of molecular techniques to the study of the brain. Salient environmental stimuli that modify neural processes are ultimately mediated via cellular and molecular changes within specific brain areas. Sensitive techniques allow one to identify the expression of messenger RNA in particular parts of the brain. A certain class of genes, referred to as the immediate-early genes, have been found to be expressed in specific parts of the brain in close temporal proximity with the experience of a particular stimulus. These immediate-early genes encode proteins that then induce the subsequent expression of late-response genes. Clayton and colleagues (e.g., Clayton et al. 1988; Mello Vicario and Clayton 1992) have identified genes that are specifically expressed in the avian forebrain. They have found that the vocal control nuclei exhibit unique patterns in the expression of many of these genes. They have also found that the playback of song specifically induces the expression of an immediate-early gene named ZENK in a part of the brain called NCM that receives a projection from the auditory area Field L and projects to the High Vocal Center (Mello et al. 1992). Thus, the experience of song sets in motion a change in cell functioning in a specific part of the brain. Identifying the consequences of this change for later behavioral activity will be an exciting avenue to pursue in the future.

These biochemical measures of cell activation, especially when combined with electrophysiological investigations, provide important information concerning the neural pathways mediating responses to acoustic stimulation. This information will ultimately lead to the description of the pathways by which auditory information results in a change in endocrine activity. However, especially for field workers, endocrine measures that are "downstream" from these neural pathways provide a relatively easy and non-invasive way to assess the salience of an auditory cue that provides powerful converging information along with behavioral measures.

5
Summary and Future Directions

The past decade has been characterized by major advances in field techniques to record and play back sounds to animals. Many of these advances are highlighted in this Volume. Simultaneous with this development has been the rise of *field endocrinology* approaches to studies of environmental physiology. As is illustrated by some of the examples referred to above, these developments can be combined relatively easily and would mutually benefit workers in both fields. By collecting blood samples and measuring hormones levels one can obtain independent measures concerning the salience of a signal. Hormonal changes provide converging evidence along with behavioral responses to allow one to assess the significance of a vocal signal. In past studies, hormones have been generally employed to assess whether a given class of vocalizations (e.g., courtship vocalizations) has any endocrine consequences. In the future more work should be done assessing whether variation in acoustic quality can influence hormonal secretions.

Sexual selection and the associated competition for mates is clearly the major ultimate reason shaping the structure of many vocalizations (Searcy and Andersson 1986; Andersson 1994). Many bioacousticians have been testing whether variation in the quality of a vocalization influences mate choice or competition among the members of one sex for access to the other sex. By elucidating the endocrine consequences of vocal behavior an important link in the proximate chain of events that mediate mate preference and mate competition can be addressed, and insights about the significance of an acoustic signal to a receiver can be attained.

References

Abraham GE (1969) Solid-phase radioimmunoassay of estradiol-17β. J Clin Endocrinol Metab 29: 866–870

Andersson M (1994) Sexual selection. Princeton University Press, Princeton

Bailey RE (1953) Surgery for sexing and observing gonad condition in birds. Auk 70: 497–499

Ball GF (1993) The neural integration of environmental information by seasonally breeding birds. Am Zool 33: 185–199

Ball GF, Wingfield JC (1987) Changes in plasma levels of luteinizing hormone and sex steroid hormones in relation to multiple-broodedness and nest-site density in male starlings. Physiol Zool 60: 191–199

Bern HA (1990) The "new" endocrinology: its scope and its impact. Am Zool 30: 877–885

Bernstein S, Lenhard RH (1953) The absorption spectra of steroids in concentrated sulfuric acid. I. Method and Data. J Org Chem 18: 1146–1165

Brauth SE (1990) Investigation of central auditory nuclei in the budgerigar with cytochrome oxidase histochemistry. Brain Res 508: 142–146

Brockway BF (1962) The effects of nest-entrance positions and male vocalizations on reproduction in budgerigars. Living Bird 1: 93–101

Brockway BF (1965) Stimulation of ovarian development and egg laying by male courtship vocalizations in budgerigars (Melopsittacus undulatus) Anim Behav 13: 575–578

Brown JB (1955) Chemical method for the determination of estriol, estrone, and estradiol in human urine. Biochem J 60: 185–193

Brown R (1994) An introduction to neuroendocrinology. Cambridge University Press, New York

Burger JW (1953) The effect of photic and psychic stimuli on the reproductive cycle of the male starling, Sturnus vulgaris. J Exp Zool 124: 227–239

Cheng MF (1986) Female cooing promotes ovarian development in ring doves. Physiol Behav 37: 371–374

Cheng MF (1992) For whom does the female dove coo? A case for the role of vocal self-stimulation. Anim Behav 43: 1035–1044

Cheng MF, Follett BK (1976) Plasma luteinizing hormone during the breeding cycle of the female ring dove. Horm Behav 7: 199–205

Cheng MF, Desiderio C, Havens M, Johnson A (1988) Behavioral stimulation of ovarian growth. Horm Behav 22: 388–401

Clayton DF, Huecas ME, Sinclair-Thompson EY, Nastiuk KL, Nottebohm F (1988) Probes for rare mRNAs reveal distributed cell subsets in canary brain. Neuron 1: 249–261

Crews D, Garrick LD (1980) Methods of inducing reproduction in captive reptiles. In: Murphy J, Collins JT (eds) The reproductive biology and diseases of captive reptiles. Society for the Study of Amphibians and Reptiles, Lawrence, Kansas, p 49

Dawson A (1983) Plasma gonadal steroid levels in wild starlings (Sturnus vulgaris) during the annual cycle and in relation to the stages of breeding. Gen Comp Endocrinol 49: 286–294

Dawson A, Goldsmith AR (1983) Plasma prolactin and gonadotrophins during gonadal development and the onset of photorefractoriness in male and female starlings (Sturnus vulgaris) on artificial photoperiods. J Endocrinol 97: 253–260

Delville Y, Sulon J, Hendrick JC, Balthazart J (1984) Effect of the presence of females on the pituitary-testicular activity in male Japanese quail (Coturnix coturnix japonica). Gen Comp Endocrinol 55: 295–305

Delville Y, Sulon J, Balthazart J (1985) Hormonal correlates of gonadal regression and spontaneous recovery in Japanese quail exposed to short day lengths. Arch Int Physiol Biochim 93: 123–133

Dittami JP (1981) Seasonal changes in the behavior and plasma titers of various hormones in barheaded geese, Anser indicus. Z Tierpsychol 55: 289–324

Dodd JM (1986) The ovary. In: Pang PKT, Schreibman MP (eds) Vertebrate endocrinology: fundamentals and biomedical implications, vol 1. Morphological considerations. Academic Press, New York, p 351

Dufty AM Jr, (1982) Responses of brown-headed cowbirds to simulated conspecific intruders. Anim Behav 30: 1043–1052

Dufty AM Jr Wingfield JC (1986a) The influences of social cues on the reproductive endocrinology of male brown-headed cowbirds: field and laboratory studies. Horm Behav 20: 222–234

Dufty AM Jr, Wingfield JC (1986b) Temporal patterns of circulating LH and steroid hormones in a brood parasite, the brown-headed cowbird, Molothrus ater. I. Males. J Zool (Lond) 208: 191–203

Dufty AM Jr, Wingfield JC (1990) Endocrine response of captive male brown-headed cowbirds to intrasexual social cues. Condor 92: 613–620

Etches RJ, Cunnigham FJ (1977) The plasma concentrations of testosterone and LH during the ovulation cycle of the hen (Gallus domesticus). Acta Endocrinol 84: 357–366

Erlanger BF, Borek F, Beiser SM, Lieberman S (1957) Steroid-protein conjugates. I. Preparation and characterization of conjugates to bovine serum albumin with testosterone and cortisone. J Biol Chem 228: 713–727

Everett JW (1988) Pituitary and hypothalamus: perspective and overview. In: Knobil E, Neil J (eds) The physiology of reproduction. Raven Press, New York, p 1143

Feder HH, Storey A, Goodwin D, Reboulleau C, Silver R (1977) Testosterone and "5-alpha-dihydrotesto-sterone" levels in peripheral plasma of male and female ring doves (Streptopelia risoria) during the reproductive cycle. Biol Reprod 16: 666–677

Ficken RW, van Tienhoven A, Ficken MS, Sibley FC (1960) Effect of visual and vocal stimuli on breeding in the budgerigar (Melopsitacus undulatus). Anim Behav 8: 104–106

Finkelstein M (1952) Fluorometric determination of micro amounts of oestrone-oestradiol and oestriol in urine. Acta Endocrinol 10: 149–166

Follett BK, Hinde RA, Steel E, Nicholls TJ (1973) The influence of photoperiod on nest-building, ovarian development and LH secretion in canaries (Serinus canarius). J Endocrinol 59: 151–162

Friedmann H (1929) The cowbirds, a study in the biology of social parasitism. Thomas, Springfield, Illinois

Friedman MB (1977) Interactions between visual and vocal courtship stimuli in the neuroendocrine response of female doves. J Comp Physiol Psychol 91: 1408–1416

Garnier DH (1971) Variations de la testostrone du plasma priphrique chez le canard Pkin au cours du cycle annuel. C R Acad Sci (Paris) 272: 1665–1668

Goldzieher JW (1963) Fluorescence spectra. In: Engel LL (ed) Physical properites of the steroid hormo-nes. MacMillan, New York, p 288

Green S, Marler P (1979) The analysis of animal communication. In: Marler P, Vandenbergh JG (eds) Handbook of behavioral neurobiology, vol 3. Social behavior. Plenum Press, New York p 73

Gwinner H, Gwinner E, Dittami J (1987) Effects of nest boxes on LH, testosterone, testicular size, and reproductive behavior of male European starlings in spring. Behaviour 103: 68–82

Haase E, Paulke E, Sharp PJ (1976) Effects of seasonal and social factors on testicular activity and hormone levels in domestic pigeons. J Exp Zool 197: 81–88

Hännsler I, Prinzinger R (1979) The influence of the sex hormone testosterone on body temperature and metabolism of Japanese quail. Experientia 35: 1037–1043

Hauser MD (1996) The evolution of communication. MIT Press, Cambridge

Hinde RA, Steel E (1978) The influence of day length and male vocalizations on the estrogen-dependent behavior of female canaries and budgerigars, with discussion of data from other species. Adv Study Behav 8: 39–73

Högstad O (1987) It is expensive to be dominant. Auk 104: 333–336

Inman B (1986) Female vocalizations and their role in the avian breeding cycle. Ann NY Acad Sci 474: 44–52

Jallageas M, Assenmacher I, Follett BK (1974) Testosterone secretion and plasma luteinizing hormone concentration during a sexual cycle in the Pekin duck, and after thyroxine treatment. Gen Com Endocrinol 22: 428–432

Kelley DB, Brenowitz EA (1992) Hormonal influences on courtship behaviors. In: Becker JB, Breedlove SM, Crews D (eds) Behavioral endocrinology. MIT Press, Cambridge, p 187

Kerlan JT, Jaffe RB (1974) Plasma testosterone levels during the testicular cycle of the red-winged blackbird (Agelaius phoeniceus). Gen Comp Endocrinol 2: 428–432

King JA, Millar RP (1991) Gonadotropin-releasing hormones. In: Pang P, Schreibman M (eds) Vertebrate endocrinology: fundamentals and biomedical implications, vol 4, part B. Academic Press, New York p 1

Kroodsma DE (1976) Reproductive development in a female songbird: Differential stimulation by quality of male song. Science 192: 574–575

Lam F, Farner DS (1976) The ultrastructure of the cells of Leydig in the white-crowned sparrow (Zonotrichia leucophrys gambelii) in relation to plasma levels of luteinizing hormone and testo-sterone. Cell Tissue Res 169: 93–109

Lehrman DS (1959) Hormonal responses to external stimuli in birds. Ibis 101: 478-296

Lott DF, Brody PN (1966) Support of ovulation in the ring dove by auditory and visual stimuli. J Comp Physiol Psycho 62: 311–313

Lott DF, Scholz SD, Lehrman DS (1967) Exteroceptive stimulation of the reproductive system of the female ring dove (Streptopelia risoria) by the mate and by the colony milieu. Anim Behav 15: 433–437

Marler P, Mundinger P, Waser MS, Lutjen A (1972) Effects of acoustical stimulation and deprivation on song development in red-winged blackbirds (Agelaius phoeniceus). Anim Behav 20: 586–606

Marshall AJ (1959) Internal and environmental control of breeding. Ibis 101: 456–478

McComb K (1987) Roaring by red deer stags advances the date of oestrus in hinds. Nature 330: 648–649

Mello CV, Vicario DS, Clayton DF (1992) Song presentation induces gene expression in the songbird forebrain. Proc Natl Acad Sci USA 89: 6818–6822

Midgley AR Jr, Niswender GD, Ram JS (1969) Hapten radioimmunoassay: a general procedure for the estimation of steroidal and other haptenic substances. Steroids 13: 731–737

Moore MC (1982) Hormonal response of free-living male white-crowned sparrows to experimental manipulation of female sexual behavior. Horm Behav 16: 323–329

Morton ML, Pereyra ME, Baptista LF (1985) Photoperiodically induced ovarian growth in the white-crowned sparrow (Zonotrichia leucophrys gambelii) and its augmentation by song. Comp Biochem Physiol 80A: 93–97

Muller SC, Scheich H (1985) Functional organization of the avian auditory field L. J Comp Physiol A 156: 1–12

Murton RK, Westwood NJ (1977) Avian breeding cycles. Oxford University Press, New York

Nagahama Y (1986) Testis. In: Pang PKT, Schreibman MP (eds) Vertebrate endocrinology: Fundamentals and biomedical implications, vol 1. Morphological considerations. Academic Press, New York, p 399

Nelson RJ (1995) An introduction to behavioral endocrinology. Sinauer, Sunderland

Nicholls TJ (1974) Changes in plasma LH levels during a photoperiodically controlled reproductive cycle in the canary (Serinus canarius). Gen Comp Endocrinol 24: 442–445

Niswender GD, Midgley AR Jr (1969) Hapten-radioimmunoassay for testosterone. 51st Proc Endocrinol Soc Abstr 22 New York

Nowaczynski WJ, Steyermark PR (1955) Absorption spectra of steroids in "100 %" phosphoric acid. Arch Biochem Biophys 58: 453–460

O'Connell ME, Reboulleau C, Feder HH, Silver R (1981) Social interactions and androgen levels in birds. I. Female characteristics associated with increased plasma androgen levels in the male ring dove (Streptopelia risoria). Gen Comp Endocrinol 44: 454–463

Papi F, Keeton WT, Brown AI, Benvenuti S (1978) Do American and Italian pigeons rely on different homing mechanisms? J Comp Physiol 128: 303–317

Pearlman WH, Crepy O (1967) Steroid-protein interaction with particular reference to testosterone binding by human serum. J Biol Chem 242: 182–189

Preedy JRK, Aitken EH (1961) Determination of estrone, 17β-estradiol, and estriol in urine and plasma with column partition chromatography. J Biol Chem 236: 1300–1311

Risser AC (1971) A technique for performing laparotomy on small birds. Condor 73: 376–379

Rowan W (1925) Relation of light to bird migration and development changes. Nature 115: 494–495

Roy EJ, Brown JB (1960) A method for the estimation of oestriol, oestrone and oestradiol-17 in the blood of pregnant women and of the foetus. J Endocrinol 21: 9–23

Runfeldt S, Wingfield JC (1985) Experimentally prolonged sexual activity in female sparrows delays termination of reproductive activity in their untreated mates. Anim Behav 33: 403–410

Sachs BD (1967) Photoperiodic control of the cloacal gland of the Japanese quail. Science 157: 201–203

Searcy WA, Andersson M (1986) Sexual selection and the evolution of song. Annu Rev Ecol Syst 17: 507–533

Seibert G (1983) Monoclonal antibodies in hormone analysis. In: Twelfth training course on hormonal assay techniques. The Endocrine Society, Bethesda, p 114

Shields KM, Yamamoto JT, Millam JR (1989) Reproductive behavior and LH levels of cockatiels (Nymphicus hollandicus) associated with photostimulation, nest-box presentation and degree of mate access. Horm Behav 23: 68–82

Silver R (1992) Environmental factors influencing hormone secretion. In: Becker JB, Breedlove SM, Crews D (eds) Behavioral endocrinology. MIT Press, Cambridge, p 401

Silverman HI (1978) Effects of different levels of sensory contact upon reproductive activity of adult male and female sarotherodon (Tilapia mossambicus); Pisces; chichlidae. Anim Behav 26: 1081–1090

Temple SA (1974) Plasma testosterone titers during the annual reproductive cycle of starlings (Sturnus vulgaris). Gen Comp Endocrinol 22: 470–479

van Tienhoven A (1983) Reproductive physiology of vertebrates. Cornell University Press, Ithaca

Vanden Heuvel WJA, Sweeley CC, Horning EC (1960) Separation of steroids by gas chromatography. J Am Chem Soc 82: 3481–3482

Vaugien L (1951) Ponte induite chez la Perruche ondulee maintenue à l'obscurité et dans l'ambience des voliéres. C R Acad Sci (Paris) 232: 1706–1708

Vermeulen A (1976) Estimation of steroid hormones by competitive protein-binding techniques. In: Antoniades HN (ed) Hormones in human blood. Harvard University Press, Cambridge, p 720

Wilczynski W, Allsion JD, Marler CA (1993) Sensory pathways linking social and environmental cues to endocrine control regions of amphibian forebrains. Brain Behav Evol 42: 252–264

Wingfield JC (1980a) Sex steroid-binding proteins in vertebrate blood. In: Ishii S, Hirano T, Wada M (eds) Hormones, adaptation, and evolution. Japan Scientific Societies Press/Springer, Berlin Heidelberg New York, p 135

Wingfield JC (1980b) Fine temporal adjustment of reproductive functions. In: Epple A, Stetson MH (eds) Avian endocrinology. Academic Press, New York, p 367

Wingfield JC (1983) Environmental and endocrine control of avian reproduction: an ecological approach, In: Mikami S, Homma K, Wada M (eds) Avian endocrinology: environmental and ecological perspectives. Japan Scientific Societies Press/Springer-Verlag, Berlin Heidelberg New York, p 265

Wingfield JC (1984) Environmental and endocrine control of reproduction in the song sparrow, *Melospiza melodia*. II. Agonistic interactions as environmental information stimulating secretion of testosterone. Gen Comp Endocrinol 56: 417–424

Wingfield JC (1985) Short-term changes in plasma levels of hormones during establishment and defense of a breeding territory in male song sparrows, *Melospiza melodia*. Horm Behav 19: 174–187

Wingfield JC, Farner DS (1975) The determination of five steroids in avian plasma by radioimmunoassay and competitive protein binding. Steroids 26:311–327

Wingfield JC, Farner DS (1976) Avian endocrinology – field investigations and methods. Condor 78: 570–572

Wingfield JC, Farner DS (1978a) The endocrinology of a natural breeding population of the white-crowned sparrow (*Zonotrichia leucophrys pugetensis*). Physiol Zool 51: 188–205

Wingfield JC, Farner DS (1978b) The annual cycle of plasma irLH and steroid hormones in feral populations of the white-crowned sparrow, *Zonotrichia leucophrys gambelii*. Biol Reprod 19: 1046–1056

Wingfield JC, Farner DS (1993) Endocrinology of reproduction in wild species. In: Farner DS, King J R, Parkes K (eds) Avian biology, vol IX. Academic Press, New York, p 163

Wingfield JC, Wada M (1989) Changes in plasma levels of testosterone during male-male interactions in the song sparrow, *Melospiza melodia*: time course and specificity of response. J Comp Physiol A 166: 189–194

Wingfield JC, Whaling CS, Marler PR (1994) Communication in vertebrate aggression and reproduction: the role of hormones. In: Knobil E, Neil JD (eds) The physiology of reproduction, 2nd edn. Raven Press, New York, p 303